1st International Workshop towards the Giant Liquid Argon Charge Imaging Experiment 2010

Journal of Physics: Conference Series, Volume 308

Tsukuba, Japan
29-31 March 2010

ISBN: 978-1-61839-128-5
ISSN: 1742-6588

Printed from e-media with permission by:

Curran Associates, Inc.
57 Morehouse Lane
Red Hook, NY 12571

Some format issues inherent in the e-media version may also appear in this print version.

Printed by Curran Associates, Inc. (2011)

For permission requests, please contact the Institute of Physics
at the address below.

Institute of Physics
Dirac House, Temple Back
Bristol BS1 6BE UK

Phone: 44 1 17 929 7481
Fax: 44 1 17 920 0979

techtracking@iop.org

Additional copies of this publication are available from:

Curran Associates, Inc.
57 Morehouse Lane
Red Hook, NY 12571 USA
Phone: 845-758-0400
Fax: 845-758-2634
Email: curran@proceedings.com
Web: www.proceedings.com

Preface

Introduction

Special session

Main goals of Giant Liquid Argon Charge Imaging Experiments

Lessons on Liquid Argon Charge Imaging technology from ongoing developments

Lessons from Xe based Liquids Imaging detectors

Studies on physics performance

Ways to improve the Liquid Argon Charge Imaging technology

1st International Workshop towards the Giant Liquid Argon Charge Imaging Experiment

Atsuto Suzuki and Koichiro Nishikawa

PREFACE

"Neutrino physics is largely an art of learning a great deal by observing nothing" (Haim Harari, 1988) was our general understanding of the field for the ~25 years previous. A new neutrino era was abruptly brought from outer space by a burst of SN1987A neutrinos. The detection of neutrinos from SN1987A gave a new impetus to neutrino research. As we know, new discoveries of neutrinos have since been made. Neutrinos were no longer mysterious, but attained particle citizenship.

Giant liquid argon charge imaging experiments have the prospect of opening the door to the second new era in neutrino physics. The coming era would provoke not evolution, but revolution in particle physics. However, paving the way for the new era requires not evolutionary, but revolutionary detector developments. I hope this workshop will be conducive to reaping a rich harvest from its activities.

In 1993, Professor Carlo Rubbia presented "The Renaissance of Experimental Neutrino Physics" in which he discussed various possibilities of shooting neutrino beams from CERN towards Gran Sasso, Super-Kamiokande at Kamioka and DUMAND in Hawaii. Now KEK hopes to shoot neutrino beams from J-PARC to Kamioka, Okinoshima, Korea and Gran Sasso.

Atsuto Suzuki
Director General, KEK

J-PARC has moved into a new phase of operation. The commissioning of the accelerator complex and experiment facilities has begun, and it is urgent to attain initial design performance as soon as possible. For the immediate future, KEK has a 5 year plan. The plan includes the upgrade of the J-PARC accelerator to a multi-Mega-Watt facility, and detector R&Ds to form the basis for a next step in the neutrino experiment. One of the main issues of the future neutrino experiment will be the search for CP violation in neutrino oscillation, which demands much more precision than studying neutrino oscillation or non-zero theta13. This naturally requires a very massive detector with higher precision than presently available technologies. In addition, such a detector will also be an excellent detector for the further search for proton decay.

The sensitivity of the various methods of extracting CP violation should be carefully evaluated. At the 4th International Workshop on Nuclear and Particle Phyics at J-PARC (NP08) in March 2008, the J-PARC MR power upgrade and the future neutrino experiment were discussed.

At the workshop, a comparison of first and second maximum in electron neutrino appearance in muon neutrino beam was discussed. This method is free from systematic errors due to ambiguities of neutrino and anti-neutrino cross sections. On the other hand, this method requires a very massive detector with energy reconstruction capability for a wide range of neutrino energy. Beside the statistical precision, which requires a high power proton accelerator, there are many strict requirements on the detector. Most of the requirements are common to a detector for neutrino physics and to the search for proton decay.

We need new ways of looking for new phenomena at a level far beyond on-going experiments. Future experiments should be designed based on new detector technology, improved accelerator technology and the results of present on-going experiments.

We are looking forward seeing the outcome.

Koichiro Nishikawa
Director, IPNS KEK

Introductory remark on the First International Workshop towards the Giant Liquid Argon Charge Imaging Experiment (GLA2010)

Takuya Hasegawa
High Energy Accelerator Research Organization (KEK), Tsukuba, Ibaraki 305-0801, Japan
E-mail: takuya.hasegawa@kek.jp

Abstract. Based on the strong scientific demand to realize Giant Liquid Argon Time Projection Chamber, which makes widely spread scientific exploration possible, the First International Workshop towards the Giant Liquid Argon Charge Imaging Experiment (GLA2010) is held on March 29-31 2010, in Tsukuba Japan. Here, introductory remark on the workshop which includes the motivation and the skeleton of the workshop is described.

1. Introduction

Giant Liquid Argon Time Projection Chamber with 100 kton size will open a fascinating opportunity to explore wide spread scientific subjects. It would be ideal for the next generation accelerator based neutrino research approaching the lepton sector CP symmetry breaking, would extend the search for the proton decay via modes favored by the supersymmetric grand unified models (e.g. $p \rightarrow \nu K^+$) up to 10^{35} years. Moreover, it would cover wide range of neutrino physics with astrophysical and terrestrial sources (e.g. solar and atmospheric neutrinos, neutrinos from stellar collapse and the neutrinos from Dark Matter annihilation.).

2. New generation accelerator based neutrino experiment and proton decay experiment

One of the primary motivation of neutrino physics research recently in the world is to improve the sensitivity to the $\nu_\mu \rightarrow \nu_e$ conversion phenomenon in the atmospheric regime. In Japan, as the first experiment utilizes KEK/J-PARC (Japan Proton Accelerator Research Complex) neutrino facility, T2K (Tokai to Kamioka Long Baseline Neutrino Experiment) starts operation to explore above subject. The similar project, NOvA is under preparation in USA.

The final goal for T2K is to accumulate an integrated proton power on target of $0.75\ \text{MW} \times 5 \times 10^7$ seconds. As is shown in Figure 1, within a few years of run, critical information, which will guide the future direction of the neutrino physics, will be obtained based on the data corresponding to about 1 to 2 $\text{MW} \times 10^7$ seconds integrated proton power on target (roughly corresponding to a 3σ discovery at $\sin^2 2\theta_{13} > 0.05$ and 0.03, respectively).

If a significant $\nu_\mu \rightarrow \nu_e$ conversion signal were to be observed at T2K, an immediate step forward to a next generation experiment aimed at the discovery of CP violation in the lepton sector would be recommended with high priority. Compared with presently available experimental conditions, lepton sector CP violation discovery requires

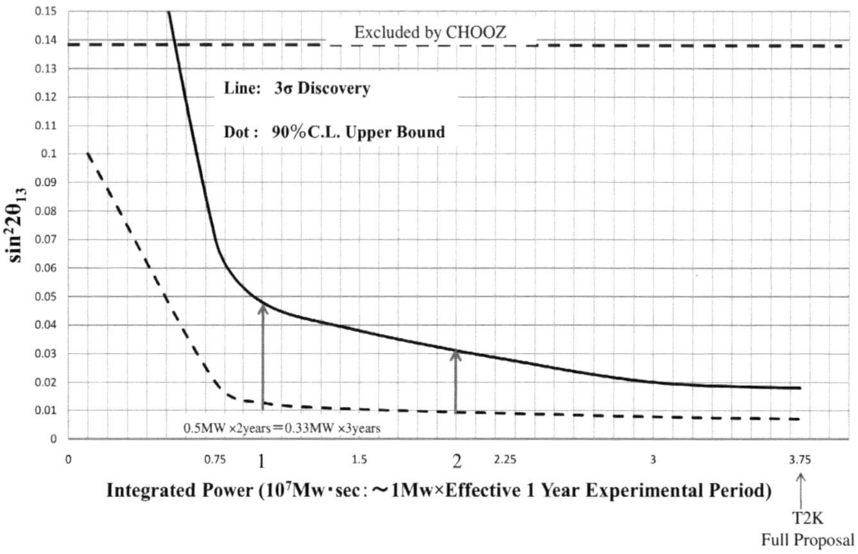

Figure 1. T2K discovery potential on $\nu_\mu \rightarrow \nu_e$ as a function of integrated proton power on target.

- improved neutrino beam intensity;
- an improved main far neutrino detector.

Detector improvements include

- detector technology;
- its volume;
- its baseline and off-axis angle with respect to the neutrino source.

Naturally, next generation far neutrino detectors for lepton sector CP violation discovery will be very massive and huge. As a consequence, the same detector will give us the rare and important opportunity to discover proton decay. A total research subject would be, to address a long standing puzzle of our physical world, the "Quest for the Origin of Matter Dominated Universe" (see e.g. [1]), with exploration of

- the lepton sector CP violation by precise testing of the neutrino oscillation processes;
  - measure precisely the CP phase in lepton sector (δ_{CP}) and the mixing angle θ_{13};
  - examine matter effect in neutrino oscillation process and possibly conclude the mass hierarchy of neutrinos.
- proton decay:
  - Search for p $\rightarrow \nu$ K^+ and p $\rightarrow e^+\pi^0$ in the life time range 10^{34} to 10^{35} years,

with assuming non-equilibrium environment in the evolution of universe.

Even in case that $\sin^2 2\theta_{13}$ is below T2K sensitivity, it is still worth while trying to improve neutrino beam intensity and far detector performance to open the way to explore $\nu_\mu \rightarrow \nu_e$ conversion phenomenon with by an order of magnitude better sensitivity.

3. Far detector options: How to approach lepton sector CP violation

The effects of lepton sector CP phase δ_{CP} appear either

- in the energy spectrum shape of the appearance oscillated ν_e charged current events (sensitive to all the non-vanishing δ_{CP} values including 180°);
- as a difference between ν and $\bar{\nu}$ behaviors (this is sensitive to the CP-odd term which vanishes for $\delta_{CP} = 0$ or 180°).

It should be noted that if one precisely measures the ν_e appearance energy spectrum shape (peak position and height for 1st and 2nd oscillation maximum and minimum) with high resolution, CP effect could be investigated with neutrino run only. One more aspect which should be noted is that antineutrino beam conditions are known to be more difficult than those for neutrinos (lower beam flux due to leading charge effect in proton collisions on target, small antineutrinos cross-section at low energy, etc.). Moreover, the systematic uncertainty for the neutrino mode experiment and antineutrino mode experiment is different each other and not much cancellation is foreseen.

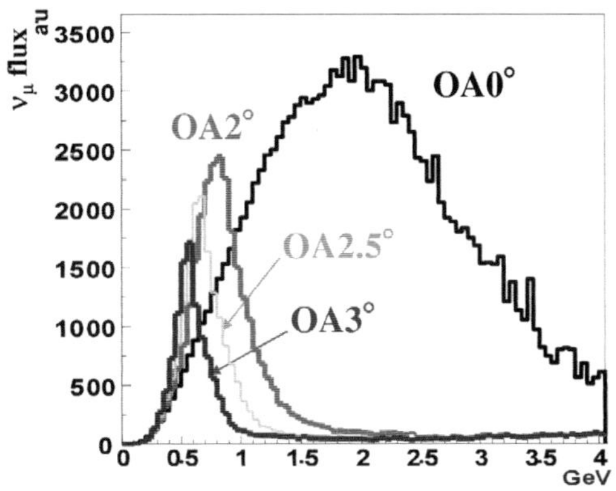

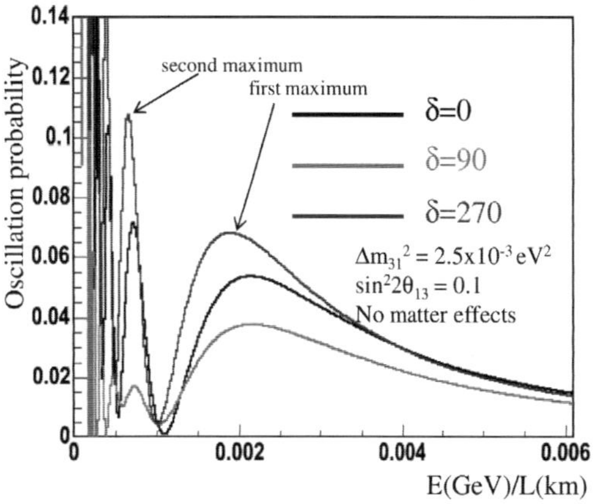

Figure 2. Neutrino flux for various off-axis angle (left) and probability for $\nu_\mu \rightarrow \nu_e$ oscillations as a function of the E(GeV)/L(km) for various δ_{CP} (right).

Figure 2 (left) shows neutrino flux for various off-axis angles. If one selects on-axis setting, 1) wide energy coverage is foreseen which is necessary to cover the 1st and 2nd maximum simultaneously, and 2) measurement suffers from severe π^0 background originated from high energy neutrino which requires the detector with high performance discrimination ability between π^0 and electron. On the other hand, if one selects off-axis setting, 1) requirement for π^0 background discrimination is soft, and 2) measurement is essentially counting experiment at the 1st oscillation maximum.

Figure 2 (right) shows the oscillation probability as a function of the E(GeV)/ L(km). If the distance between source and detector is fixed, the curves can be easily translated to that for the expected neutrino energy spectrum of the oscillated events. As can be seen, if the neutrino energy spectrum of the oscillated events could be reconstructed with sufficiently good resolution in order to distinguish first and second maximum, useful information to extract the CP phase would be available even only with a neutrino run. If baseline is set to be long, 1) energy of 2nd oscillation maximum gets measurable. 2) statistical significance may get worse, and 3) measurement is affected by large matter effect. On the other hand, if baseline is set to be short, 1) it is impossible to extract 2nd oscillation maximum information, 2) statistical significance may get better, and 3) measurement is less affected by matter effect.

In addition to above consideration on the measurement of lepton sector CP phase δ_{CP}, discovery potential for proton decay and reality to realize huge detector are also the essential issues to be taken into account, to define far detector option.

4. Optimal configuration for the investigation of lepton sector CP Phase δ_{CP} with Liquid Argon Time Projection Chamber

Optimal experimental set up such as, the length of the baseline, the angle with respect to the neutrino beam axis and the detector technology, highly depends on the way to extract the effects of CP phase as discussed in the previous section. Since the Liquid Argon TPC has an excellent energy resolution for the neutrino energy measurement and event reconstruction capability from sub GeV to a few GeV and from single prong to high multiplicity configuration, it is suitable for spectrum measurement with wide energy coverage. Moreover, since the Liquid Argon TPC has an excellent discrimination capability of π^0 and electron, wide band on-axis beam is tolerable. It should be seriously taken into account the fact that there is an inevitable boundary condition from available accelerator facility, namely, the finite neutrino intensity and the energy range of the neutrino beam (30 GeV 1.66MW KEK/J-PARC MR for Japanese case). Also it is important to obtain experimental result within reasonable time scale of about five years. Thus, the optimal choice for the Liquid Argon TPC for the investigation of lepton sector CP symmetry breaking is the "Measurement of the energy spectrum shape of the appearance oscillated ν_e charged current events (emphasis on the 1st and 2nd oscillation maximum) with on-axis beam with 5 years neutrino beam running". After this first phase measurement, next experimental configuration (e.g. antineutrino mode running) will be decided.

5. Possible scenarios in Europe

The FP7 Design Study of a pan-European Infrastructure for Large Apparatus studying Grand Unification and Neutrino Astrophysics (LAGUNA) [2] is a collaborative project involving 21 beneficiaries, composed of academic institutions from Denmark, Finland, France, Germany, Poland, Spain, Switzerland, the United Kingdom as well as industrial beneficiaries specialized in civil and mechanical engineering and rock mechanics. LAGUNA brings together on one hand the scientific community interested in this kind of research, and on the other hand the industrial and technical experts able to help assess the feasibility of this infrastructure.

The principal goal of LAGUNA is to assess the feasibility of a new international research infrastructure in Europe able to host the next generation, very large mass, deep-underground neutrino observatory, with a volume in the range of 100,000 to 1,000,000 m^3. Europe currently hosts four world-class national deep underground laboratories –with high-levels of technical expertise– that are located in Boulby (UK), Canfranc (Spain), Gran Sasso (Italy), and Modane (France). However, none of these is large enough for the detector envisioned by LAGUNA. Therefore, in addition to the possible extensions of the existing underground laboratories, the LAGUNA consortium has been studying the creation of new laboratories in the region of Umbria (Italy), Pyhäsalmi (Finland), Sierozsowice (Poland) and Slanic (Romania).

The LAGUNA design study considers three different underground neutrino detector technologies located at the several potential underground sites cited above, in order to identify the scientifically and technically most appropriate and cost-effective strategy for future large-scale underground detectors in Europe. The first detector option (MEMPHYS [3]) is based on the well-established Water Cerenkov Imaging technology developed by IMB in USA and Kamiokande/SuperKamiokande in Japan. The second option (LENA [4]) considers a very large non-segmented liquid scintillator detector. The third is based on the next generation liquid argon time projection chamber technique (GLACIER [5]). At this stage, the three gliquidh detectors are studied in parallel, exploiting synergies and common issues.

After three years of detailed studies that led among others to the publication of seven reports describing the feasibility of the LAGUNA infrastructure in each of the pre-selected site the Pyhäsalmi site has been identified by the LAGUNA consortium as a high-priority potential site together with the Fréjus site. Both sites will be further investigated in the context of the recently approved LAGUNA-LBNO study (2011-2014) [6].

LAGUNA is high priority project of the Roadmap of Astroparticle Physics (ASPERA) which states gWe recommend that a new large European infrastructure is put forward as a future international multi-purpose facility on the 100-1000 ktons scale for improved studies of proton decay...h and gThe three detection techniques being studied for such large detectors in Europe, Water Cherenkov, Liquid Scintillator and Liquid Argon, should be evaluated in the context of a common design study which should also address the underground infrastructure and the possibility of an eventual detection of future accelerator neutrino beams.h. At the same time, the possibility to couple the far detector to a high intensity neutrino beam from CERN clearly places the LAGUNA project at the interface with the CERN European Strategy Update to be delivered by the end of 2012. As such the LAGUNA project constitutes a high astro-particle physics priority to be discussed within the CERN strategy update process.

6. Possible scenario in USA

There is a proposal named the Long-Baseline Neutrino Experiment (LBNE) [7] which utilize a high intensity neutrino source at Fermi National Accelerator Laboratory (Fermilab) aiming at 20 kt Liquid Argon TPC located at the proposed Deep Underground Science and Enginnering Laboratory (DUSEL) 1300km away from Fermilab.

7. Possible scenario in Japan: J-PARC to Okinoshima long baseline neutrino experiment

The study of possible new generation discovery experiments with KEK/J-PARC neutrino beam was initiated at the 4th International Workshop on Nuclear and Particle Physics at J-PARC (NP08) [8]. With the same configuration as T2K (2.5° off-axis angle), the center of the neutrino beam will go through underground beneath SK (295 km baseline), and will automatically reach the Okinoshima island region (658 km baseline) with an off-axis angle 0.76° (almost on-axis).

7.1. KEK/J-PARC and Main Ring Synchrotron

KEK/J-PARC (Figure 3) is a KEK-JAEA joint facility for MW-class high intensity proton accelerator. It provides unprecedented high flux of various secondary particles, such as neutrons, muons, pions, kaons, and neutrinos, which are utilized for elementary particle physics and material and life science.

In the accelerator complex, H^- ions are accelerated to 181 MeV with LINAC, fed into Rapid Cycling Synchrotron (RCS) with stripping out electrons and are accelerated to 3 GeV. At final stage, proton beam goes into Main Ring Synchrotron (MR) and accelerated to 30 GeV. For the neutrino experiment, accelerated protons are kicked inward to neutrino beam facility by single turn with fast extraction devices. Main characteristics of MR is described in Figure 4.

7.2. KEK/J-PARC neutrino beam facility

The proton beam from MR run through KEK/J-PARC neutrino beam facility and producing intense muon neutrinos toward the west direction. KEK/J-PARC neutrino beam facility is

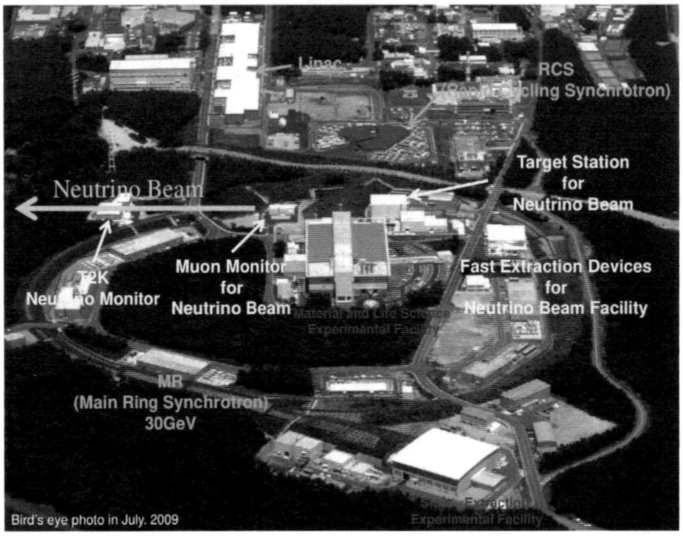

Figure 3. KEK/J-PARC accelerator and experimental facility.

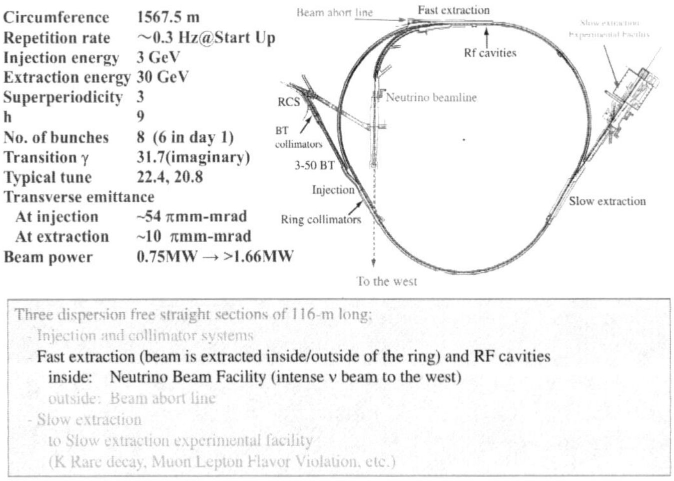

Figure 4. The overview of the KEK/J-PARC MR.

composed of following parts with their functionalities (Figure 5).

- Preparation section: Match the beam optics to the arc section.
- Arc section: Bend the beam $\sim 90^\circ$ toward the west direction with superconducting combined function magnet.

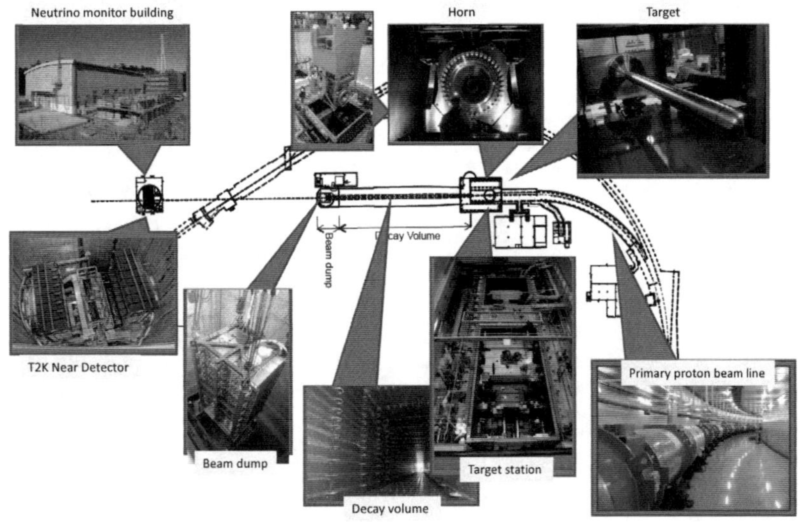

Figure 5. KEK/J-PARC neutrino beam facility.

- Final focus section: Match the beam optics to target both in position and in profile. Level of mm control is necessary which corresponds to 1 mrad ν direction difference, also not to destroy target.
- Graphite target and horn magnet: Produce intense secondary π's and focus them to the west direction. (3 horns system with 320 kA pulse operation)
- Muon monitor: Monitor μ direction (= ν direction), pulse to pulse, with measuring center of muon profile.
- On-axis neutrino monitor (INGRID): Monitor ν direction and intensity.

This facility is designed so that up to ~2 MW beam power is tolerable. The limitation is due to temperature rise and thermal shock for the components such as Al horn, graphite target, and Ti vacuum window. Since everywhere suffers from high radiation, careful treatment of radioactive water and air (~10 GBq/3weeks) is required. Moreover, maintenance scenario of radio active components has to be seriously planned.

7.3. KEK/J-PARC neutrino beam upgrade plan

As for the neutrino beam intensity improvement, MR power improvement scenario toward MW-class power frontier machine, KEK Roadmap plan, is analyzed and proposed by the KEK/J-PARC accelerator team as shown in Table 1.

The plan will be realized with the combination of high repetition cycle and high beam density. Items to be modified from start up toward high intensity are listed as following.

- Number of bunches in MR should be increased from 6 to 8. For this purpose, fast rise time extraction kicker magnet have to be prepared. Its installation is foreseen in 2010 summer.
- Repetition cycle of MR has to be improved from 3.5 seconds to <1.92 seconds. For this purpose RF and magnet power supply improvement is necessary.
- RCS operation with harmonic number 1 has to be conducted. This is to make the beam bunch to be longer in time domain to decrease space charge effect. For this purpose RF improvement is necessary. When RCS is operated with harmonic number 2, beam is injected

Table 1. MR power improvement scenario toward MW-class power frontier machine (KEK Roadmap).

	Achieved(March2011)	Next Step	**KEK Roadmap**
Power (MW)	0.145	0.45	**>1.66**
Energy (GeV)	30	30	**30**
Rep. Cycle (sec.)	3.04	2.2	**1.92 ~ 0.5**
No. of Bunches	8	8	**8**
Particles/Bunch	1.2×10^{13}	2.5×10^{13}	$\mathbf{(4.1 \sim 8.3)\times10^{13}}$
Particles/Ring	9.2×10^{13}	2.0×10^{14}	$\mathbf{(3.3 \sim 6.7)\times10^{14}}$
LINAC (MeV)	181	181	**400**
RCS[a]	h=2	h=2	**h=2 or 1**

[a] Harmonic number of RCS

to MR with 2 bunches × 4 cycles. On the other hand, when RCS is operated with harmonic number 1, beam is injected to MR with single bunch with doubled number of protons × 8 cycles.

- LINAC 400 MeV operation is required to avoid severe space charge effect at RCS injection. Construction of necessary component is already approved and started.

7.4. J-PARC to Okinoshima long baseline neutrino experiment
The scenario is depicted in Figure 6 [9][10]. In order to cover a wider energy range, detector

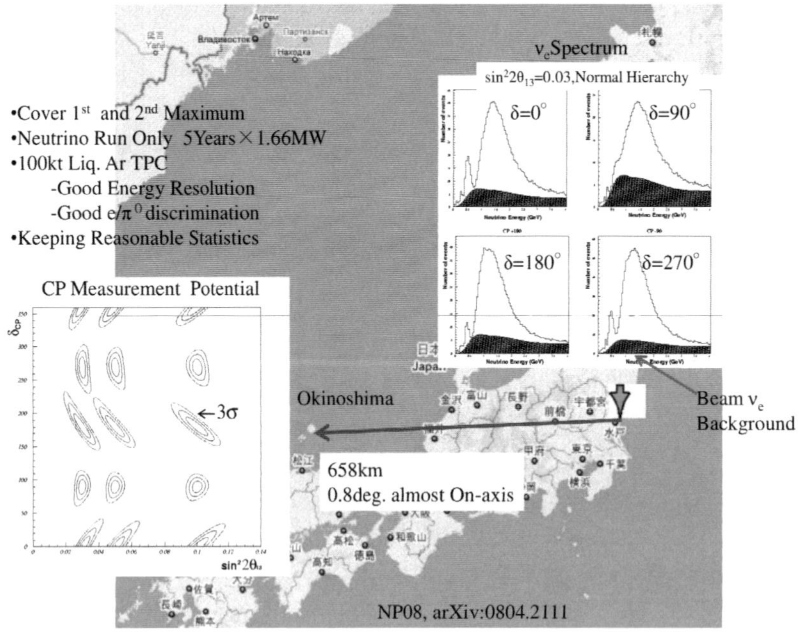

Figure 6. J-PARC to Okinoshima long baseline neutrino experiment.

location which is near on-axis is favored. If one assumes that the second oscillation maximum has to be located at an energy larger than about 400 MeV, the baseline should be longer than about 600 km. In addition, in order to collect enough statistics, baseline should not be too much longer than above stated. Taking into account all of the above mentioned conditions, the Okinoshima region (658 km baseline and almost on-axis (0.76° off-axis) configuration) turns out to be ideal.

Analysis here based on the assumption using a neutrino run only during five years to be reasonable time duration for the single experiment (10^7 seconds running period/year is assumed), under the best KEK/J-PARC beam assumption. An anti-neutrino beam (opposite horn polarity) might be considered in a second stage in order to cross-check the results obtained with the neutrino run (in particular for mass hierarchy problem). Detector is assumed to be a 100 kton Liquid Argon Time Projection Chamber. This type of detector is supposed to provide higher precision than other huge detectors to separate the two peaks in energy spectrum. In addition, the π^0 background is expected to be highly suppressed thanks to the fine granularity of the readout, hence the main irreducible background will be the intrinsic ν_e component of the beam.

The right hand side plot in Fig. 6 shows the energy spectra of electron neutrino at the cases of δ_{CP} equal 0°, 90°, 180°, 270°, respectively. Shaded region is common for all plots and it shows the background from beam ν_e. Here perfect resolution is assumed. As shown, the value of δ_{CP} varies the energy spectrum, especially the first and the second oscillation peaks (heights and positions), therefore comparison of the peaks determine the value δ_{CP}, while the value of $\sin^2 2\theta_{13}$ changes number of events predominantly. Allowed regions in the perfect resolution case are shown in left hand side of Fig. 6. Twelve allowed regions are overlaid for twelve true values, $\sin^2 2\theta_{13}$=0.1, 0.05, 0.02, and δ_{CP}=0°, 90°, 180°, 270°, respectively. The δ_{CP} sensitivity is 20∼30° depending on the true δ_{CP} value [9][10].

8. World wide consensus and the motivation of the workshop

There is no question about importance of 100 kt Liquid Argon Time Projection Chamber. The question is how to realize it. The first International Workshop towards the Giant Liquid Argon Charge Imaging Experiment (GLA2010) is the place to discuss how to realize 100 kt Liquid Argon Time Projection Chamber.

9. The aim of the GLA2010

The aim of the GLA2010 is to bring together researchers having common interest in realizing a giant neutrino observatory based on the liquid Argon time projection chamber technology combining next-generation searches for proton decay and neutrino physics with natural and artificial sources. The workshop will review the current worldwide efforts towards large liquid Argon detectors and aims at fostering collaborations on the medium and long time scales.

10. The skeleton of the GLA2010

The skelton of the GLA2010 program is as follows.

- Learn experiences thoroughly, and integrate presently available knowledge
  - Special session the ICARUS Liquid Argon TPC
  - Lessons on Liquid Argon Charge Imaging technology from ongoing developments
  - Lessons from Xe based Liquids Imaging detectors
  - Studies on physics performance
- Push present efforts forward coherently
  - Ways to improve the Liquid Argon Charge Imaging technology
- Toward realization of Giant Liquid Argon TPC
  - Localization studies

  - Future steps towards the realization of Giant Liquid Argon Charge Imaging detectors
- Needless to say, based on physics argument
  - Main goals of Giant Liquid Argon Charge Imaging Experiments

The workshop starts with the talk on the pioneering and world leading experiences and efforts by the ICARUS Liquid Argon project presented by Professor Carlo Rubbia.

References

[1] G. Steigman Ann.Rev.Astron.Astrophys, 14(1976)339; A.D. Sakharov Pisma Zh.ETF 5(1967)32.
[2] Coordinated by A.Rubbia, "Design of a pan-European Infrastructure for Large Apparatus Studying Grand Unification and Neutrino Astrophysics,"Design Study proposal (LAGUNA) for FP7 INFRASTRUCTURE, http://www.laguna-science.eu/.
[3] A. de Bellefon et al., arXiv:hep-ex/0607026.
[4] M.Wurm et al., arXiv:1104.5620.
[5] A.Rubbia, arXiv:hep-ph/0402110.
[6] Coordinated by A.Rubbia, "Design of a pan-European Infrastructure for Large Apparatus Studying Grand Unification, Neutrino Astrophysics and Long Baseline Neutrino Oscillations,"Design Study proposal (LAGUNA-LBNO) for FP7 INFRASTRUCTURE.
[7] http://lbne.fnal.gov/.
[8] Talk given at the 4th International Workshop on Nuclear and Particle Physics at J-PARC (NP08). Slides available at http://j-parc.jp/NP08/.
[9] A. Badertscher, T. Hasegawa, T. Kobayashi, A. Marchionni, A. Meregaglia, T. Maruyama, K. Nishikawa and A. Rubbia, " A Possible Future Long Baseline Neutrino and Nucleon Decay Experiment with a 100 kton Liquid Argon TPC at Okinoshima using J-PARC Neutrino Facility," arXiv:0804.2111.
[10] A. Badertscher et al., P32: Towards a Long Baseline Neutrino and Nucleon Decay Experiment with a next-generation 100 kton Liquid Argon TPC detector at Okinoshima and an intensity upgraded JPARC Neutrino beam, presented at the 9th Program Advisory Committee (PAC) for Nuclear and Particle Physics Experiments at the J-PARC 50GeV Proton Synchrotron (J-PARC PAC), see http://j-parc.jp/NuclPart/pac 1001/pdf/KEK J-PARC-PAC2009-10.pdf.

The ICARUS Liquid Argon TPC: A neutrino "bubble chamber" after Gargamelle

Carlo Rubbia

CERN, European Laboratory for Particle Physics, CH-1211 Geneve 23, Switzerland

Laboratori Nazionali del Gran Sasso dell'INFN, I-67010 Assergi (AQ), Italy

E-mail: carlo.rubbia@cern.ch

Abstract. The success of the bubble chamber in high energy physics was due to two fundamental characteristics: (1) it provided a massive target (2) with complete imaging and reconstruction of the events. In this context, the Liquid Argon Time Projection Chamber can be considered as a modern fully electronic "bubble chamber". In this talk, we review the developments that led to the ICARUS T600 detector.

The need of visual detectors of high target density

- The success of the bubble chamber as a main tool in high energy fixed target physics is due to two characteristics:
  - It provides a massive target, of substantial density
  - It provides complete imaging and reconstruction of the events in itself
- This technology has permitted in the past very substantial advances based on :
  - Single events with complete reconstruction (e.g. discovery of Ω-)
  - Surprise events, i.e. topologies not a priori expected (e.g. Gargamelle neutral currents)
- Technology is costly and complicated
  - It requires high pressures and mechanical expansion
  - Its sensitivity is limited to about a few milliseconds
  - Optics limits viewing of large volumes

Rubbia, KEK_March2010 Slide: 2

A Gargamelle neutrino event

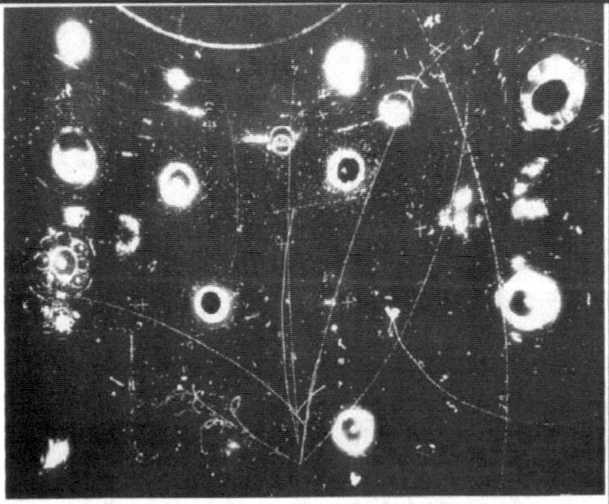

Figure 1 (a) Photograph of an event having e^+V^0 in the final state. The positron is unambiguously identified by bremsstrahlung.

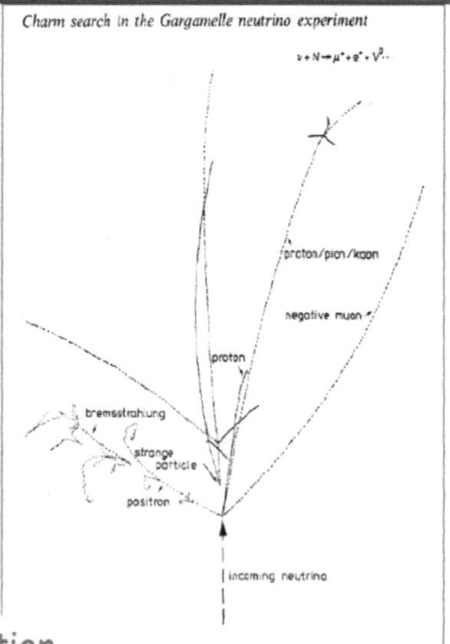

(b) Sketch of the event shown in figure 1(a).

Charm production in a neutrino interaction
The total visible energy is 3.58 GeV.

Rubbia, KEK_March2010 Slide: 3

Journal of Physics: Conference Series **308** (2011) 012002 doi:10.1088/1742-6596/308/1/012002

The ICARUS initial idea

- A multi-kton detector based on a new powerful detection technique:
  - the Liquid Argon Time Projection Chamber
    - [C. Rubbia: CERN-EP/77-08 (1977)]
  - first proposed to INFN in 1985
    - [ICARUS: Imaging Cosmics And Rare Underground Signals: INFN/AE-85/7]
  - capable of providing a 3D imaging of any ionizing event ("electronic bubble chamber") with in addition:
    - continuously sensitive
    - Operating without pressure, eventually underground
    - high granularity (~mm)
    - self triggering
    - excellent calorimetric properties
    - particle identification (through dE/dx vs range)

Rubbia, KEK_March2010 Slide: 4

Noble liquid as tracking medium

- Ideal material for detection of ionizing radiation:
  - Dense (≈ g/cm^3), homogeneous, target and detector
  - Does not attach electrons (long drift paths in liquid phase)
  - High electron mobility (≈quasi-free drift electrons)
  - Commercially easy to obtain (in particular, liquid Argon)
  - Can be made very pure, with many impurities freezing out
  - Inert, not flammable

Argon constitutes 0.9 % of air content

Type	Density (ρ/cm^3)	Energy loss dE/dx (MeV/cm)	Radiation length X_0 (cm)	Collision length λ (cm)	Boiling point @ 1 bar (K)	Thermal electron mobility (cm^2/Vs)	
Neon	1.2	1.4	24	80	27.1	high&low	
Argon	1.4	2.1	14	80	87.3	500	€
Krypton	2.4	3.0	4.9	29	120	1200	€€
Xenon	3.0	3.8	2.8	34	165	2200	€€€

Rubbia, KEK_March2010 Slide: 5

Processes induced by charged particles in liquid Argon

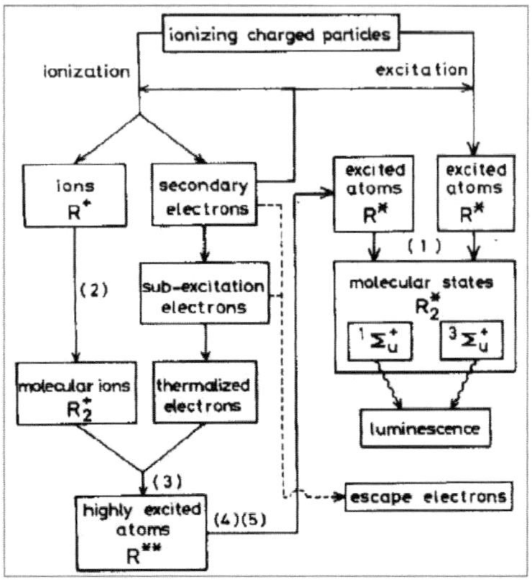

M. Suzuki et al., NIM 192 (1982) 565

When a charged particle traverses medium:

- Ionization process
- Scintillation (luminescence)
  - UV spectrum (λ=128 mn)
  - Not energetic enough to further ionize, hence, argon is transparent
  - Rayleigh-scattering
- Cerenkov light (if fast particle)

→ *UV light*

→ *Charge*

→ *Cerenkov light (if $\beta > 1/n$)*

The pioneering work of T.DOKE must be acknowledged

Rubbia, KEK_March2010 Slide: 6

Signals from dielectric liquids

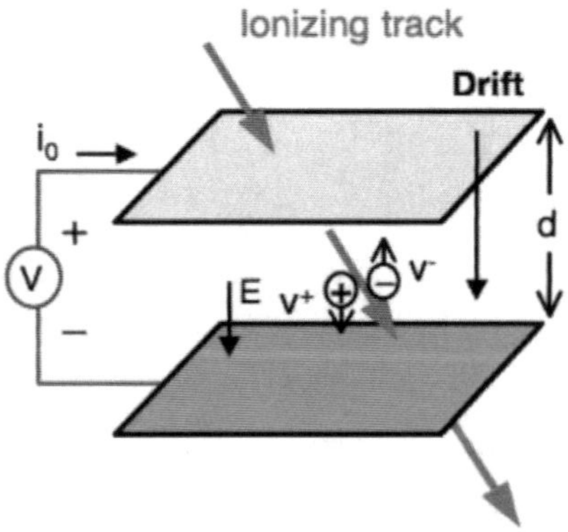

The "work" performed by the power supply which puts electrons and ions in movement is given by:

$$dW = eE(v^+ + v^-)dt = Vi_0 dt$$

from which the current and charge are:

$$i_0 = e\,(v^+ + v^-)/d$$

$$Q^{\pm} = e(d - x^{\pm})/d$$

since $v^- / v^+ \approx 10^5$, electron current is dominant.

If a grid is inserted on the path of the electron, the work changes sign under traversal and so does the current, with a total charge collected equal to zero

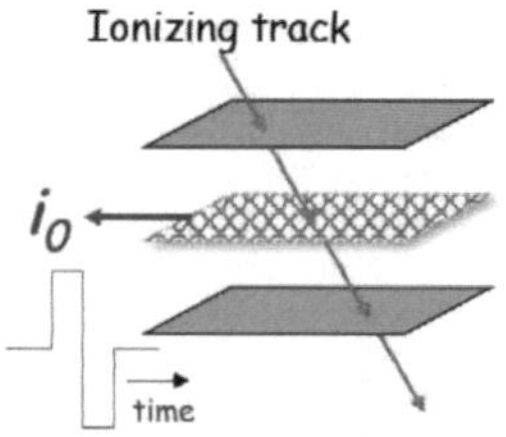

Rubbia, KEK_March2010 Slide: 7

Journal of Physics: Conference Series **308** (2011) 012002 doi:10.1088/1742-6596/308/1/012002

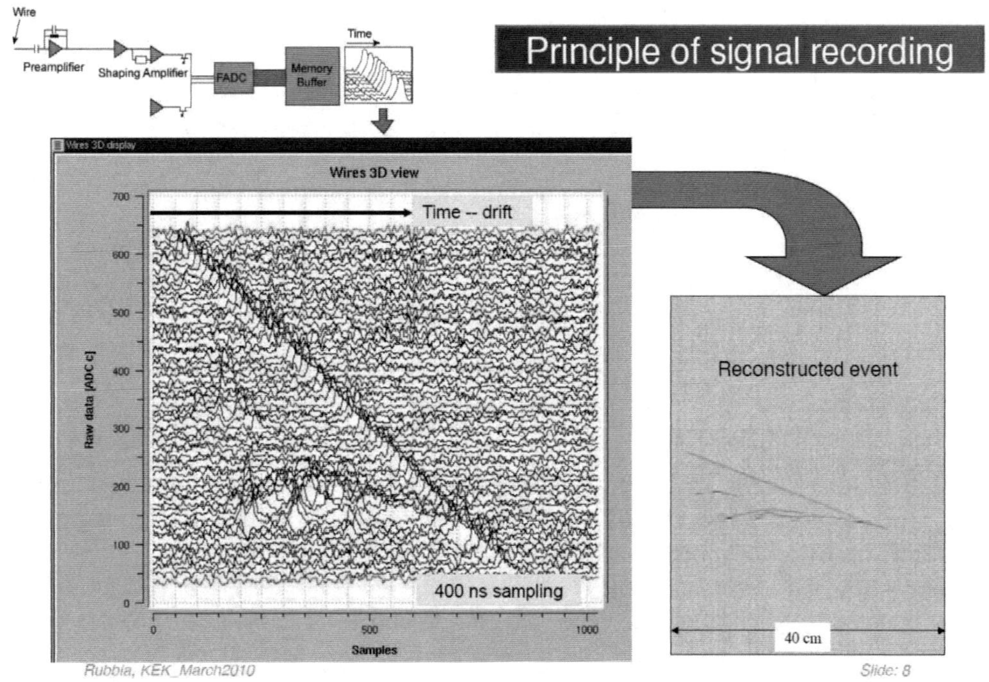

Signal extraction procedure

Collection plane wire analysis: charge = signal area

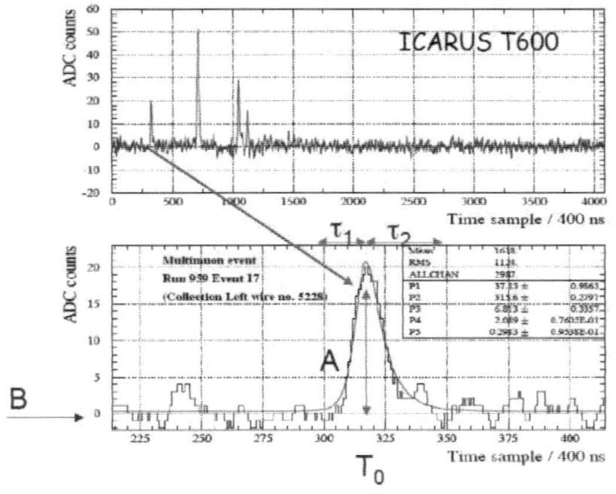

$$f(t) = B + A \frac{e^{-\frac{t-T_0}{\tau_2}}}{1 + e^{-\frac{t-T_0}{\tau_1}}}$$

B = baseline
A = amplitude
τ_1 , τ_2 = rise and fall time
T_0 = peak position

The same empirical function used to fit Ionization hits and test pulse hits (calibration)

Rubbia, KEK_March2010 Slide: 9

Non distructive readout: the induction signals (3D)

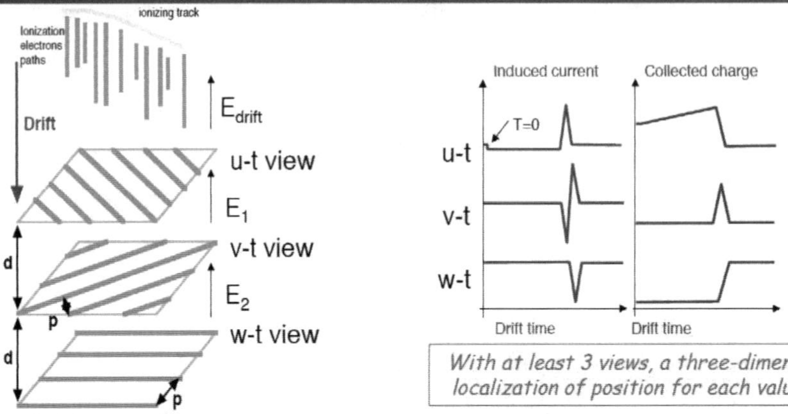

With at least 3 views, a three-dimensional localization of position for each value of t

The drifting electron is traversing several transparent wires arrays oriented in the direction of the required view. The ionizing track records in each of them a *triangular induction signal*.

Finally the electron charge is *collected* by the collection plane.

The generated topological view of the event is the one seen by a camera at infinity with the optical axis in the direction of the wires.

Rubbia, KEK_March2010 Slide: 10

Grid shielding and transparency

- Best signal localization and charge collection is realized if each grid perfectly shields the adjacent electrodes.
  - Shielding inefficiency σ is a pure geometrical effect:

$$\sigma = \frac{p}{2\pi d}\ln\left(\frac{p}{2\pi r}\right) \quad p = 3mm; d = 3mm; r = 75\mu m \Rightarrow \sigma = 18\%$$

    - Depends on pitch p, wire radius r and distance from other grids d.
  - Full transparency is required not to loose signals along the path before reaching the collecting grid

$$\frac{E_1}{E_2} > \frac{1+\rho}{1-\rho} \qquad \rho = \frac{2\pi r}{p} \qquad \frac{E_1}{E_2} > 1.4 \Rightarrow \text{Full transparency}$$

    - Depends on E-field ratio across the grid
    - For equal fields loss is twice the geometrical cross-sect

Rubbia, KEK_March2010 Slide: 11

Free electron drift velocity in LAr

- How fast can free electron drift in Liquid Argon ?

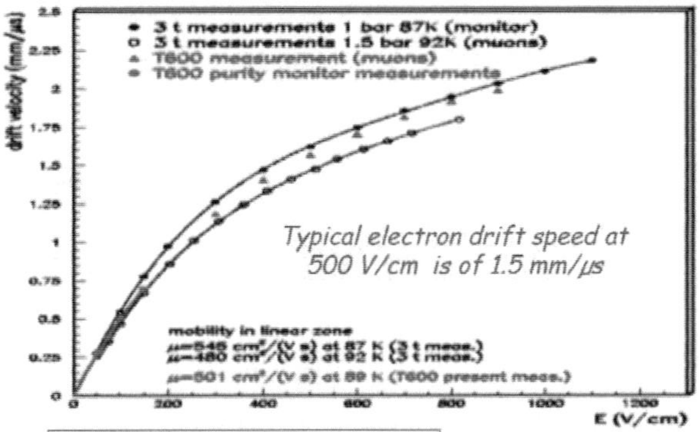

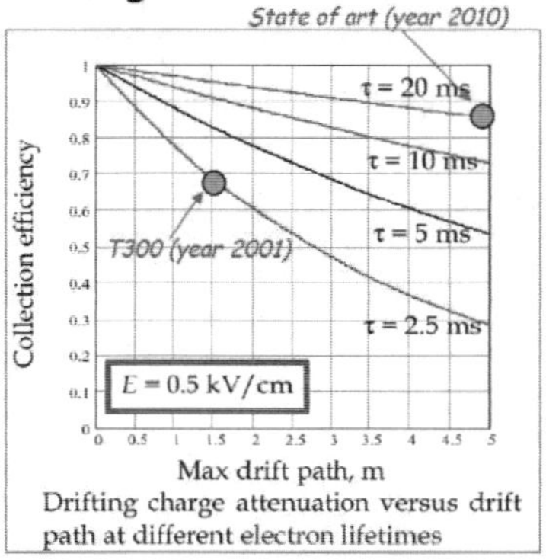

Drifting charge attenuation versus drift path at different electron lifetimes

Fit Function: $v_D = \dfrac{\mu \cdot E}{1+\beta \cdot E}$ ← Electron mobility, μ = 542 cm²/V/s

$\mu \approx T^{-3/2}$

$$v_{DRIFT} \approx \frac{Ee}{m_e} \cdot \frac{\lambda}{v_T}$$

λ → Mean free path; v_T → Maxwellian velocity

Velocity saturation factor $\beta^{-1} \approx$ 450 V/cm
Accounts for loss of linearity when electron is no longer thermalized

Rubbia, KEK_March2010 Slide: 12

Free electron yield in LAr

- Strong e⁻-ion recombination due to comparable thermalization distance (~140 nm) and e⁻-ion separation.
- Both geminate (Onsager) and columnar (Jaffe) active depending on dE/dx and E.F.
- Observed dE/dx linearity (particle i.d.) and energy resolution
- Additional scintillation light is generated in the recombination process

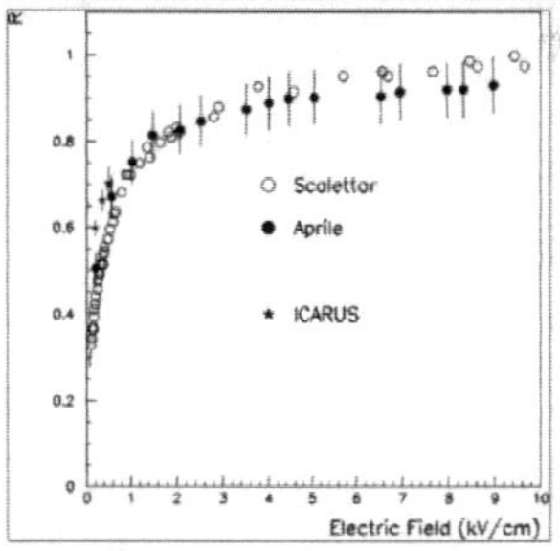

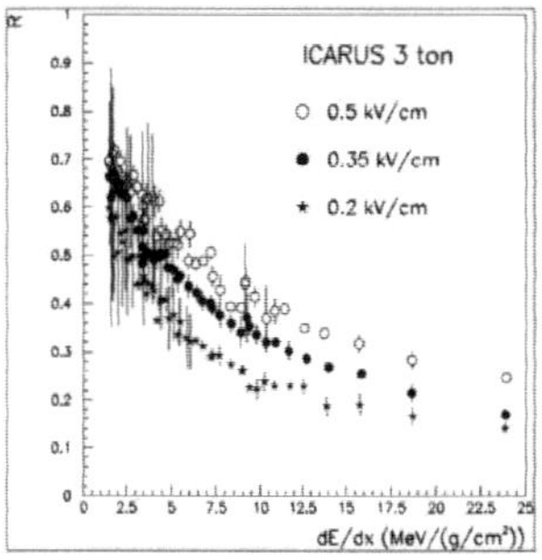

Relative scintillation and electron charge for electrons (m.i.) and alpha particles

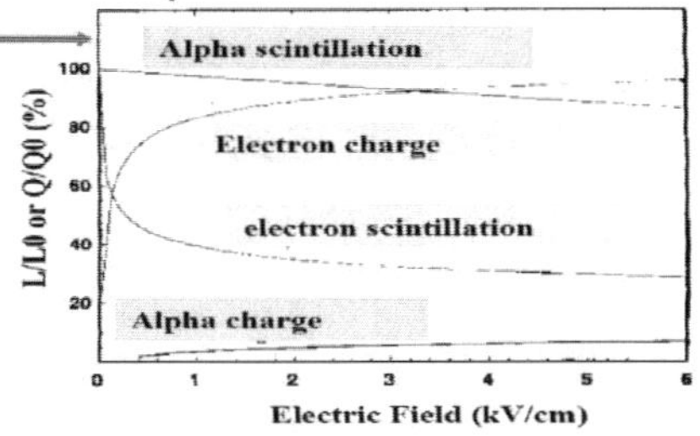

Rubbia, KEK_Marc Slide: 13

Free electron lifetime in LAr

- The density of the Liquid Argon is about 1000 times larger than the one ofArgon gas at pnt. Therefore the mean free path for free electron drift is correspondingly shorter.
- The presence of electron trapping polar impurities can be expressed in equivalent Oxygen molar densities

$$\tau \approx 300\mu s \times \frac{1\,ppb}{N(O_2)}$$

- Because of temperature (87 K) most of the contaminants freeze out spontaneously.
- Main residual contaminants are O_2, H_2O, CO_2 and N_2.
- The goal: 10 ms lifetime for a 30 ppt (t=trillion !) of Oxygen equivalent !

A. Bettini et al., NIM A305 (1991) 177

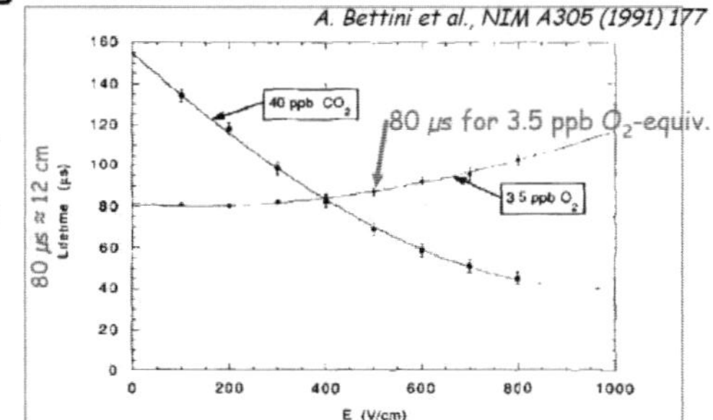

Fig. 12. Electric field dependence of the electron lifetime τ in purified LAr doped with 40 ppb CO_2 and 3.5 ppb O_2. Lines are drawn to guide the eye.

Rubbia, KEK_March2010 Slide: 14

Free electron diffusion in LAr

- At 500v/cm, a 5m drift length corresponds to a drift time of 3.1 ms (1.6 m/ms).
- The intrinsic bubble size (rms diffusion) is given by

$$\sigma_D[mm] = 0.9\sqrt{T_D[ms]}$$

- The values for 5m drift are $\langle\sigma_D\rangle \approx$ 1.1 mm and $\sigma_{max} \approx$ 1.6 mm , tiny with respect to the wire pitch (≥ 2mm).

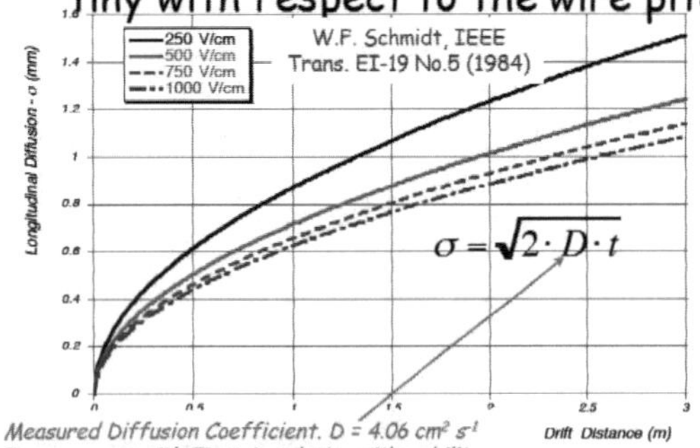

Measured Diffusion Coefficient. D = 4.06 cm² s⁻¹
In agreement with Einstein relation with mobility for quasi thermal electrons.

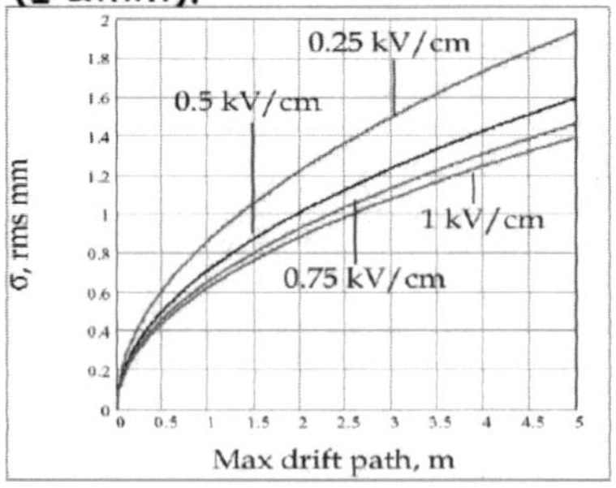

Lateral diff. ~ longitudinal diff.

Rubbia, KEK_March2010 Slide: 15

15

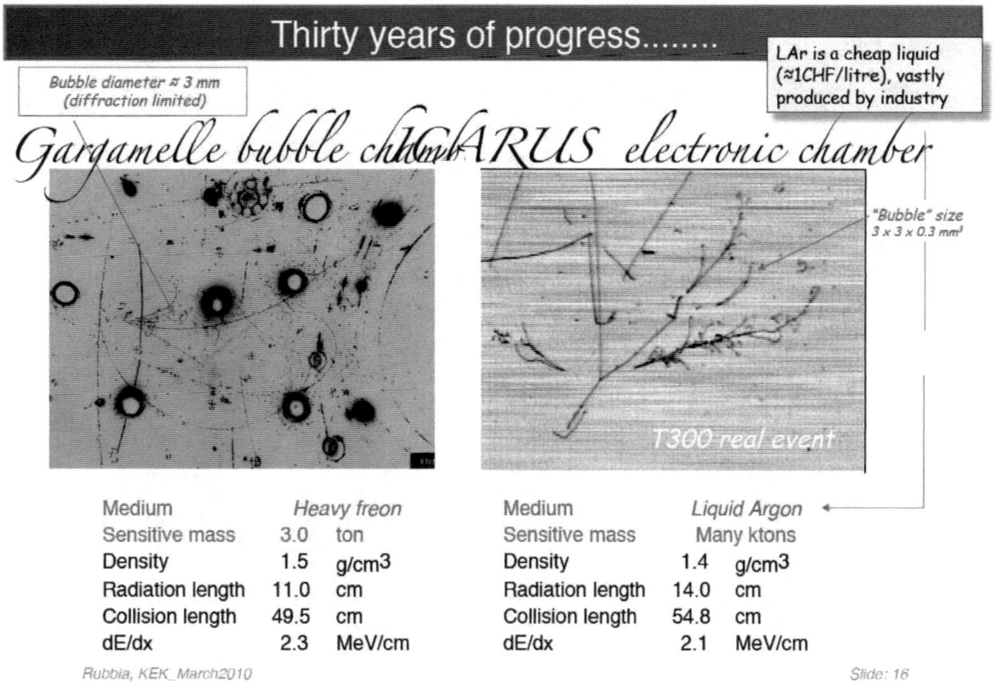
Thirty years of progress........
LAr is a cheap liquid (≈1CHF/litre), vastly produced by industry
Bubble diameter ≈ 3 mm (diffraction limited)
Gargamelle bubble chamber
ICARUS electronic chamber
"Bubble" size 3 x 3 x 0.3 mm³
T300 real event
Medium Heavy freon
Sensitive mass 3.0 ton
Density 1.5 g/cm3
Radiation length 11.0 cm
Collision length 49.5 cm
dE/dx 2.3 MeV/cm
Medium Liquid Argon
Sensitive mass Many ktons
Density 1.4 g/cm3
Radiation length 14.0 cm
Collision length 54.8 cm
dE/dx 2.1 MeV/cm
Rubbia, KEK_March2010
Slide: 16

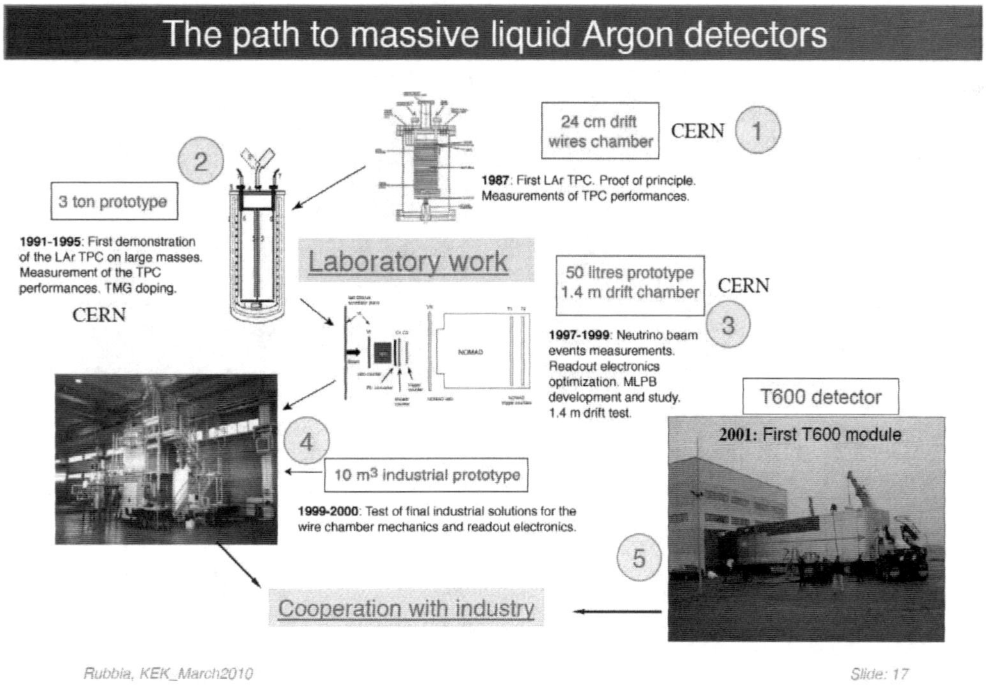
The path to massive liquid Argon detectors
24 cm drift wires chamber
CERN
1
1987: First LAr TPC. Proof of principle. Measurements of TPC performances.
2
3 ton prototype
1991-1995: First demonstration of the LAr TPC on large masses. Measurement of the TPC performances. TMG doping.
CERN
Laboratory work
50 litres prototype 1.4 m drift chamber
CERN
3
1997-1999: Neutrino beam events measurements. Readout electronics optimization. MLPB development and study. 1.4 m drift test.
T600 detector
2001: First T600 module
4
10 m3 industrial prototype
1999-2000: Test of final industrial solutions for the wire chamber mechanics and readout electronics.
5
Cooperation with industry
Rubbia, KEK_March2010
Slide: 17

First T300 cryostat during construction (2001)
≈ 270'000 liters LAr
= T300
Rubbia, KEK_March2010
Slide: 18

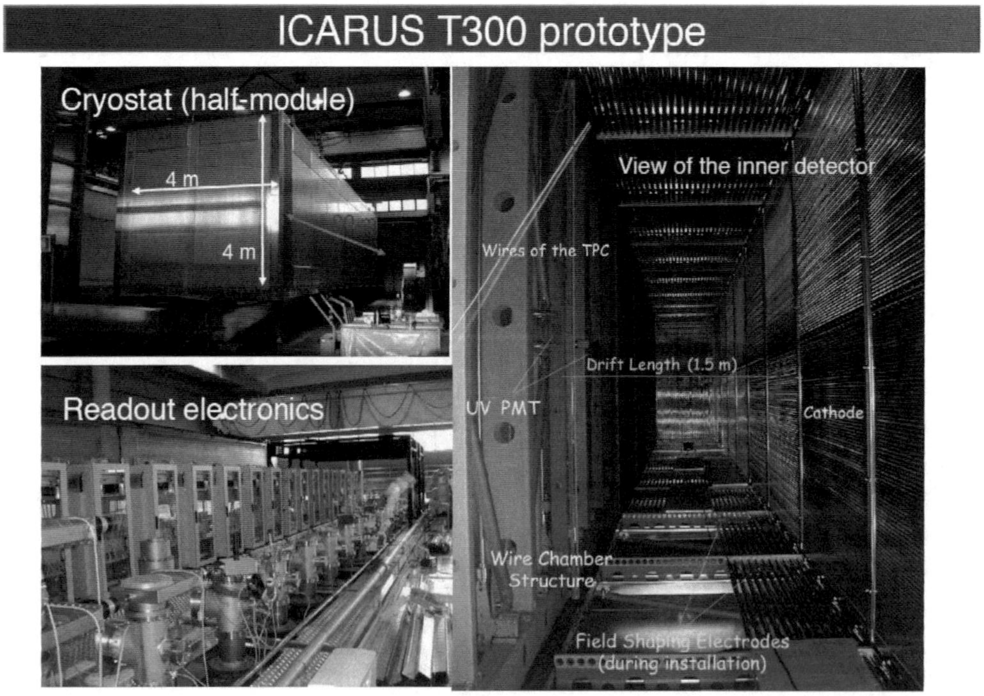
ICARUS T300 prototype
Cryostat (half-module)
4 m
4 m
Readout electronics
View of the inner detector
Wires of the TPC
Drift Length (1.5 m)
UV PMT
Cathode
Wire Chamber Structure
Field Shaping Electrodes (during installation)

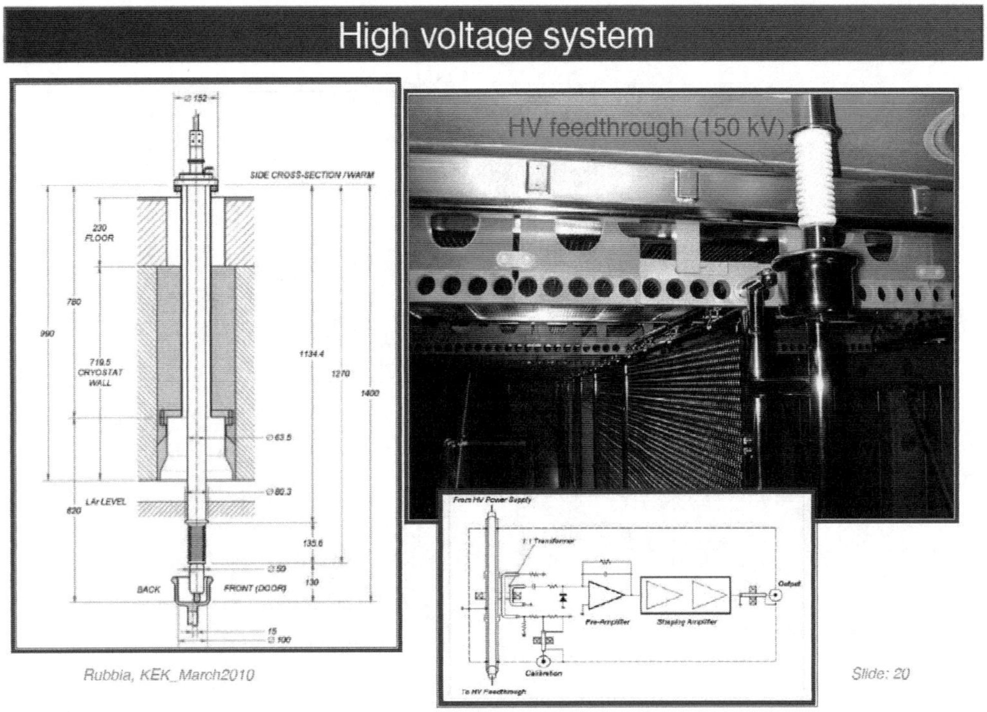
High voltage system
SIDE CROSS-SECTION / WARM
230
FLOOR
780
990
719.5
CRYOSTAT
WALL
LAr LEVEL
820
1134.4
1270
1400
Ø 152
Ø 63.5
Ø 80.3
135.6
Ø 50
130
BACK
FRONT (DOOR)
15
Ø 100
HV feedthrough (150 kV)
Rubbia, KEK_March2010
Slide: 20

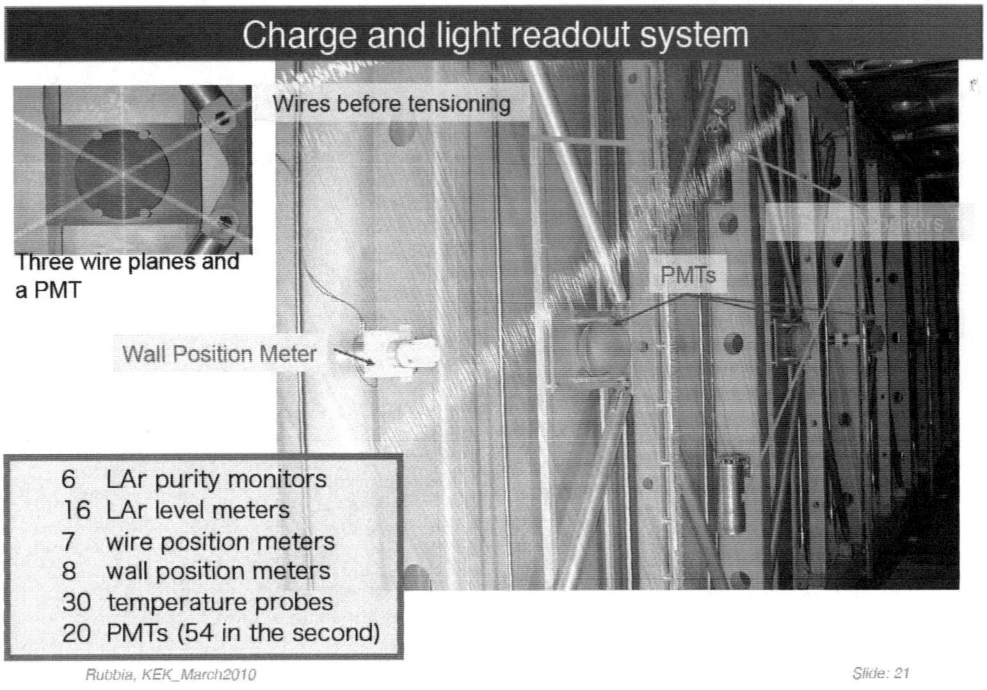
Charge and light readout system
Wires before tensioning
Three wire planes and
a PMT
PMTs
Wall Position Meter
6 LAr purity monitors
16 LAr level meters
7 wire position meters
8 wall position meters
30 temperature probes
20 PMTs (54 in the second)
Rubbia, KEK_March2010
Slide: 21

Electronic racks (54000 ch.)
Rubbia, KEK_March2010
Slide: 22

Technical run in Pavia
Walking on the detector...
August 2001
...and observing cosmic ray events
Rubbia, KEK_March2010
Slide: 23

Filling procedures

- Filling procedure towards the ultra pure (ttb) LAr:
  - Use ultra high vacuum standards for detector components design, construction, cleaning and assembly;
  - Removal of air and outgassing of surfaces by evacuating the argon container volume to the molecular vacuum level ($\approx 10^{-5}$ mbar);
  - Fast cooling (to minimize out-gassing) and filling with ultra-purified LAr;
  - Recirculation/purification of the gas phase to block the diffusion of the impurities from the hot parts of the detector and from micro-leaks on the openings (typically located on the top of the device) in the bulk liquid;
  - Recirculation/purification of the bulk liquid volume to further reduce the impurities concentration.

Rubbia, KEK_March2010 Slide: 24

Purification in the liquid and in the surrounding gas

- According to the kinetic theory, the densities of the gaseous contaminating impurities are approximately the same in the liquid and in the gas regions.
- Since the density of the cold Argon gas is much smaller than in the liquid, the *fractional density* of impurities is correspondingly larger in the gas.
- For ICARUS and at the design rate of 100 Nm^3/h— the gas driven purification speed will be of the order of 1/16 $hour^{-1}$, much faster than the one with the standard liquid purification.
- Incidentally, the liquid/gas/liquid process would require zero energy in the reversible iso-temperature limit with no losses

Liquid recirculation scheme

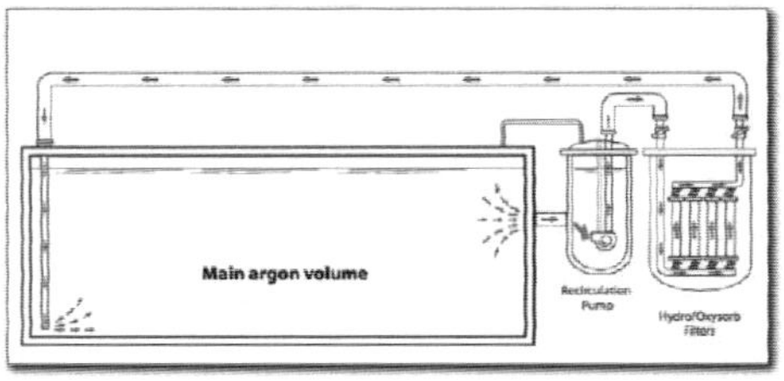

Gas recirculation scheme

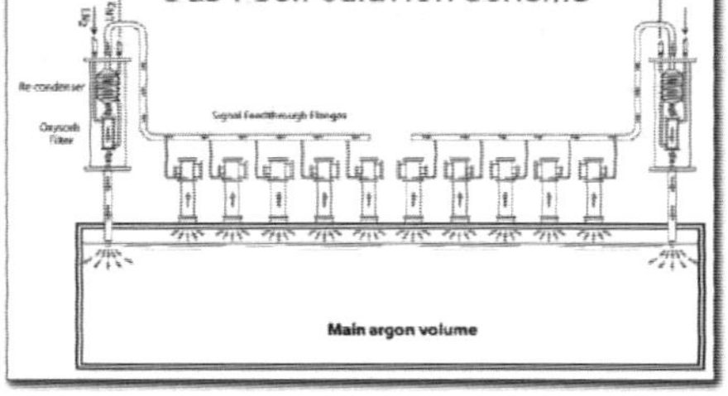

Rubbia, KEK_March2010 Slide: 25

Purification systems for the T600.

Liquid recirculation

Gas recirculation

Rubbia, KEK_March2010 Slide: 26

"Purity": state of the art in 2001

- During 2001 a first real size test of the 1/2 of the T600 detector was performed on surface in Pavia.
- In the detector the drift length was set to 1.5 m, corresponding to a drift time of about 1 ms.
- The purity performance at that time is shown here, measured both with purity monitors and muon tracks.

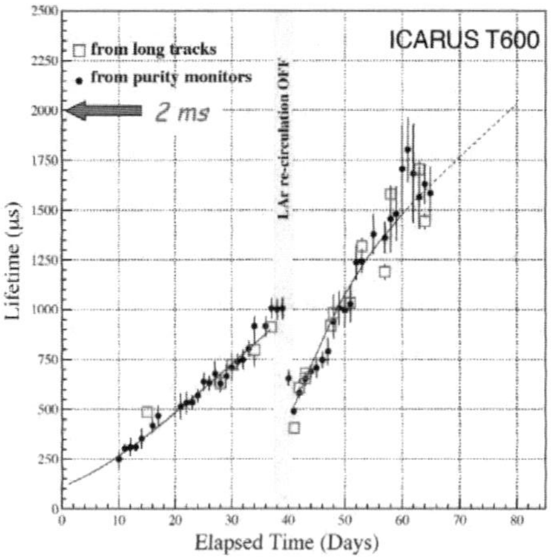

- A considerable progress over the last few years has permitted to reach industrial purification techniques which are capable of a much better performance.

Rubbia, KEK_March2010 Slide: 27

Recent progress in experimental purity achievements

- New industrial purification methods have been developed at an exceptional level, especially remnants of O2 and N2, which have to be initially and continuously purified.
- Extremely high τ_{ele} have been determined with cosmic ν's.
- The short path length used (30 cm) is compensated by the high accuracy in the observation of the specific ionization
- The result here reported is τ_{ele} ≈21 ms corresponding to ≈15 ppt, namely a ≈10^{-11} molecular impurities in Ar

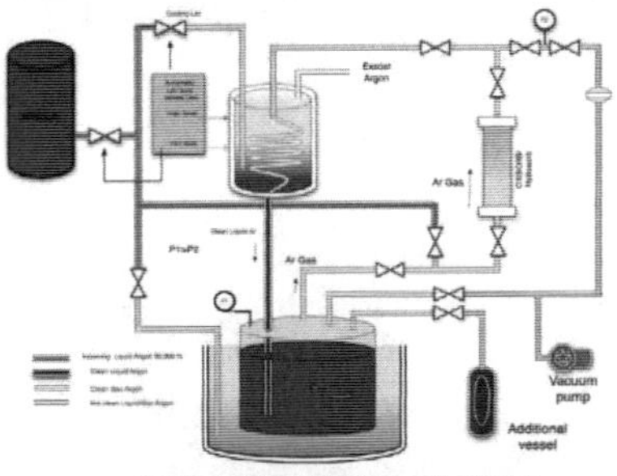

Rubbia, KEK_March2010 Slide: 28

Purity achievement

- The measured value to the experimental τ_{ele} corresponds to an attenuation of about 15 % for a longest drift of 5 meters, opening the way to exceptionally long drift distances.

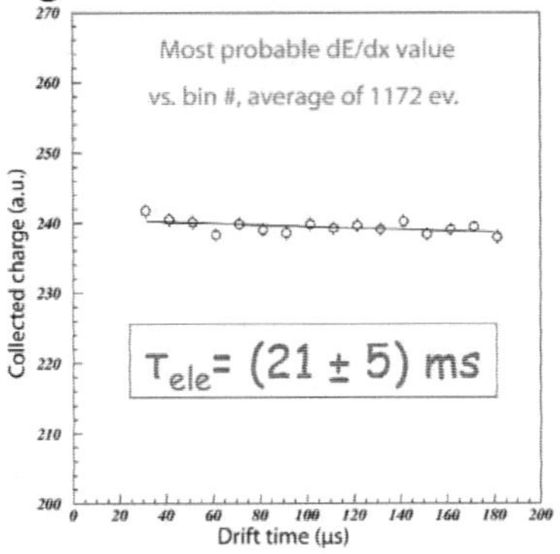

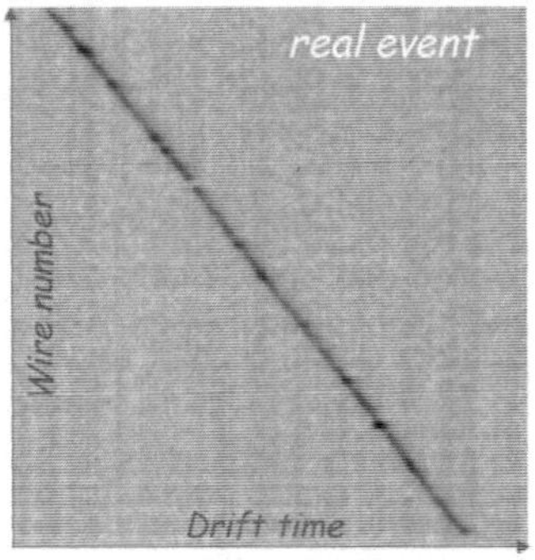

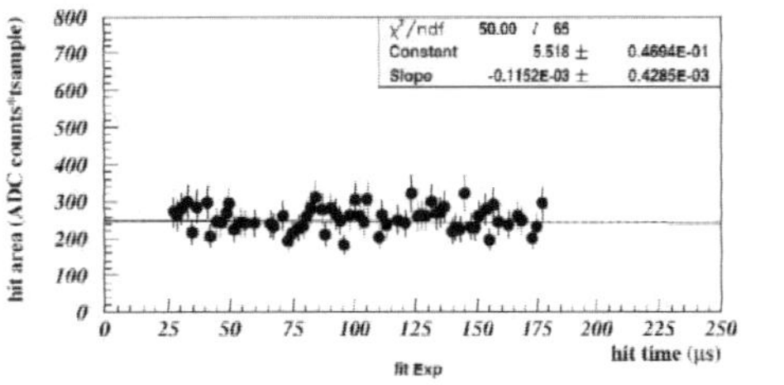

Rubbia, KEK_March2010 Slide: 29

Quasi-elastic neutrino interactions at CERN

- 50 liter LAr-TPC exposed to CERN WANF (NOMAD as muon spectrometer)
  - Full 3D event kinematics reconstruction, particle id, momentum balance, π^0 rejection

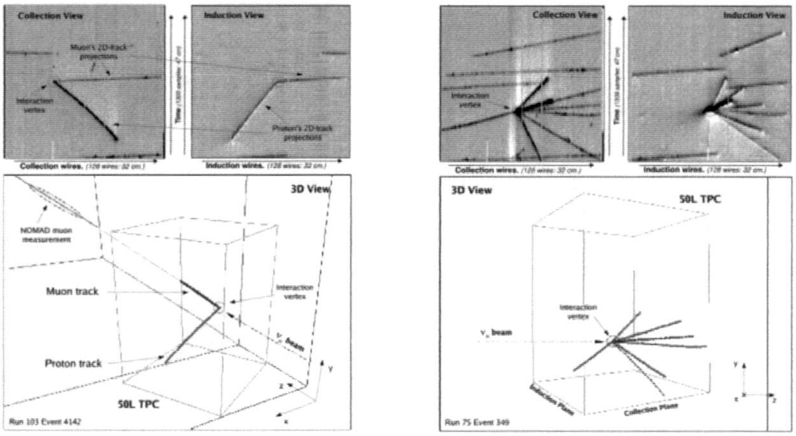

Rubbia, KEK_March2010 Slide: 30

Quasi-elastic event reconstruction at CERN

Selection of ~ 200 pure lepton-proton final state with exactly one proton T_p>50 MeV (range > 2 cm) and any number protons T_p<50 MeV

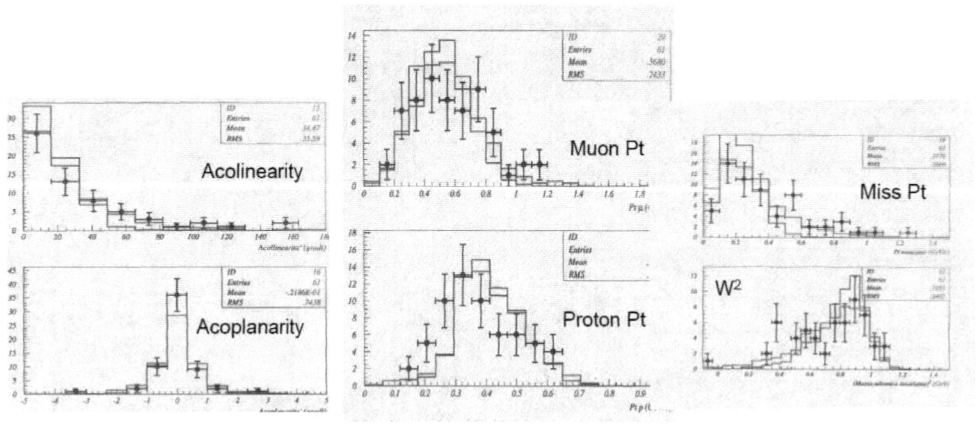

Good agreement with FLUKA expectations (Red line), accounting for Nuclear Fermi motion and re-interactions in nuclei.

Rubbia, KEK_March2010 Slide: 31

Journal of Physics: Conference Series **308** (2011) 012002 doi:10.1088/1742-6596/308/1/012002

Summary of LAr TPC performances

- Tracking device
  - Precise event topology
  - Momentum via multiple scattering
- Measurement of local energy deposition dE/dx
  - e / γ separation (2%X_0 sampling)
  - Particle ID by means of dE/dx vs range
- Total energy reconstruction of the events from charge integration
  - Full sampling, homogeneous calorimeter with excellent accuracy for contained events

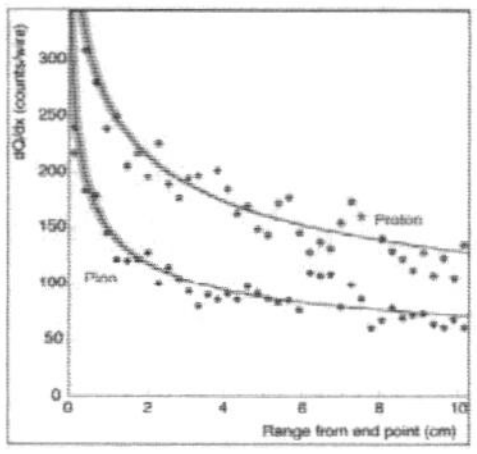

RESOLUTIONS

Low energy electrons: $\sigma(E)/E = 11\% / \sqrt{E(MeV)} + 2\%$

Electromagn. showers: $\sigma(E)/E = 3\% / \sqrt{E(GeV)}$

Hadron shower (pure LAr): $\sigma(E)/E \approx 30\% / \sqrt{E(GeV)}$

Rubbia, KEK_March2010 Slide: 32

Long muon tracks

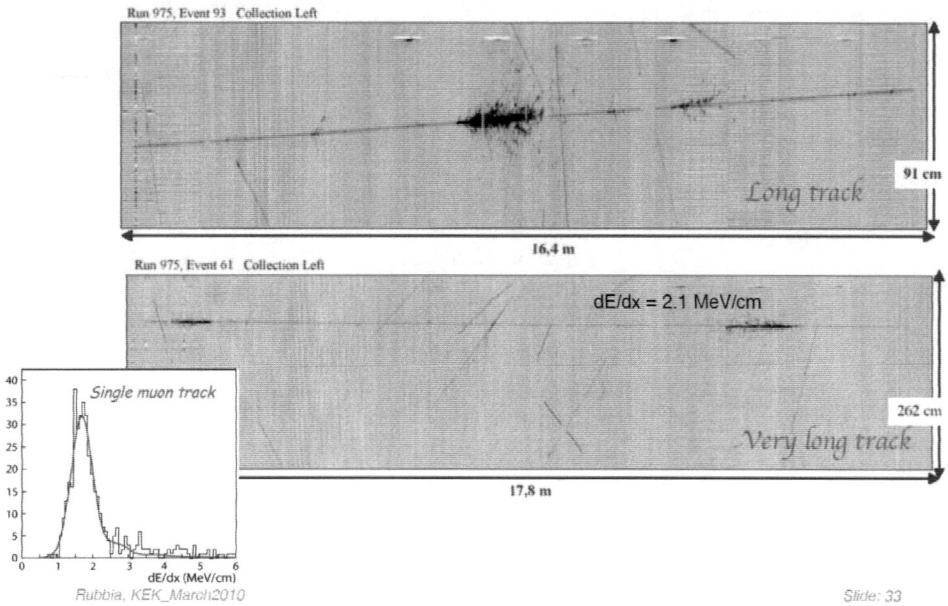

Rubbia, KEK_March2010 Slide: 33

Muon momentum measurement with multiple scattering

In case of non-contained events

Multiple Scattering is the key tool to measure momentum, essential for atmospheric and CNGS ν

- Kalman filter method applied to segmented tracks (L_{seg} = segment length)
- Momentum p extracted from measurement of deflection angle θ and from χ^2 of the fit

Deflection angle contributions:

$$\begin{cases} \theta_{MS} \propto \sqrt{L_{seg}}/p & \leftarrow \text{MS angle} \\ \theta_{det} \propto L_{seg}^{-3/2} & \leftarrow \text{detector resolution} \end{cases}$$

- procedure validated on cosmic ray data (stopping μ's) - extended to higher energy with MC calculations
- Resolution $\Delta p/p$ depends mainly on track length : on μ's from CNGS ν's $\Delta p/p$ ~16%

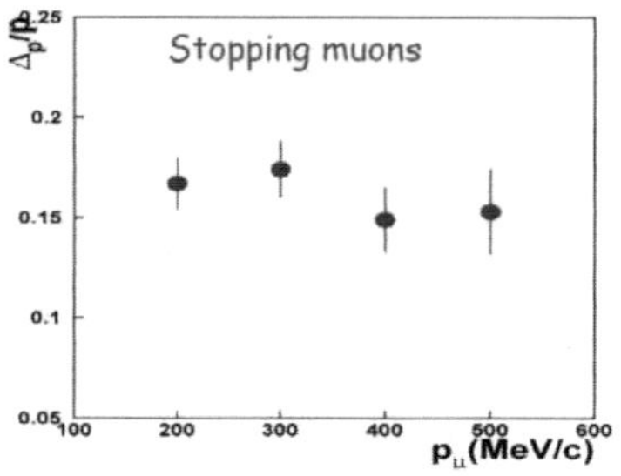

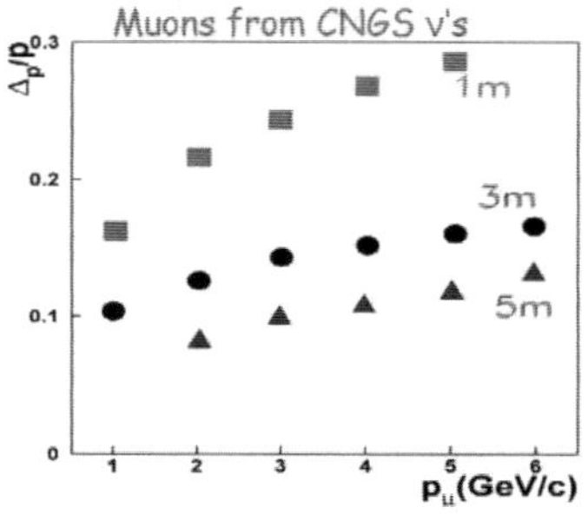

Rubbia, KEK_March2010 Slide: 34

Particle Identification ($k^+ \rightarrow \mu^+ \rightarrow e^+$)

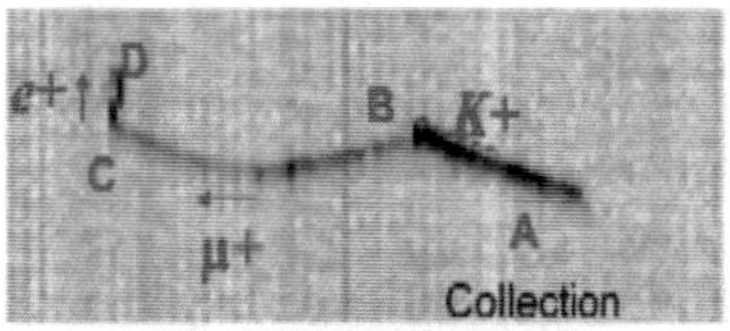

$K^+[AB] \rightarrow \mu^+[BC] \rightarrow e^+[CD]$

T300 real event

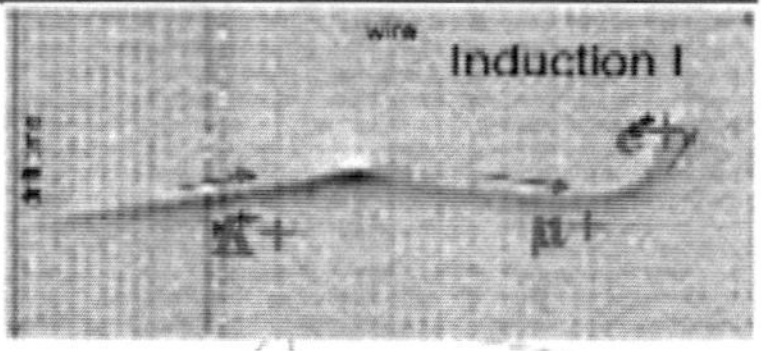

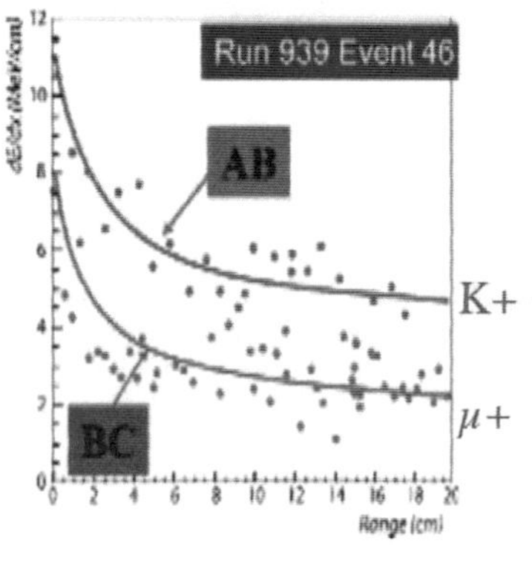

Efficient particle identification due to precise 3D reconstruction, dE/dx, range measurement:

- stopping power;
- recognition of secondary particle production after decay/interaction.

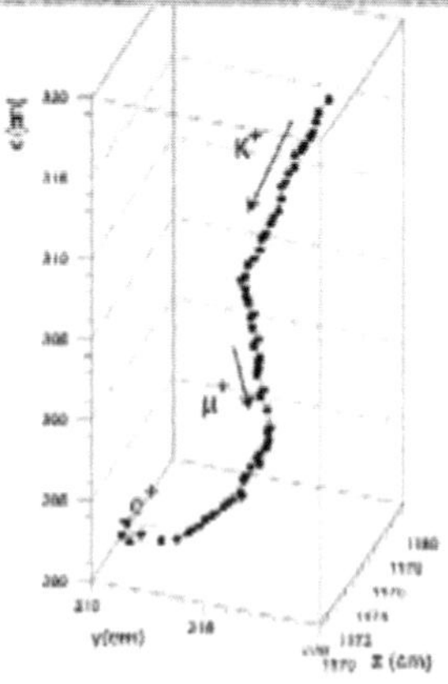

Rubbia, KEK_March2010

3D reconstruction (stopping μ)

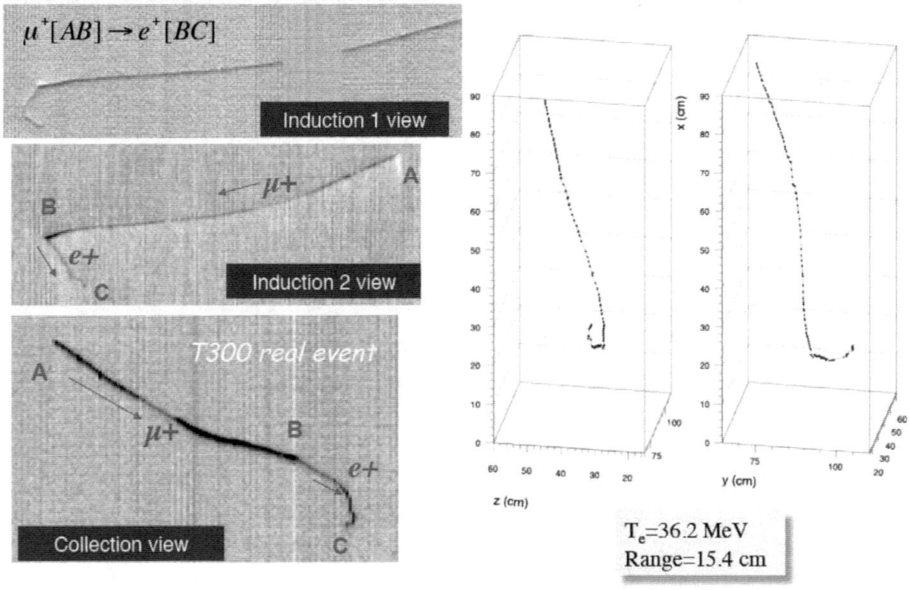

Rubbia, KEK_March2010 Slide: 36

Energy resolution from μ-decays

Michel electrons (μ - e decays)

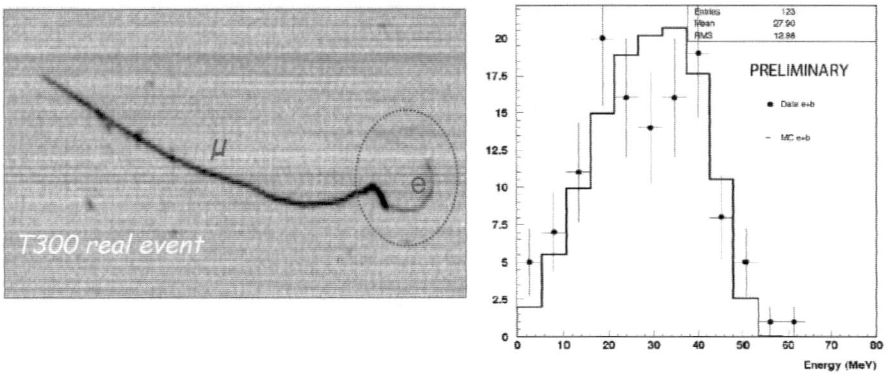

Energy resolution: $$\frac{\sigma}{E} = \frac{(13 \pm 2)\%}{\sqrt{E(MeV)}} \oplus (1.8 \pm 0.3)\%$$

Rubbia, KEK_March2010 Slide: 37

Journal of Physics: Conference Series **308** (2011) 012002 doi:10.1088/1742-6596/308/1/012002

Bremsstrahlung+ Pair-production

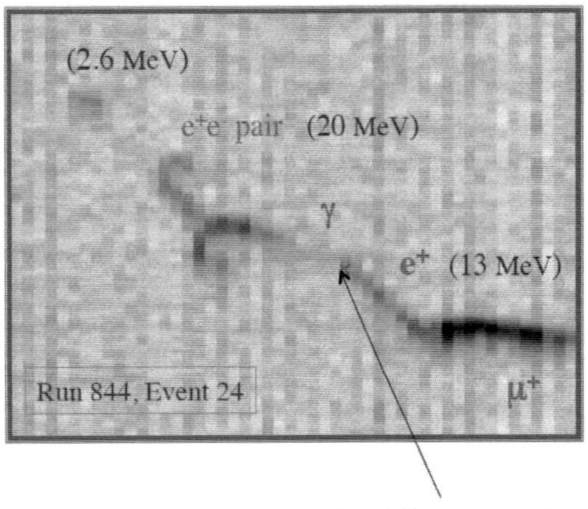

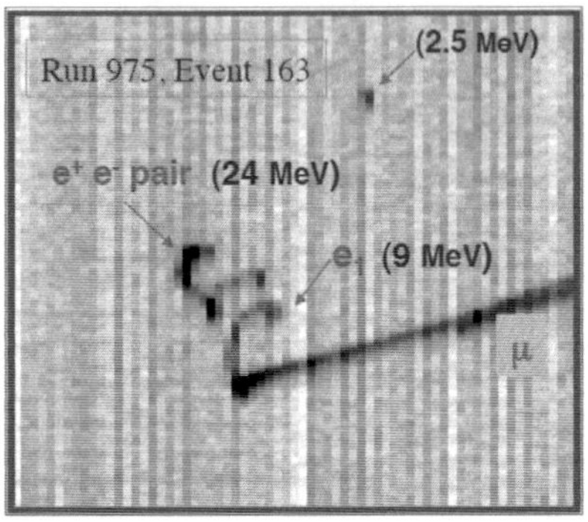

≈20% of positron from μ decays expected to annihilate before stopping

Rubbia, KEK_March2010 Slide: 38

Low energy detail

Single wire performance

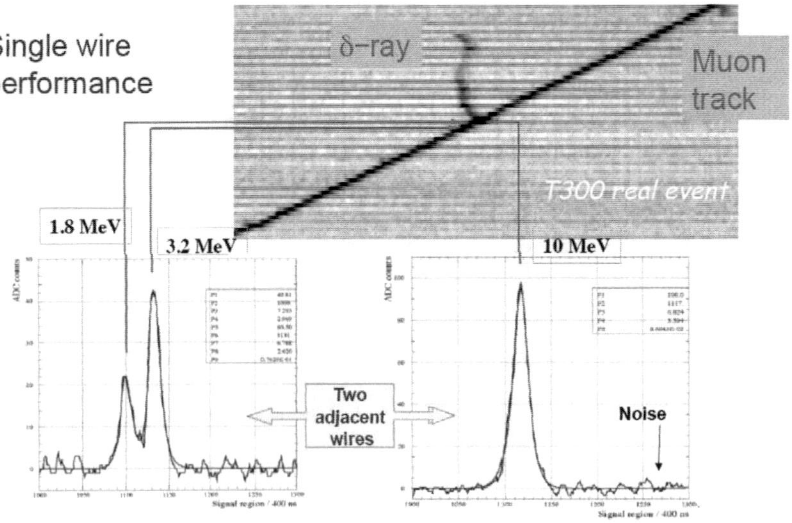

Rubbia, KEK_March2010 Slide: 39

Pi-zero identification

•Reconstruction of π^0-showers

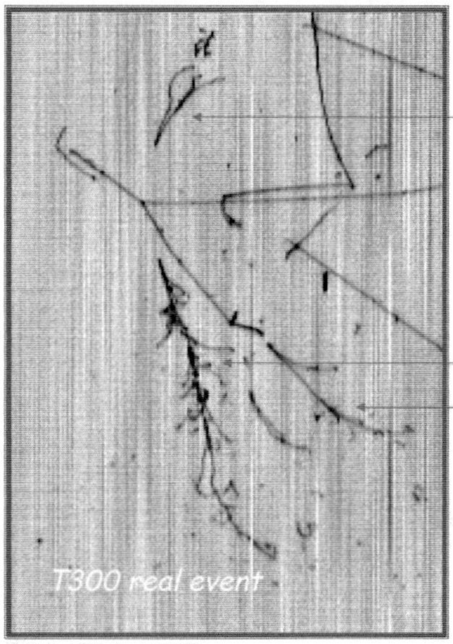

$$M_{inv} = \sqrt{2E_1E_2\cos(\theta_{12})}$$

158 MeV

752 MeV

140 MeV

θ = 141°

M_{inv} = 650 MeV

θ = 25°

M_{inv} =140 MeV

Rubbia, KEK_March2010 Slide: 40

π^0 reconstruction

230 hadronic interactions with $\pi^0 \to \gamma\,\gamma$ candidates have been selected from ICARUS T300 Pavia run

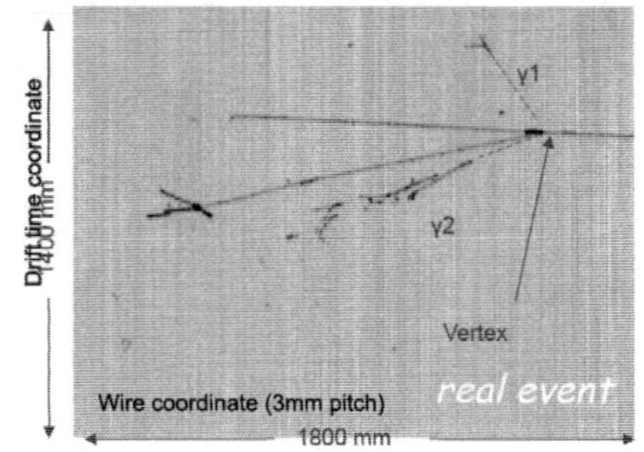

The average (γ,γ) invariant mass is in agreement with the π^0 mass hypothesis (m_{π^0} = 135 MeV/c²);

$$m_{\gamma\gamma} = 133.4 \pm 3.0(stat) \pm 4.0(sys)\mathrm{MeV/c^2}$$

The systematic error is mostly due to calibration

The measured photon radiation length is

$$X_{\gamma,meas} = (17.4 \pm 0.8)\mathrm{cm}$$

in agreement with expectations:

$$X_{\gamma,\exp} = \frac{9}{7}\cdot 14\mathrm{cm} = 18\mathrm{cm}$$

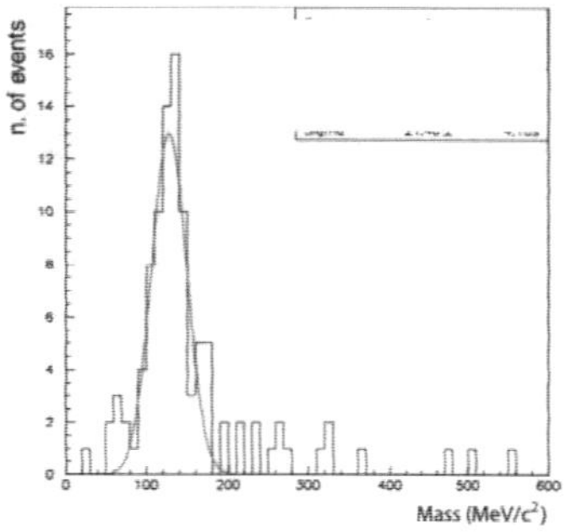

Rubbia, KEK_March2010 Slide: 41

Electron – π^0 separation

- NC in LAr suppressed by:
  - topology (γ conversion from vertex)
  - reconstruction of π^0 mass
  - electron/photon separation (dE/dx)
- Electron identification eff. = 90 %
- Residual misidentification < 0.1%

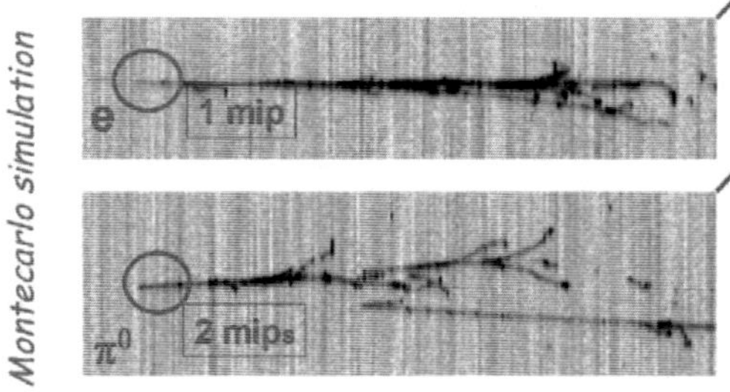

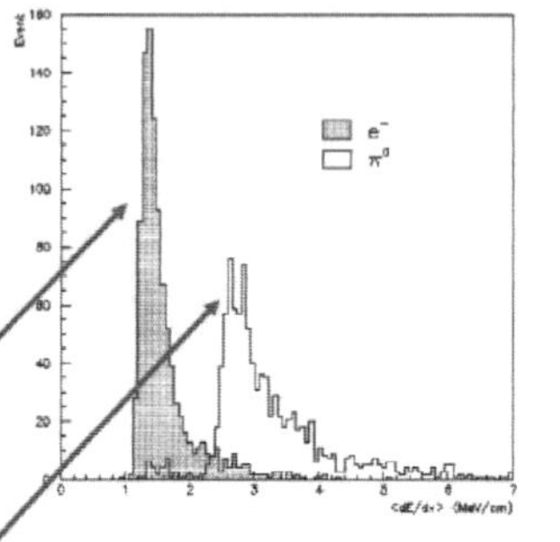

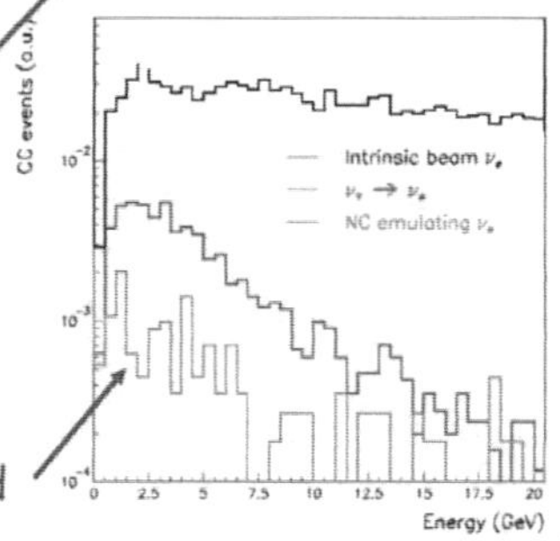

NC-e background

Rubbia, KEK_March2010 42

The ICARUS Collaboration in CNGS2

A. Ankowski[1], M. Antonello[2], P. Aprili[3], F. Arneodo[3], B. Baibussinov[4], M. Baldo Ceolin[4], G. Battistoni[5], L. Barzè[6], P. Benetti[6], A. Borio[6], E. Calligarich[6], M. Cambiaghi[6], F. Carbonara[7], F. Cavanna[2], S. Centro[4], A. Cesana[5], K. Cieslik[8], D. B. Cline[17], A. G. Cocco[7], A. Dabrowska[8], R.Dolfini[6], C. Farnese[4], A. Fava[4], A. Ferrari[5], G. Fiorillo[7], S. Galli[2], D. Gibin[4], A. Gigli Berzolari[6],A.Giuliano[9],K.Graczyk[1],S. Gninenko[10], A. Guglielmi[4],C. Juszczak[1], J. Holeczek[11], D. Kielczewska[12], M. Kirsanov[10], J. Kisiel[11], T. Kozlowski[13], N. Krasnikov[10],M. Lantz[5], G. Mannocchi[9], M. Markiewicz[8], Ch. Matthey[17], V. Matveev[10], A. Menegolli[6], G. Meng[4],C. Montanari[6], S. Muraro[5], J. Nowak[1], S. Otwinowski[17], O. Palamara[3], L. Periale[9], G. PianoMortari[2], A. Piazzoli[6], P. Picchi[9], F. Pietropaolo[4], W. Polchlopek[14], M. Posiadala[12], P. Przewlocki[13], A. Rappoldi[6],.G. L. Raselli[6], E. Rondio[13], M. Rossella[6], C. Rubbia[3], P. Sala[5], L. Satta[9], D. Scannicchio[6], Y. Seo[17], E. Segreto[3],F. Sergiampietri[15], J. Sobczyk[1], D. Stefan[8], J, Stepaniak[13], R. Sulej[16], M. Szarska[8], M. Terrani[5], G. Trinchero[9], F. Varanini[4], S. Ventura[4], C. Vignoli[6], T. Wachala[8], A. Zalewska[8], K. Zaremba[16] B. Lisowski[17], H. Wang[17], X. Yang[17]

[1] Wroclaw University of Technology, Wroclaw, Poland
[2] Dipartimento di Fisica e INFN, Università di L'Aquila, Via Vetoio, I-67100
[3] Laboratori Nazionali del Gran Sasso dell'INFN, Assergi (AQ), Italy
[4] Dipartimento di Fisica e INFN, Università di Padova, Via Marzolo 8, I-35131
[5] Dipartimento di Fisica e INFN, Università di Milano, Via Caloria 2, I-20123
[6] Dipartimento di Fisica Nucleare, Teorica e INFN, Università di Pavia, Via Bassi 6, I-27100
[7] Dipartimento di Scienza Fisiche, INFN e Università Federico II, Napoli, Italy
[8] H. Niewodniczanski Institute of Nuclear Physics, Krakow, Poland
[9] Laboratori Nazionali di Frascati (INFN), Via Fermi 40, I-00044
[10] INR RAS, prospekt 60-letiya Oktyabrya 7a, Moscow 117312, Russia
[11] University of Silesia, 12 Bankowa st., 40-007 Katowice, Poland
[12] Warsaw Univeristy, Krakowskie Przedmiescie 26/28, 00-927 Warszawa, Poland
[13] A. Soltan Institute for Nuclear Studies, 05-400 Swierk/Otwock, Poland
[14] AGH University of Science and Technology, Al. Mickiewicza 30, 30-059 Krakow, Poland
[15] Dipartimento di Fisica, Università di Pisa, Largo Bruno Pontecorvo 3, I-56127
[16] Univeristy of Technology, Pl. Politechniki 1, 00-661 Warsaw, Polanda
[17] Department of Physics and Astronomy, University of California, Los Angeles, USA

Rubbia, KEK_March2010 Slide: 43

ICARUS (CNGS2): the first large scale LAr experiment

- ICARUS represents a major milestone in the practical realization of a large scale LAr detector. Successfully operated on surface in Pavia in 2002, will soon be operational in the underground HallB of LNGS.
- The T600 at LNGS will collect simultaneously "bubble chamber like" neutrino events events of different nature
- Cosmic ray events
  - ≈ 100 ev/year of unbiased atmospheric CC neutrinos.
  - Solar neutrino electron rates >5 MeV. ~1-2 ev/day
  - Supernovae neutrinos.
  - A zero background proton decay with 3 x 10^{32} nucleons for "exotic" channels.
- CERN beam associated events: 1200 ν_μ CC ev/y and 7-8 ν_e CC ev/year
  - Observation of neu-tau events in the electron channel (with sensitivity comparable to OPERA
  - A search fo sterile neutrinos
- Other unexpected phenomena

Rubbia, KEK_March2010 Slide: 44

The ICARUS detector in underground Hall B of LNGS

Rubbia, KEK_March2010 Slide: 45

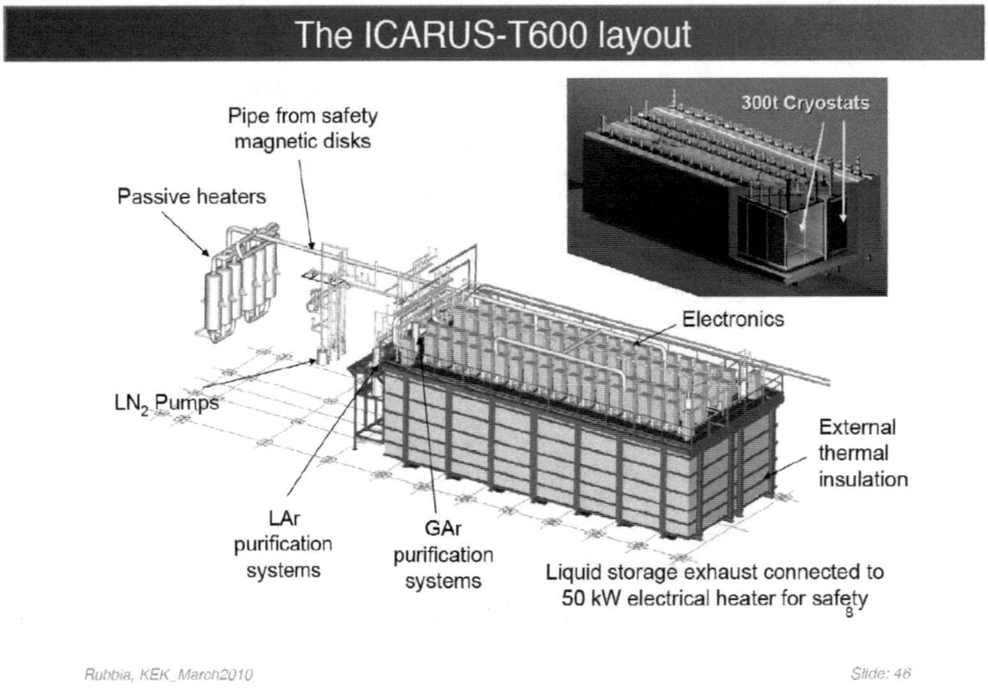
The ICARUS-T600 layout
300t Cryostats
Pipe from safety magnetic disks
Passive heaters
Electronics
LN2 Pumps
External thermal insulation
LAr purification systems
GAr purification systems
Liquid storage exhaust connected to 50 kW electrical heater for safety
8
Rubbia, KEK_March2010
Slide: 46

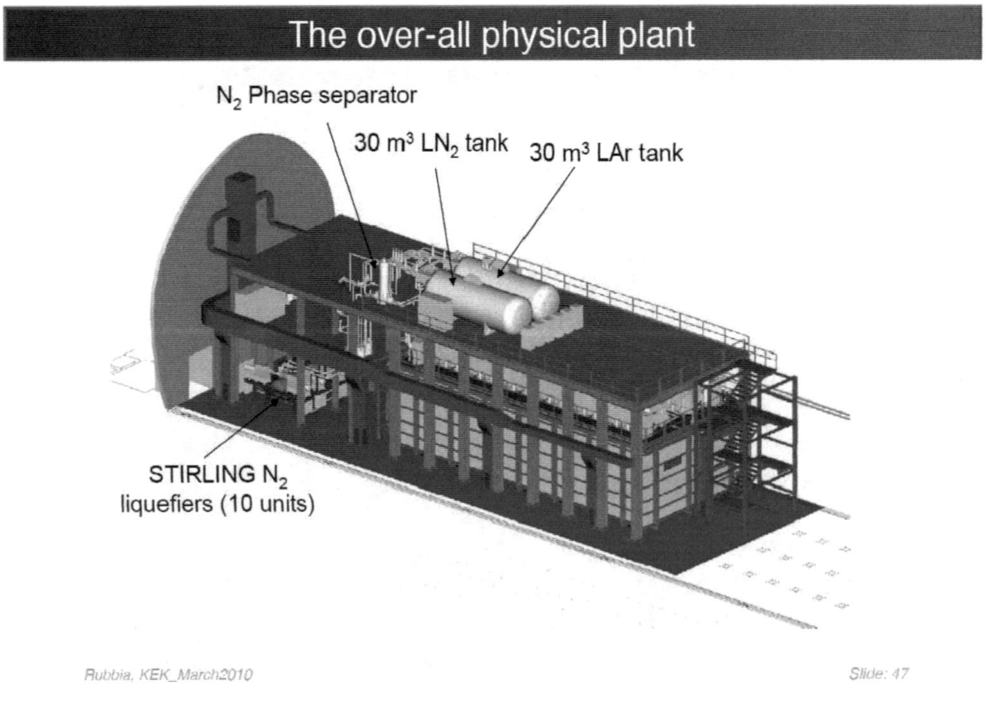
The over-all physical plant
N2 Phase separator
30 m3 LN2 tank
30 m3 LAr tank
STIRLING N2 liquefiers (10 units)
Rubbia, KEK_March2010
Slide: 47

T600 in hall B (CNGS2-2010)

Rubbia, KEK_March2010 Slide: 48

ICARUS: CNGS2 searching for the ν_τ signature.

- CERN beam associated events: 1200 ν_μ CC ev/yr and 8 ν_e CC ev/yr for 4.5 x 10^{19} pot/yr beam intensity
- The electron decay channel is a quite significant goal also for the present T600 mass, uniquely characterized by a large transverse momentum unbalance due to the emission of the two neutrinos.
- Several background free ν_τ events are expected in 3 yrs data taking.

CNGS ν_τ interaction, E_ν=18.7 GeV

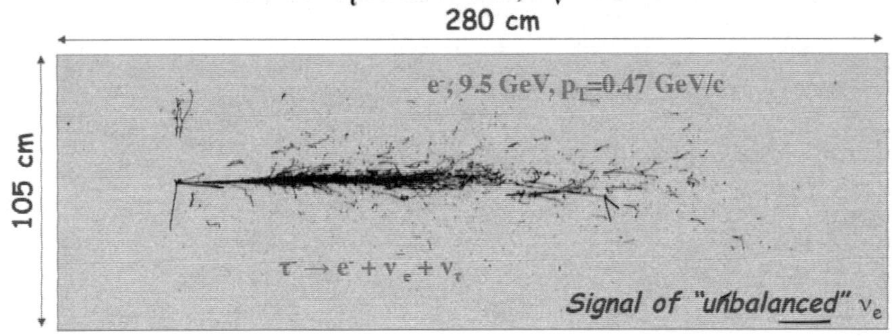

Rubbia, KEK_March2010 Slide: 49

Sterile neutrino search with CNGS2 (ICARUS)

- The $\sin^2(2\theta)$-Δm^2 explored region covers most of LNSD allowed areas and extends to lower values of Δm^2
- Two indicated points are reference values of MiniBooNE proposal and of previous slides
- Data taking will be enough to exclude $\sin^2(2\theta)$ values $> 5\ 10^{-3}$ at $3\ \sigma$ with ν (!)
- Smaller $\sin^2(2\theta)$ are not explored. *An additional LAr experiment at PS is proposed for ν and ν-bar.*

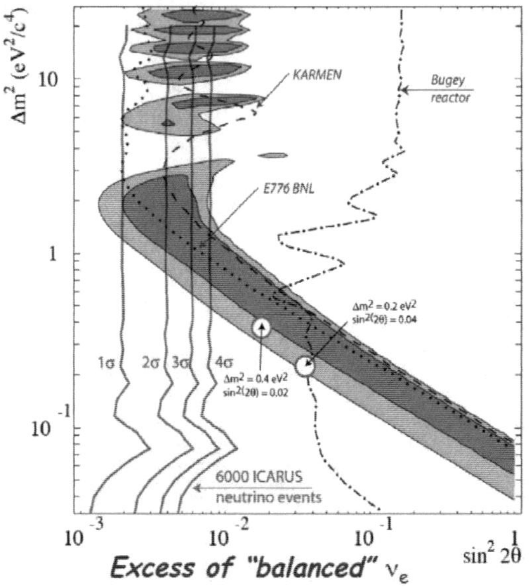

Rubbia, KEK_March2010 Slide: 50

Triggering "low energy events"

- ICARUS-T600 is a self-triggering detector exploiting both scintillation light and TPC wire signals.
- Wire signals are used for local trigger purposes to select isolated low energy event (solar ν, SN ν), atmospheric ν's and p-decays. Well suited also for very large LAr-TPC's.

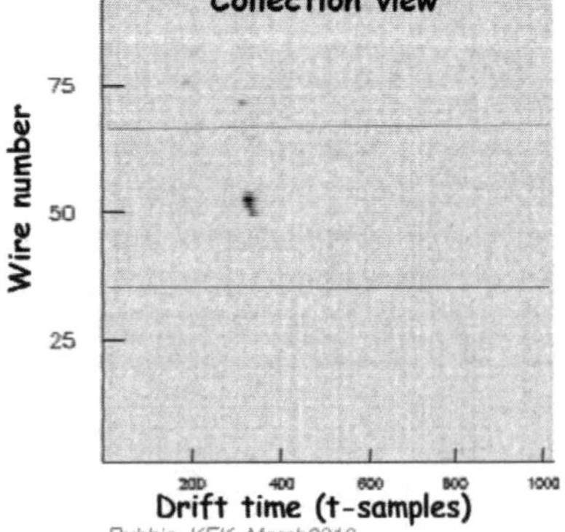

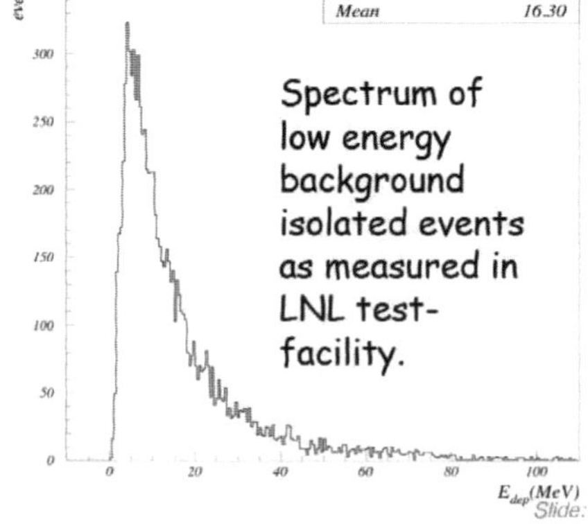

Rubbia, KEK_March2010 Slide: 51

Neutrinos !?

- Neutrinos have been the origin of an impressive number of "Surprises". It has been demonstrated that the sum of the strengths of the coupling of different ν is very close to 3. But it is only assuming that neutrinos, in similarity to charged leptons, have unitary strengths that the *resulting number of neutrinos is 3.* The situation may be altered by the additional presence of sterile neutrinos.
- The experimentally measured weak coupling strengths are only rather poorly known, leaving room for many other alternatives.
- *Are neutrino a simple carbon copy repetition of quarks?*
- Important discoveries may be ahead:
  - CP violation in the lepton sector
  - Sterile neutrino and other "surprises"
  - Majorana or Dirac ν's;ν-less β-decay,ν-masses

Rubbia, KEK_March2010 Slide: 52

The LSND Experiment: antineutrino oscillations ?

LSND has observed an excess of☒ν_e events in a☒ν_μ beam, 87.9 ± 22.4 ± 6.0 **(3.8σ)**

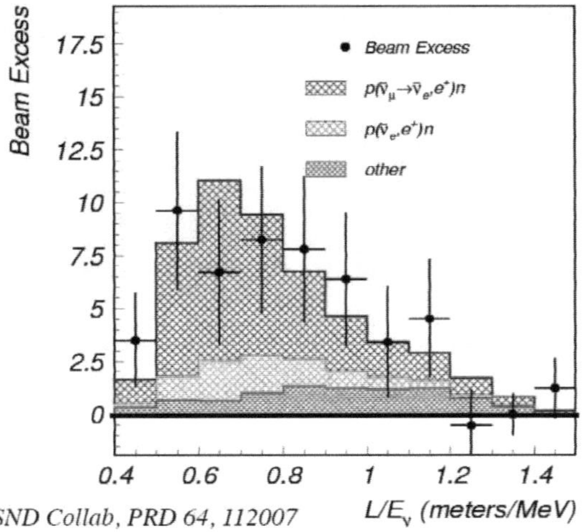

LSND Collab, PRD 64, 112007

Points -- LSND data
Signal (blue)
Backgrounds (red, green)

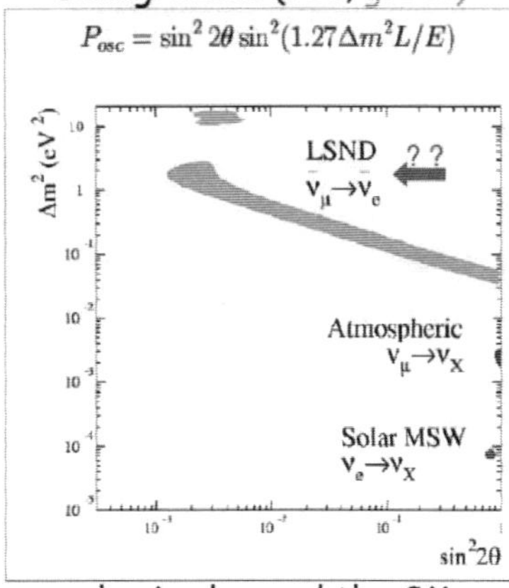

- 3 oscillation signals, if confirmed, require new physics beyond the SM

Rubbia, KEK_March2010 Slide: 53

Many theoretical hypothesis.....

3+2 Sterile Neutrinos — Sorel, Conrad, & Shaevitz (PRD70(2004)073004)

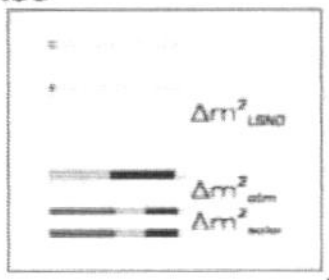

← Additional, sterile neutrinos ?

← Standard neutrinos

MaVaNs & 3+1 — Hung (hep-ph/0010126)

Sterile Neutrino — Kaplan, Nelson, & Weiner (PRL93(2004)091801

CPT Violation & 3+1 Sterile Neutrino — Barger, Marfatia, & Whisnant (PLB576(2003)303)

Quantum Decoherence — Barenboim & Mavromatos (PRD70(2004)093015)

Lorentz Violation — Kostelecky & Mewes (PRD70(2004)076002)
Katori, Kostelecky, Tayloe (hep-ph/0606154)

Extra Dimensions — Pas, Pakvasa, & Weiler (PRD72(2005)095017)

Sterile Neutrino Decay — Palomares-Ruiz, Pascoli, & Schwetz (JHEP509(2005)48)

And so on....

Rubbia, KEK_March2010 Slide: 54

MiniBooNE experiment: antineutrino data

- MiniBooNE result for anti-neutrino events, the direct analog of LSND, is based on 3.39 x 10^{20} POT.
- The result is inconclusive with respect to the LNSD result.
- LSND is still alive and well.

MiniBooNE limit at 90 %

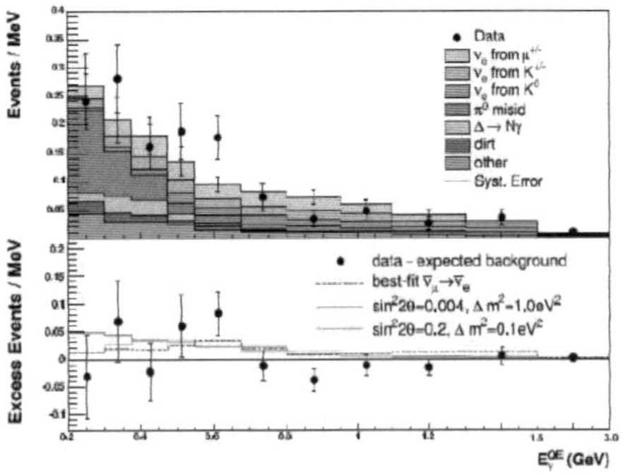

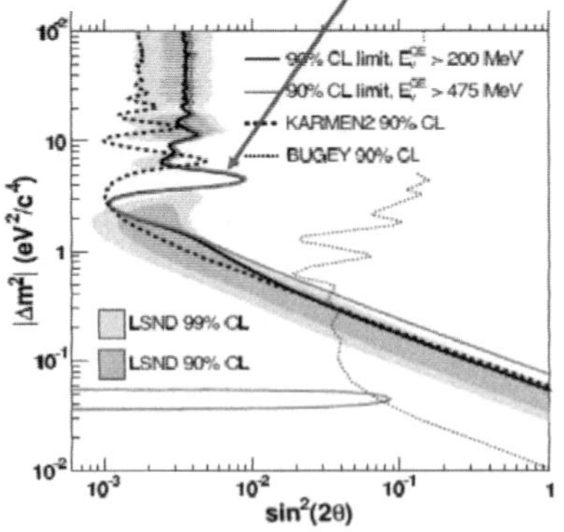

Rubbia, KEK_March2010 Slide: 55

MiniBooNE experiment: neutrino data

- MiniBooNE result for neutrino events is based on 6.46×10^{20} POT, corresponding to 1.5×10^5 $\nu\mu$ CC-QE events
- We expect 375 ν e CC-QE intrinsic background events. with a possible LSND signal of ~200 ν e CC-QE events
- 96±17±20 events above background, for $300 < E_\nu^{QE} < 475$ MeV:

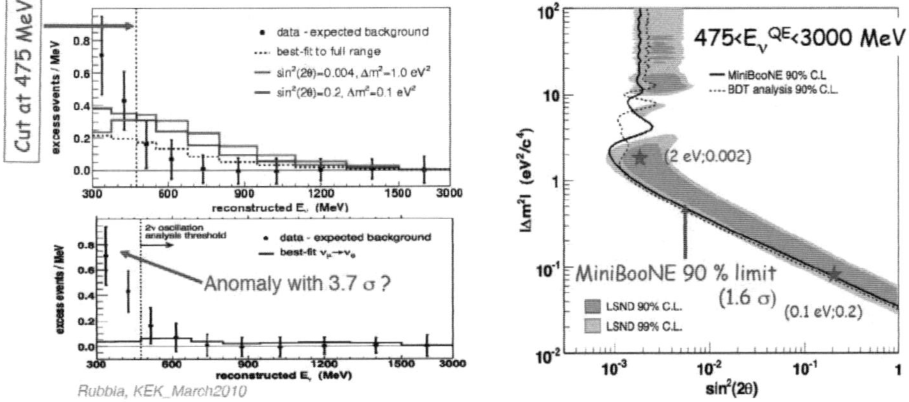

Rubbia, KEK_March2010

ICARUS after CNGS2: The CERN- PS neutrino beam

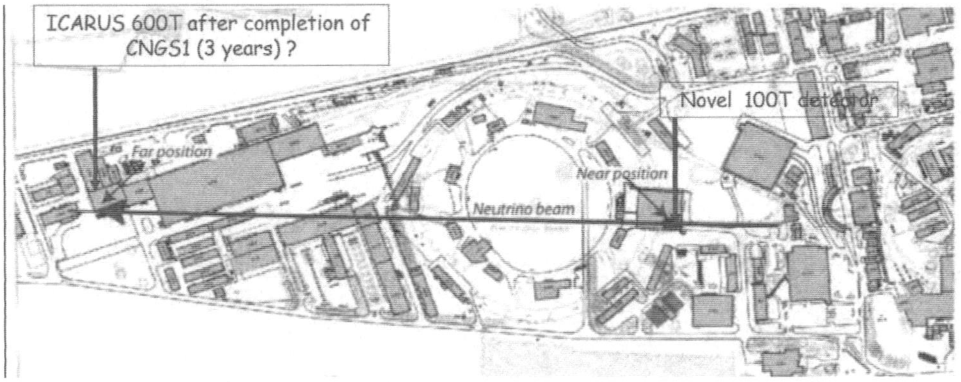

- The PS proton beam at 19.2 GeV/c is extracted via TT2, TT1 and TT7.
- The magnetic horn is designed to focus particles of momentum≈2GeV/c.
- The decay tunnel is about 50 m long, followed by an iron beam stopper

Two positions are foreseen for the detection of the neutrinos
The far (main) location at 850 m from the target (600T);
The near location at a distance of 127 m from the target (100T).

Rubbia, KEK_March2010 Slide: 57

Unique features of the CERN SPS beam

- The present proposal is a search for spectral differences of electron like specific signatures in *two identical detectors* but at two different neutrino decay distances.
- *In absence of oscillations* the two ν_e observed spectra are a precise copy of each other, independently of the specific experimental event signatures and without any Monte Carlo comparisons.

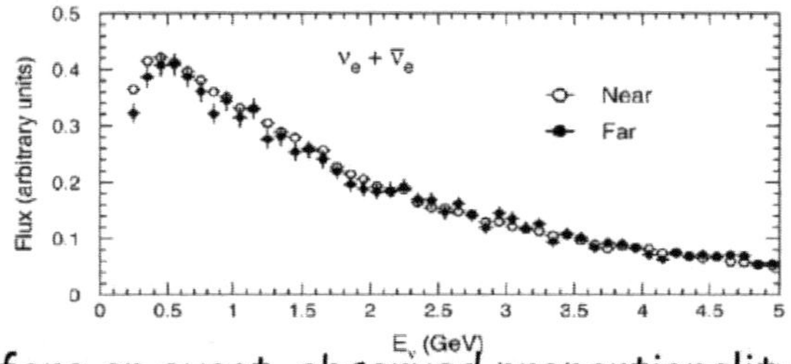

The two ν_e spectra in the far (850 m) and near (127 m locations are nearly identical

- Therefore an exact, observed proportionality between the two ν_e spectra implies directly the absence of neutrino oscillations over the measured interval of L/E.

Rubbia, KEK_March2010 Slide: 58

New features of the CERN proposal

- It appears that the present proposal, unlike LNSD and MiniBooNE, can *determine both the mass difference and the value of the mixing angle.*
- Very different and clearly distinguishable patterns are possible depending on the actual values in the (Δm^2 - $\sin^2 2\theta$) plane.
- The intrinsic ν_e background due to the beam contamination is also shown.
- The magnitude of the LNSD expected oscillatory behaviour, for the moment completely unknown, is in all circumstances well above the background, also considering the very high statistical impact and the high resolution of the experimental measurement.

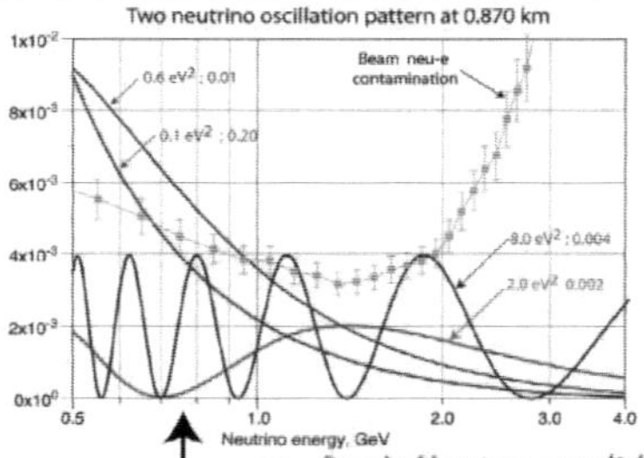

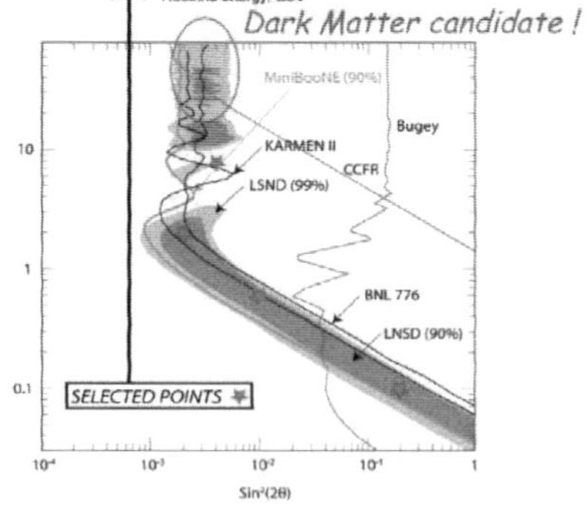

Rubbia, KEK_March2010 Slide: 59

Comparing sensitivities

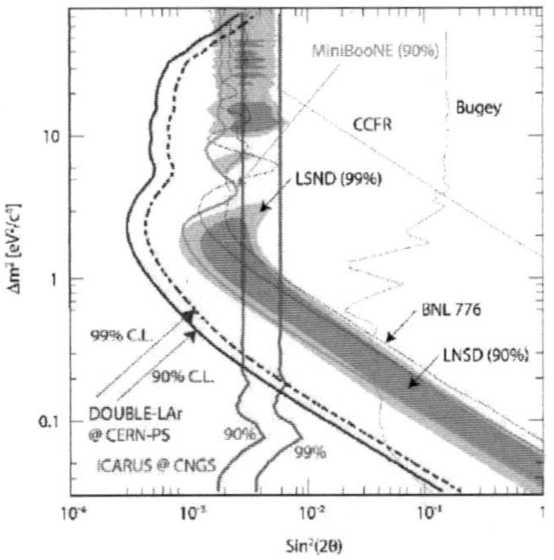

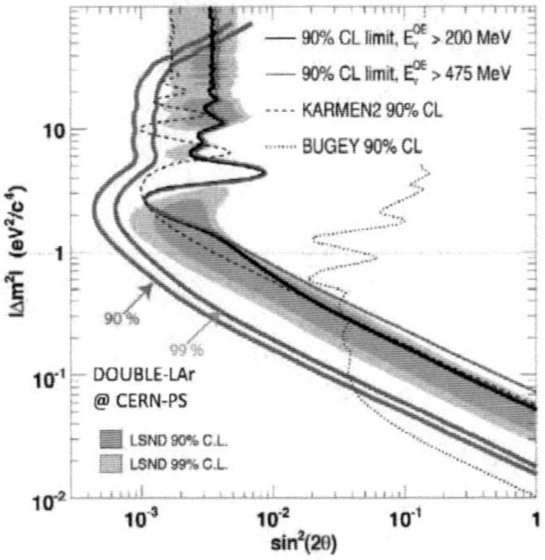

Expected sensitivity for the proposed experiment exposed at the CERN-PS neutrino beam (left) and anti-neutrino (right) for 2.5 10^{20} pot. The LSND allowed region is fully explored both for neutrinos. In the neutrino case, the expectations from CNGS2/ICARUS T600 at LNGS are also shown.

Rubbia, KEK_March2010 Slide: 60

From T600 to multi-kton LAr-TPC: a MODULAr approach

- The MODULAr proposal is focused on the following three main items:
  - a new neutrino beam derived from the existing horn focussing and the proton beam line from the 400 GeV SPS, eventually with an increased intensity. Relatively modest changes in the neutrino beam focussing of CERN will produce a nearly optimal beam configuration.
  - A new experimental area, eventually enlarged in future phases, to be realised about 10 km off-axis from the main laboratory, away from the protected area of the Gran Sasso National Park. The high event rejection power of the LAr-TPC detector will ensure the absence of backgrounds not only for the neutrinos from the CNGS but also for proton decay and cosmic neutrinos if a shallow depth location (1.2 km of equilavent water depth) is provided.
  - A new LAr-TPC Imaging underground detector made of several identical units, each of about 10 kt fiducial mass. As a first step, a total of about 20 kt will be realised along the lines of the vast R&D work carried out over the last decades. This programme may eventually be further improved with additional modules, depending of the developments of the programmes with and without accelerators.

Rubbia, KEK_March2010 Slide: 61

The MODULAr concept

- Indicative cross section of the T600 "clone" in the dedicated "swimming pool like" underground hall.
- It is made of two twin separate LAr containers made of Aluminum extruded structures, thermally stabilized with forced N2 circulation.
- Outside the dual structure, an about 1.5 m thick perlite wall provides spontaneous, passive heat insulation.
- The region on top of the "swimming pool" is accessible to auxiliary equipments.
- *Total L Ar mass for a detector 60 m long (3 x ICARUS) is about 10'000 tons. Same gaps as for ICARUS, but 4 m drift.*

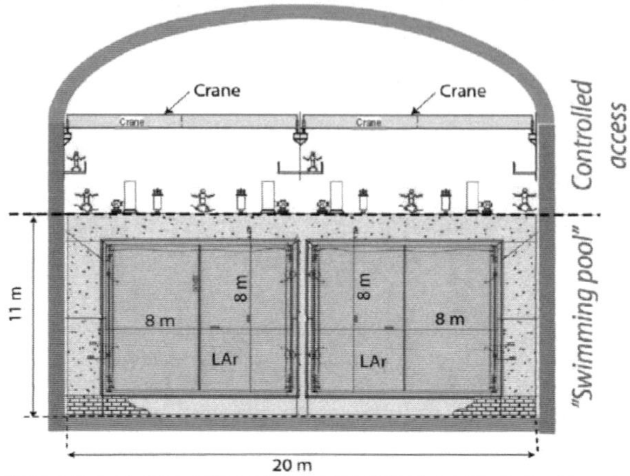

The volume of each of the two gaps should be about 8 x 8 m^2 and 60 m long, corresponding to 3840 m3 of liquid or 5370 t of LAr. The drift length is 4 m. A reasonable three-plane wire pitch should be of the order of 6 mm, twice the value of the T600. The full detector is made of two such dewars.

Rubbia, KEK_March2010 Slide: 62

3σ MODULAr sensitivity to θ_{13} and δ_{CP}

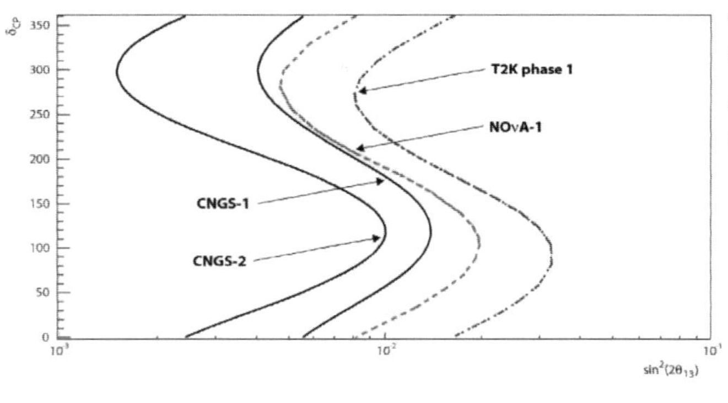

CNGS-1: 1.2×10^{20} pot/y
CNGS-2: 4.3×10^{20} pot/y
$\Delta m_{23}^2 = 2.5\times10^{-3} eV^2$
normal mass hierarchy

GLoBES calculation:
5 % beam systematics
$\Delta E/E$ = 15 %

Rates for 20 kt mass, 5 y and 1.2×10^{20} pot/y →
$\sin^2(2\theta_{13})$ = 0.1

	ν_μ CC	bg	signal	$S/\sqrt{bg}$
MODULAr	13000	70	390	47
NOνA (20kt)		19.5	142	32
T2K		23	103	23

Rubbia, KEK_March2010

The future of the LAr

- A number of possible future experiments with masses much larger than T600 have been discussed by a number of authors and are a subject of discussion for the long range future of neutrino physics and perhaps proton decay.
- These authors have presented masses of 5'000 and up to 100'000 tons, namely between 10x and 200x the today'sT600.
- But containing the LAr is not enough: for instance purity levels of 10^{-11} O_2 equivalent must be created and maintained.
- In our view, before translating more or less generic R&D into a detector of such an enormous magnitude, intermediate steps must be performed, consolidating realistically the physics already possible with detectors of the present size.
- Gargamelle has already shown that remarkable results may be obtained with a very sensitive detector even if much smaller than the one of larger and coarser calorimeters of that time.
- There may be a similar opportunity in the future, paving on the same time the way to the much larger ultimate facilities.

Rubbia, KEK_March2010 Slide: 64

Thank you !

Rubbia, KEK_March2010 Slide: 65

Journal of Physics: Conference Series **308** (2011) 012003 doi:10.1088/1742-6596/308/1/012003

Results from massive underground detectors on solar and atmospheric neutrino studies and proton decay searches

Takaaki Kajita
ICRR and IPMU, University of Tokyo,
Kashiwa-no-ha 5-1-5, Kashiwa, Chiba 277-8582, JAPAN
E-mail: kajita@icrr.u-tokyo.ac.jp

Abstract. Massive underground detectors have been playing important roles in particle and astro-particle physics. Results from massive underground detectors on solar and atmospheric neutrino studies and proton decay searches are reviewed.

1. Introduction

Massive underground detectors have long history. In the mid. 1960's, a solar neutrino experiment was proposed to study the interior of the Sun [1], [2]. The proposed detector was constructed at 4400 m.w.e. underground in the Homestake gold mine in the late 1960's. It might be fair to mention that this detector was the first massive underground detector, since the total mass of the target material (C_2Cl_4) was approximately 600 tons. This detector observed solar neutrinos for the first time confirming that the energy generation in the Sun is due to nuclear fusion reactions. However, the observed flux [3] by this detector was approximately 1/3 of the predicted flux by the standard solar model (SSM). This problem was called the "solar neutrino problem". Because the information available by this detector was the event rate only, various massive underground detectors were required to understand the cause of the problem.

In the 1970's, the idea of Grand Unified Theories emerged [4], [5]. These theories predicted nucleon decays with a typical lifetime of 10^{30} years. If the nucleon lifetime is 10^{30} years, it is possible to observe many proton and bound neutron decays with a massive underground detector. In the early 1980's, several underground detectors with the masses ranging 100 tons to several ktons were constructed to discover nucleon decays. These experiment did not discover nucleon decays. However, some of these detectors observed neutrinos from a supernova SN1987A [6], [7], discovered the atmospheric ν_μ deficit [8], and confirmed the solar neutrino problem [9]. These results required more studies with the next generation massive underground detectors.

In the following sections, recent results from massive underground detectors on solar and atmospheric neutrino studies and proton decay searches are discussed.

2. Solar neutrino experiments

In 1989, Kamiokande observed the ^8B solar neutrino flux by neutrino electron scattering, $\nu e \rightarrow \nu e$ [9]. This experiment confirmed that neutrinos are coming from the Sun and that the solar neutrino flux was significantly lower than the standard solar model (SSM) prediction. Subsequently, solar neutrino experiments that used 30 to 60 tons of Gallium were carried out [10], [11]. These experiments were unique in the sense that about half of the calculated event rate was due to fundamental *pp* solar neutrino reactions. These low energy solar neutrino experiments also observed that the solar neutrino flux was lower than the SSM prediction. Although the solar neutrino deficit was clearly observed in these experiments, the cause of the deficit was not firmly identified, partly because no experiment observed the ν_e and $(\nu_\mu+\nu_\tau)$ fluxes separately. MSW mechanism [12], [13] was a serious possibility, but was not uniquely identified as the solution to this problem.

The situation changed drastically in 2001 and 2002. In 2001, the SNO experiment, which used 1 kton of heavy water (D_2O) measured the ^8B solar neutrinos by $\nu_e + D \rightarrow e^- + p + p$ (CC) [14], i.e., by charged-current ν_e interactions. In the mean time, Super-Kamiokande (Super-K) measured the ^8B neutrino flux precisely by neutrino-electron scattering [15]. By comparing these results, it was found that the flux measured by Super-K (assuming all the neutrinos are electron-neutrinos) was 3.3 σ higher than that measured by SNO [14]. This discrepancy was concluded as evidence for solar neutrino oscillations, since ν_μ's and ν_τ's generated by neutrino oscillations contribute only to the neutrino-electron scattering.

In 2002, the evidence was substantially strengthened by the measurement of the total $(= \nu_e + \nu_\mu + \nu_\tau)$ neutrino flux in SNO. The total flux was observed by $\nu_x + D \rightarrow \nu_x + p + n$ (NC) [16]. The observed total flux was consistent with the SSM prediction, confirming the solar neutrino oscillation at the 5.5 standard deviation level [16]. Since the CC/NC ratio is approximately equal to $\sin^2\theta_{12}$ in the LMA region, the measured solar neutrino fluxes strongly constrain $\sin^2\theta_{12}$.

Furthermore, a long-baseline reactor neutrino experiment, KamLAND, has been contributing to $\nu_e \rightarrow \nu_x$ oscillation studies. KamLAND is a 1 kton liquid scintillator detector located 2700 m.w.e. deep underground in the Kamioka mine. KamLAND observes reactor anti-electron neutrinos by $\overline{\nu}_e p \rightarrow e^+ n$. See Figure 1 for the latest results from KamLAND [17]. The observed spectrum distortion is regarded as very strong evidence for "oscillations". The mean flight length of the neutrinos detected by KamLAND is 180 km. The observed spectrum distortion together with the known flight length makes it possible to estimate the Δm^2_{12} parameter precisely. Combining the solar and KamLAND data, the 1-2 (i.e., solar) oscillation parameters have been measured accurately.

With the increasing accuracy of the solar and KamLAND data, it was pointed out that there might be slight tension in the suggested $\sin^2\theta_{12}$ value by the solar and KamLAND experiments [18], [19]. The suggested $\sin^2\theta_{12}$ value from KamLAND was larger than that from solar neutrino experiments. It was also pointed out that the tension can be reduced if $\sin^2\theta_{13}$ is non-zero [18], [19]. Due to this reason, updated data with a higher precision have been waited for.

In 2009, the SNO experiment published a low-energy analysis of the SNO-I and -II data [20]. With a better understanding of various background and signal distributions, it was made possible to estimate the total neutrino flux accurately by the NC events. The accuracy of the total flux has been improved by more than a factor of 2. It should also be noted that the total ^8B solar neutrino flux was measured to an accuracy of approximately 4%, which was a factor of 4 smaller than the estimated uncertainty by the SSM calculation.

As mentioned above, the accurately measured total neutrino flux improves our understanding of the θ_{12} parameter. The improved allowed region with the most updated solar neutrino data compared with the previous one is depicted in Ref. [21].

Figure 2 shows the estimated oscillation parameters with the updated solar neutrino and the KamLAND data. Since the central value of the measured total ^8B solar neutrino flux have not changed much with the 2009 SNO analysis, and since the uncertainty in $\sin^2\theta_{12}$ by the KamLAND experiment was larger than that by the solar neutrino experiments, the tension, and therefore the implication for non-zero θ_{13}, has not changed much [20].

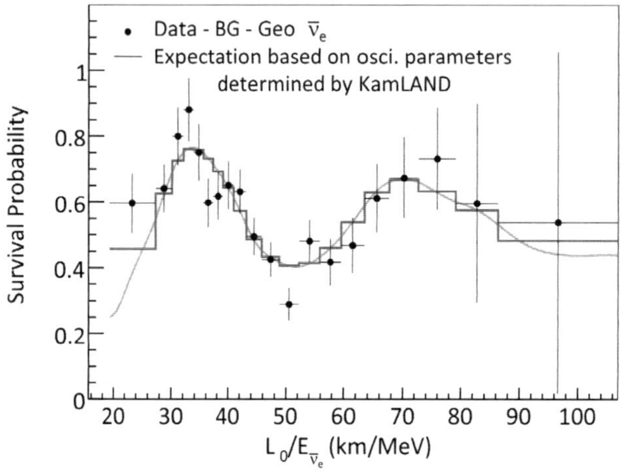

Figure 1. Ratio of the background subtracted reactor $\overline{\nu}_e$ spectrum to the no-oscillation expectation as a function of L_0/E. 2881 ton·year of the KamLAND data are used. $L_0 = 180$ km is assumed. (From Ref. [17].)

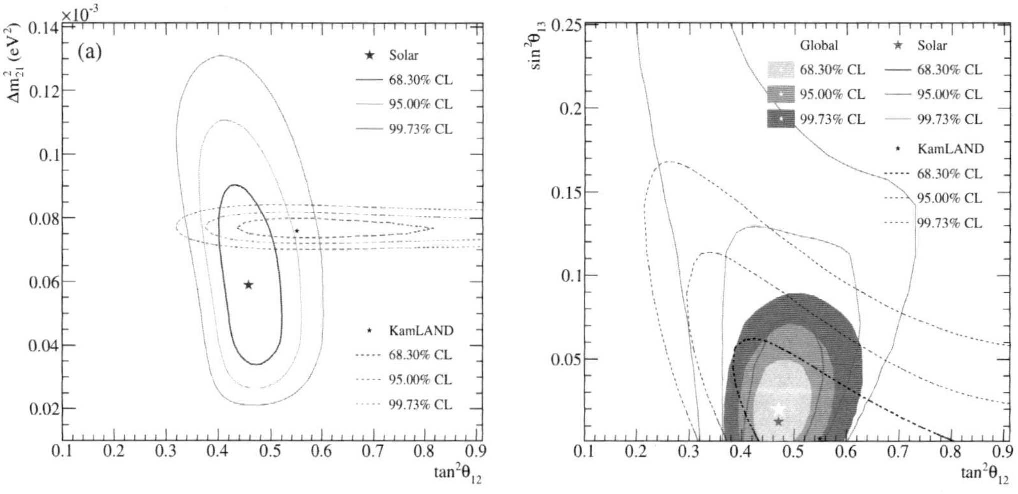

Figure 2. Left panel: Allowed regions of 12-oscillation parameters by solar neutrino and the KamLAND experiments. Right panel: Allowed $\tan^2\theta_{12}$ vs. $\sin^2\theta_{13}$ regions by the solar neutrinos experiments only, KamLAND only, and solar plus KamLAND experiments. (From Ref. [20].)

3. Atmospheric neutrino experiments

Atmospheric neutrino experiments played a fundamental role in the discovery of neutrino oscillations [8], [22], [23], [24], [25], [26]. As of this writing, the statistics of the atmospheric neutrino events are dominated by the data from Super-K. Therefore, we discuss the atmospheric neutrino results from Super-K.

Super-K have several phases. As of this writing, data up to Super-K-III have been used for the final physics analysis. The total exposure of the detector for fully-contained atmospheric neutrino analysis is 173 kton·yr. In the Super-K data, zenith-angle and energy dependent deficit of ν_μ charged current (CC) events has been observed. See Fig. 3 (left), which shows the zenith angle distributions for some sub-samples from Super-K. By fitting the zenith angle distributions [27], the oscillation parameters are accurately measured.

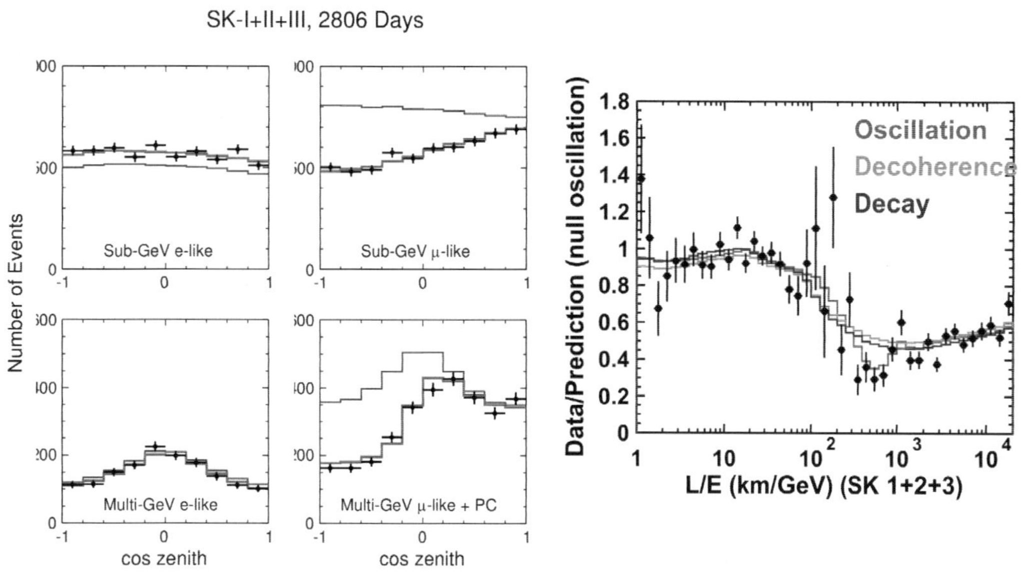

Figure 3. Left: Zenith angle distributions observed in Super-K-I+II+III (173 kton·year exposure). Top panels show fully-contained (FC)) sub-GeV e-like (left) and μ-like (right) events. The bottom panels show FC multi-GeV e-like (left) and FC multi-GeV μ-like plus partially-contained events (right). Sub- and multi-GeV are defined to be less and more than 1.33 GeV visible energy, respectively. The blue histograms show the prediction without oscillations. The red histograms show the best fit $\nu_\mu \to \nu_\tau$ expectation. Right: L/E distribution observed in Super-K-I+II+III. The red, blue and green histograms show the best-fit predictions based on neutrino oscillation, decay and decoherence models, respectively.

Although, the Super-K zenith angle data were explained well by oscillations, the sinusoidal oscillation pattern was not seen in the left panel of Fig. 3 due to the smearing. Therefore, Super-K carried out a dedicated L/E analysis using only high L/E resolution events [28]. Figure 3 (right) shows the L/E distribution from Super-K-I+II+III. The L/E distribution shows a dip around $L/E = 500$ km/GeV that corresponds to the first oscillation minimum. The distribution is used to compare the oscillation and other hypotheses, and to constrain the other hypotheses. Due to the observation of the dip, the allowed region of the oscillation parameters, especially that of Δm^2_{23}, is strongly constrained.

Figure 4 shows the estimated allowed regions of 2-flavor $\nu_\mu \to \nu_\tau$ oscillation parameters from the atmospheric neutrino and long baseline neutrino oscillation experiments. We find that the allowed regions overlap well, suggesting further that the standard neutrino oscillation scenario is valid. From this figure, we find that Δm^2_{23} is most accurately measured by the MINOS

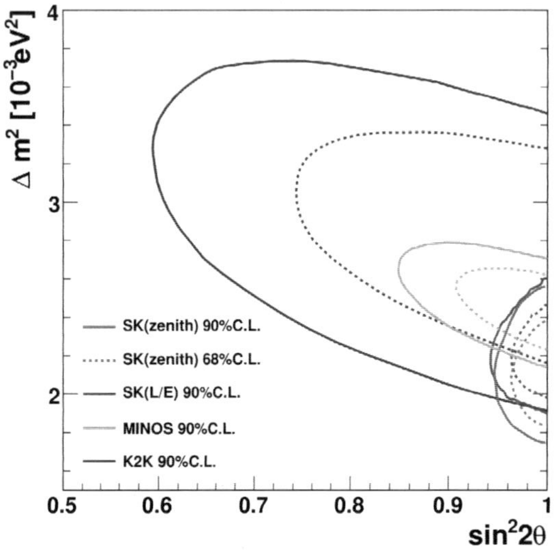

Figure 4. Estimated allowed regions of $\nu_\mu \rightarrow \nu_\tau$ oscillation parameters from atmospheric neutrino and long baseline neutrino oscillation experiments at 68%C.L. (dotted) and 90%C.L. (solid).

experiment [29], while the mixing parameter is still measured most accurately by the Super-K atmospheric neutrino experiment.

The neutrino flight length extends up to 12800 km in atmospheric neutrino experiments. The oscillation length relevant to Δm^2_{12} is shorter than the diameter of the earth for E_ν below 1 GeV. Therefore the effects due to the solar oscillation terms can be observable in atmospheric neutrino experiments. It has been pointed out that these terms play a unique role in atmospheric neutrino oscillations, such as the possible measurement of $\sin^2\theta_{23}$, (i.e., the discrimination of the octant of θ_{23}) [30]. For simplicity, we assume that $\theta_{13} = 0$. The change in the atmospheric ν_e flux due to oscillations driven by the solar oscillation terms is written as [30];

$$\frac{F^{osc}_{\nu e}}{F^0_{\nu e}} - 1 = P_2(r\cos^2\theta_{23} - 1), \tag{1}$$

where $F^{osc}_{\nu e}$ and $F^0_{\nu e}$ are the atmospheric ν_e fluxes with and without oscillations, $r(\equiv F^0_{\nu\mu}/F^0_{\nu e})$ is the ratio of the un-oscillated atmospheric ν_μ and ν_e fluxes, and P_2 is the two neutrino transition probability ($\nu_e \rightarrow \nu_x$) in matter driven by the solar oscillation terms. P_2 is large for neutrinos passing through the earth with the energies below 1 GeV. Thus the sub-GeV atmospheric neutrinos play an important role in observing the solar term effect. Since the ν_μ and ν_e flux ratio (r) is approximately 2 in the sub-GeV neutrino energy region, the $F^{osc}_{\nu e}/F^0_{\nu e}$ value in Eq.1 is very close to 1 in the case of the maximal 2-3 mixing (i.e., $\theta_{23} = 45°$). However, according to Eq.1, if θ_{23} is in the first (second) octant ($\theta_{23} < 45°$ ($> 45°$)), an excess (deficit) of the sub-GeV e-like events is expected.

Figure 5 shows the χ^2 distribution as a function of $\sin^2\theta_{23}$ obtained by the Super-K experiment from an oscillation analysis that includes the solar oscillation terms [31]. With the present data, there is no indication for the deviation from $\sin^2\theta_{23} = 0.5$ even if the solar terms are taken into account.

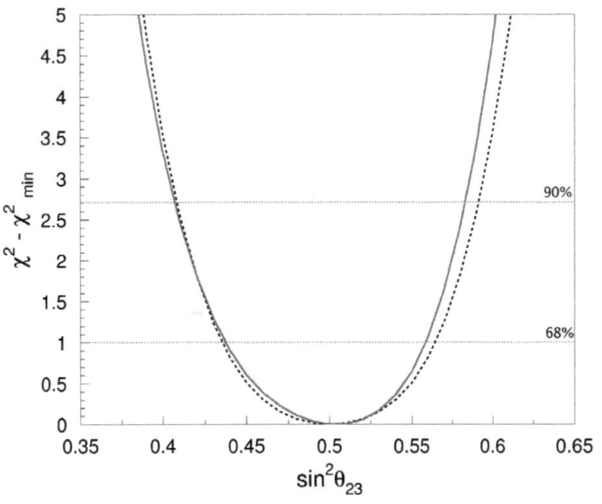

Figure 5. The $\chi^2 - \chi^2_{\rm min}$ distributions as a function of $\sin^2\theta_{23}$ for oscillations without the solar terms (dashed line) and with the solar terms (solid line) by the atmospheric neutrino data from the Super-Kamiokande experiment (173 kton·yr exposure) [31]. $\theta_{13} = 0$ is assumed.

4. Nucleon decay searches

An unambiguous discovery of nucleon decays would have a profound impact to particle physics, since the discovery implies that the electromagnetic, weak and strong forces are unified at a very high energy scale. Therefore dedicated experiments started to search for nucleon decays around 1980 [32], [33], [34], [35], [36] several years after the proposal for the Grand Unified Theories [4], [5]. Due to the importance, the searches for nucleon decays are still continuing. Since the exposure is dominated by the Super-K experiment, the latest results from Super-K are discussed in this section.

4.1. $p \rightarrow e^+\pi^0$ search

The final state for this decay channel is simple; 3 electromagnetic showers. 2 of them are due to π^0 decay into 2 gamma rays. The total invariant mass should be consistent with the proton mass, and the total momentum within the Fermi momentum. The produced π^0 in the oxygen nucleus may interact within the same nucleus, and therefore the detection efficiency is reduced. Figure 6 shows the total invariant mass versus total momentum plots for the data, atmospheric neutrino Monte Carlo, and proton decay Monte Carlo. The data are consistent with the atmospheric neutrino prediction. There has been no candidate event in the signal box. The 90% C.L. lower limit on the lifetime of proton for this decay mode is 1.01×10^{34} years. (See Ref. [37] for details of the data analysis with a slightly smaller data sample.)

4.2. $p \rightarrow \overline{\nu}K^+$ search

This decay mode is not trivial to search for in water Cherenkov detectors, since K^+ is invisible in these detectors. Therefore, several methods have been developed to identify low-energy K^+ decays, which would imply proton decays into $\overline{\nu}K^+$.

One method ("prompt γ search") searches for a 6 MeV de-excitation gamma ray accompanied with a delayed muon of 236 MeV/c from a stopped K^+ decay [38]. These signals are expected for proton decays occurring in oxygen nuclei. The gamma rays can be emitted by the excited,

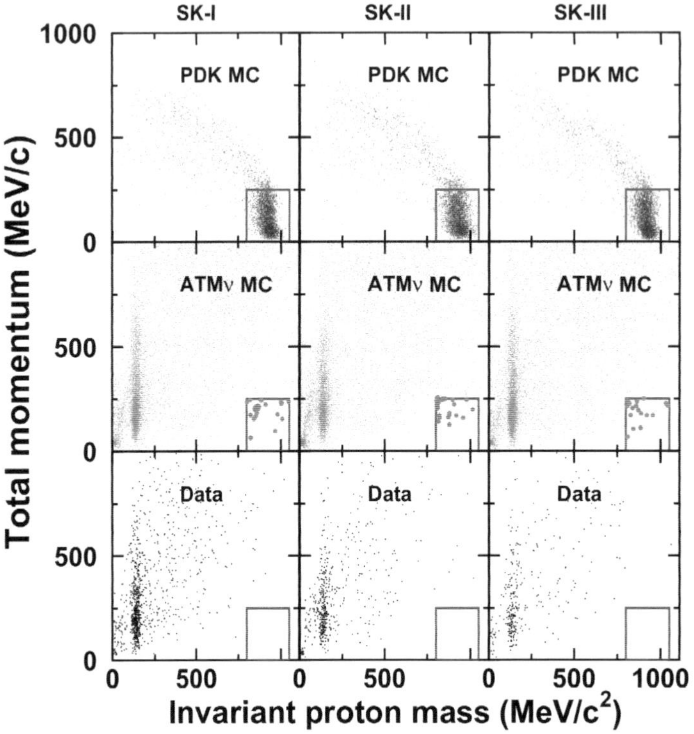

Figure 6. Search for $p \rightarrow e^+\pi^0$: Total invariant mass versus total momentum distributions for Super-K-I (left panels). Super-K-II (middle panels) and Super-K-III (right panels). The top panels show those for proton decay Monte Carlo, the middle panels for atmospheric neutrino Monte Carlo (500 years of exposure equivalent), and the bottom panels for the data. The estimated detection efficiencies are 1.6%, 43.5% and 45.2% for Super-K-I, II and III, respectively. The expected number of atmospheric neutrino background events are 0.20, 0.11 and 0.06 for the Super-K-I, II and III exposures, respectively. No candidate events has been observed.

residual nuclei ($^{15}N^*$). Another method searches for a $\pi^+ + \pi^0$ signal ("$\pi^+ + \pi^0$ search") from a stopped K^+. The third method simply searches for a single muon with the momentum consistent with 236 MeV/c. These 3 methods have been used to search for proton decays in Super-K. Table 1 shows the results for the first 2 methods. There has been no candidate event for these 2 methods. In addition, there has been no excess of muon events in the muon momentum distribution around 236 MeV/c. The limit on the proton lifetime for this decay mode is 3.3×10^{33} years at 90% C.L.. (See Ref. [39] for details of the data analysis with a slightly smaller data sample.)

5. Summary

Researches with massive underground detectors have almost half a century of history. These detectors have been contributing to particle and astro-particle physics substantially. The solar neutrino problem, which was discovered by the first generation massive underground detector,

Table 1. Search for $p \to \overline{\nu} K^+$: Summary of the detection efficiency, background and candidate events for the prompt gamma and for the $\pi^+ + \pi^0$ searches for SK-I, II and III.

Mode		SK-I(92 kt·yr)	SK-II(49 kt·yr)	SK-III(32 kt·yr)
Prompt γ search	Efficiency(%)	7.2	5.8	7.3
	Background	0.16	0.08	0.03
	Candidates	0	0	0
$\pi^+\pi^0$ search	Efficiency(%)	6.5	5.3	6.6
	Background	0.46	0.33	0.14
	Candidates	0	0	0

has been understood as due to $\nu_e \to \nu_x$ oscillations. Recent solar and reactor long baseline experiments have measured the 1-2 oscillation parameters accurately. Atmospheric neutrino experiments of the previous generation discovered the atmospheric neutrino anomaly. They have also been understood to be due to $\nu_\mu \to \nu_\tau$ oscillations. $\sin^2 2\theta_{23}$ has been measured accurately by atmospheric neutrino experiments. Proton decay has been searched for seriously for almost 30 years. No evidence for proton decay has been found. The present 90% C.L. limits are 1.0×10^{34} and 3.3×10^{33} years for $e^+\pi^0$ and $\overline{\nu}K^+$ decay modes (preliminary), respectively.

The achievements of the large underground detectors suggest that future massive underground detectors will further contribute to particle and astro-particle physics.

Acknowledgments

The author would like to thank the organizers of this workshop for their kind invitation. The author would like to thank the collaborators in Super-Kamiokande for their help in preparing the presentation. This work has been partially supported by the Grants-in-Aid for Scientific Research, JSPS.

References

[1] Bahcall J N 1964 *Phys. Rev. Lett.* **12** 300–302
[2] Davis R 1964 *Phys. Rev. Lett.* **12** 303–305
[3] Davis Jr R, Harmer D S and Hoffman K C 1968 *Phys. Rev. Lett.* **20** 1205–1209
[4] Pati J C and Salam A 1974 *Phys. Rev.* **D10** 275–289
[5] Georgi H and Glashow S L 1974 *Phys. Rev. Lett.* **32** 438–441
[6] Hirata K *et al.* (Kamiokande-II) 1987 *Phys. Rev. Lett.* **58** 1490–1493
[7] Bionta R M *et al.* 1987 *Phys. Rev. Lett.* **58** 1494
[8] Hirata K S *et al.* (Kamiokande-II) 1988 *Phys. Lett.* **B205** 416
[9] Hirata K S *et al.* (Kamiokande-II) 1989 *Phys. Rev. Lett.* **63** 16
[10] Anselmann P *et al.* (GALLEX) 1992 *Phys. Lett.* **B285** 376–389
[11] Abdurashitov D N *et al.* 1994 *Phys. Lett.* **B328** 234–248
[12] Wolfenstein L 1978 *Phys. Rev.* **D17** 2369–2374
[13] Mikheev S P and Smirnov A Y 1985 *Sov. J. Nucl. Phys.* **42** 913–917
[14] Ahmad Q R *et al.* (SNO) 2001 *Phys. Rev. Lett.* **87** 071301 (*Preprint* nucl-ex/0106015)
[15] Fukuda S *et al.* (Super-Kamiokande) 2001 *Phys. Rev. Lett.* **86** 5651–5655 (*Preprint* hep-ex/0103032)
[16] Ahmad Q R *et al.* (SNO) 2002 *Phys. Rev. Lett.* **89** 011301 (*Preprint* nucl-ex/0204008)
[17] Abe S *et al.* (KamLAND) 2008 *Phys. Rev. Lett.* **100** 221803 (*Preprint* 0801.4589)
[18] Fogli G L, Lisi E, Marrone A, Palazzo A and Rotunno A M 2008 *Phys. Rev. Lett.* **101** 141801 (*Preprint* 0806.2649)
[19] Schwetz T, Tortola M A and Valle J W F 2008 *New J. Phys.* **10** 113011 (*Preprint* 0808.2016)
[20] Aharmim B *et al.* (SNO) 2009 (*Preprint* 0910.2984)
[21] Gonzalez-Garcia M C, Maltoni M and Salvado J 2010 *JHEP* **04** 056 (*Preprint* 1001.4524)
[22] Becker-Szendy R *et al.* 1992 *Phys. Rev.* **D46** 3720–3724
[23] Fukuda Y *et al.* (Kamiokande) 1994 *Phys. Lett.* **B335** 237–245
[24] Fukuda Y *et al.* (Super-Kamiokande) 1998 *Phys. Rev. Lett.* **81** 1562–1567 (*Preprint* hep-ex/9807003)

[25] Sanchez M C *et al.* (Soudan 2) 2003 *Phys. Rev.* **D68** 113004 (*Preprint* hep-ex/0307069)
[26] Ambrosio M *et al.* (MACRO) 2004 *Eur. Phys. J.* **C36** 323–339
[27] Ashie Y *et al.* (Super-Kamiokande) 2005 *Phys. Rev.* **D71** 112005 (*Preprint* hep-ex/0501064)
[28] Ashie Y *et al.* (Super-Kamiokande) 2004 *Phys. Rev. Lett.* **93** 101801 (*Preprint* hep-ex/0404034)
[29] Adamson P *et al.* (MINOS) 2008 *Phys. Rev. Lett.* **101** 131802 (*Preprint* 0806.2237)
[30] Peres O L G and Smirnov A Y 2004 *Nucl. Phys.* **B680** 479–509 (*Preprint* hep-ph/0309312)
[31] Wendell R *et al.* (Super-Kamiokande) 2010 *Phys. Rev.* **D81** 092004 (*Preprint* 1002.3471)
[32] Krishnaswamy M R *et al.* 1981 *Phys. Lett.* **B106** 339–346
[33] Battistoni G *et al.* 1983 *Phys. Lett.* **B133** 454
[34] Blewitt G *et al.* 1985 *Phys. Rev. Lett.* **54** 22
[35] Berger C *et al.* (Frejus) 1989 *Nucl. Phys.* **B313** 509
[36] Hirata K S *et al.* (Kamiokande-II) 1989 *Phys. Lett.* **B220** 308
[37] Nishino H *et al.* (Super-Kamiokande) 2009 *Phys. Rev. Lett.* **102** 141801 (*Preprint* 0903.0676)
[38] Totsuka Y Presented at 7th Workshop on Grand Unification, ICOBAN '86, Toyama, Japan, Apr 16-18, 1986
[39] Kobayashi K *et al.* (Super-Kamiokande) 2005 *Phys. Rev.* **D72** 052007 (*Preprint* hep-ex/0502026)

Journal of Physics: Conference Series **308** (2011) 012004 doi:10.1088/1742-6596/308/1/012004

Near detectors for long baseline neutrino experiments

Tsuyoshi NAKAYA

Department of Physics, Kyoto University, Kyoto 606-8502, Japan

E-mail: t.nakaya@scphys.kyoto-u.ac.jp

Abstract. We describe near neutrino detectors in the long baseline neutrino oscillation experiments and introduce the physics results obtained with the detectors. In the evolution of the neutrino experiments, the neutrino detectors at the near site play an essential role to measure the property of the neutrino beam and to measure the neutrino cross sections. We introduce the near neutrino detectors and the physics results in the K2K, SciBooNE and T2K neutrino experiments. We also discuss the future prospects of the neutrino experiments together with a possible advanced design of near detectors.

1. Introduction

After the discovery of neutrino oscillation by the Super-Kamiokande (Super-K) experiment [1], the study of neutrino mass and mixing becomes a very interesting subject in particle physics. Several accelerator based neutrino oscillation experiments are conducted and provide the precise measurements of neutrino oscillations. The K2K experiment was the first long baseline neutrino experiment and confirmed the neutrino oscillation [2]. Then, the MINOS experiment provides accurate information of the neutrino oscillation parameters [3], especially the precise mass square difference. Recently, the OPERA experiment finds a candidate event of ν_τ appearance in $\nu_\mu \to \nu_\tau$ oscillation [4]. In 2009, the T2K experiment launched out. T2K will provide the most precise measurements of neutrino oscillation parameters in $\nu_\mu \to \nu_\tau$ oscillation and the most sensitive search for $\nu_\mu \to \nu_e$ appearance to probe into the non-measured mixing angle, θ_{13} [5]. When we find the signal of $\nu_\mu \to \nu_e$ appearance, we could enter the new era to study CP violation in neutrinos. So, it is one of the most fascinating time to study neutrinos.

In the long baseline neutrino oscillation experiments, the two detector configuration is often applied. One detector, called a near neutrino detector, is located near the neutrino source to measure the neutrino beam flux. The other detector, called the far detector, is located at the far site to measure oscillated neutrinos. The neutrino oscillation could be precisely measured by comparing the measurements between the near detector and the far detector.

With the evolution of accelerator neutrino oscillation experiments, several type of near neutrino detectors are developed to measure the neutrino beam flux more precisely. In addition, it is found that the understanding of physics of neutrino cross sections is critical and the cross sections should be measured more precisely to achieve the better sensitivity on neutrino oscillation.

2. Roles of a near neutrino detector

We design and build near neutrino detectors with the following four roles in principle.

(i) To measure the neutrino beam flux at the near site, in reality, to measure the product of the flux times the cross section, which is used to predict the neutrino beam flux at the far detector.
(ii) To monitor the neutrino beam steadily for a long running period.
(iii) To measure and study the neutrino cross sections.
(iv) To develop a new detector technology for future neutrino experiments.

2.1. Neutrino beam flux measurement

The observed rate of neutrino events at the near detector as a function of the reconstructed neutrino energy E'_ν is expressed as

$$N_{ND}(E'_\nu) = \int \Psi_{ND}(E_\nu) \times \sigma(E_\nu) \times \epsilon_{ND}(E_\nu) \times f_{ND}(E'_\nu, E_\nu) dE_\nu, \quad (1)$$

where "ND" denotes the Near Detector, E_ν is the true neutrino energy, $\Psi_{ND}(E_\nu)$ is the neutrino beam flux at the near detector, $\sigma(E_\nu)$ is the neutrino cross section, $\epsilon_{ND}(E_\nu)$ is the detection efficiency, and $f_{ND}(E'_\nu, E_\nu)$ is the energy response function that E'_ν is measured for the true energy E_ν. The similar expression is described for the rate of neutrino events at the far detector.

$$N_{FD}(E'_\nu) = \int P(E_\nu, \Delta m^2, \sin^2 2\theta) \times \Psi_{FD}(E_\nu) \times \sigma(E_\nu) \times \epsilon_{FD}(E_\nu) \times f_{FD}(E'_\nu, E_\nu) dE_\nu, \quad (2)$$

where "FD" denotes the Far Detector and $P(E_\nu, \Delta m^2, \sin^2 2\theta)$ is the probability of neutrino oscillation with the parameters of Δm^2 and $\sin^2 2\theta$. By comparing the equations 1 and 2, the neutrino oscillation parameters could be extracted. In the equations, the neutrino flux $\Psi(E_\nu)$ is usually estimated based on data measured by the hadron production experiments. In K2K, the corresponding hadron production experiment is the HARP experiment [6] and it is CERN NA61 in T2K [7]. The precision of the neutrino flux prediction is typically ~ 20 %. When we take the flux ratio $\{\Psi_{FD}(E_\nu)/\Psi_{ND}(E_\nu)\}$ between the far and near sites, the precision is often reliably improved to be 5 % level or less. So, it is one of the important motivation to measure and extract $\Psi_{ND}(E_\nu)$ in the near detector. Another large uncertainty on the prediction of the neutrino event rate is the uncertainty of neutrino cross sections, $\sigma(E_\nu)$, which is typically ~ 20 %. The total cross section $\sigma(E_\nu)$ are the sum of several interaction channels: charged-current (CC) quasi-elastic (QE) scattering, CC resonant π production, CC coherent π production, CC deep inelastic scattering, and the same types of neutral current (NC) interactions. Therefore, we need to measure several interaction channels separately to improve the precision of the neutrino cross sections. Other quantities, $\epsilon(E_\nu)$ and $f(E'_\nu, E_\nu)$, depend on the detector design. The near detectors are designed to have the similar efficiency and the response function as those of the far detectors with a common design. Sometimes, a different type of near detector is designed and constructed because a compact and high-resolution neutrino detector is applicable and functional at the near site with the intense neutrino flux. In T2K, the far detector is a Water Cherenkov detector, Super-K, and the near detector is a fine segmented detector called ND280 which has TPCs operating under the magnetic field. We show the T2K neutrino events observed in Super-K and in ND280 in Figure 1 and 2, which looks quite different.

2.2. Neutrino beam monitor

A neutrino experiment often runs for several years to accumulate enough statistics due to the very low event rate of neutrinos at the far site. The beam quality must be monitored and guaranteed to be stable for the long running period. For this monitoring purpose, the observed event rate and the energy spectrum at the near detectors could provide essential information.

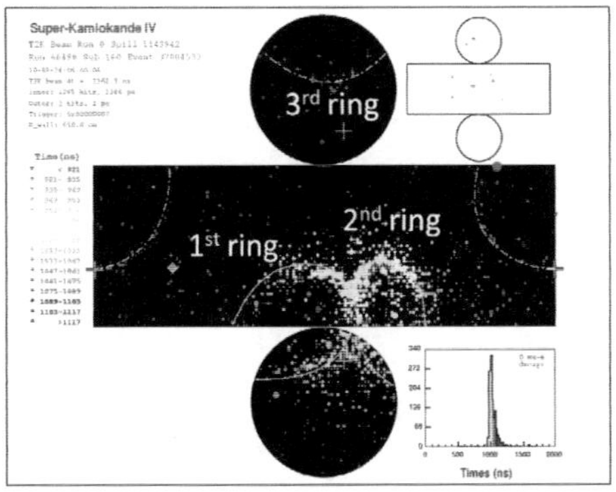

Figure 1. The First T2K Neutrino event observed in Super-K at the far site.

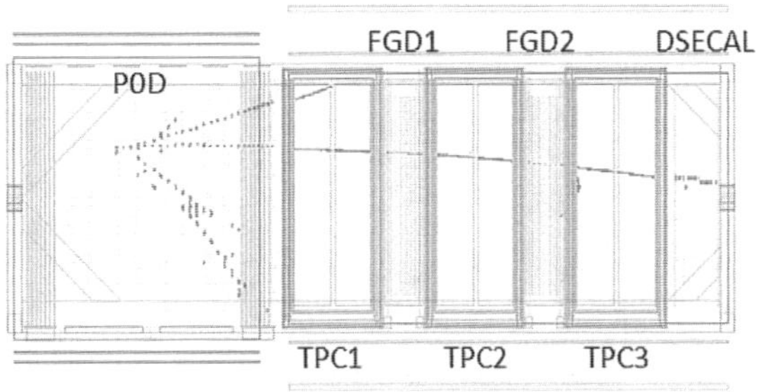

Figure 2. A T2K Neutrino event observed in ND280 at the near site.

As an example, the K2K experiment collected data from 1999 to 2004. During the data taking period, the near detector monitored the neutrino event rate, the neutrino beam direction and the neutrino energy spectrum via the measurement of the muon momentum [2]. The event rate was guaranteed to be stable within 2 %, the beam direction within 1 mrad and no change of the energy spectrum.

In T2K, the near detectors, especially the INGRID detector, are constructed to satisfy the requirements for beam monitoring. In Figure 3, the neutrino event rate normalized by the number of protons on target, the neutrino beam profile and the neutrino beam direction measured by T2K INGRID are shown.

2.3. Neutrino cross section measurement

As already introduced, the neutrino cross sections are measured in near neutrino detectors with the intense neutrino beams. The current precision of the neutrino cross sections is not as good as the demand by the future neutrino oscillation experiments. The theoretical models of neutrino interactions are not perfect, either. Then, physics of neutrino interactions provides interesting subjects of nuclear physics in the accelerator neutrino experiments.

In the K2K experiment, the cross sections of NC π^0 production [8], CC single π^+ production [9], CC coherent π^+ production [10], and CC π^0 production [11, 12] are measured. In addition, the form factor of CC-QE scattering is also studied [13].

In the SciBooNE experiment, which worked as a near detector of the MiniBooNE experiment, the CC coherent π^+ production [14], the NC single π^0 production [15] and the NC coherent π^0 productions [16] are intensively studied. In the SciBooNE experiment, the ratio of the NC to the CC coherent π cross sections is measured to be $0.14^{+0.30}_{-0.28}$ which is far away from 2 predicted by isospin symmetry. The result receives much attentions.

In future neutrino oscillation experiments, we will measure the neutrino cross sections more precisely with sophisticated neutrino detectors [17]. Based on the measurements, several models of neutrino-nucleus interactions could be well tuned, so that the understanding of neutrino interaction physics will be advanced.

2.4. Development of the new detector technology

Another important reason to construct the near neutrino detectors is to develop and test the new detector technology, which might be utilized to a novel type of neutrino detectors and spin off a new application.

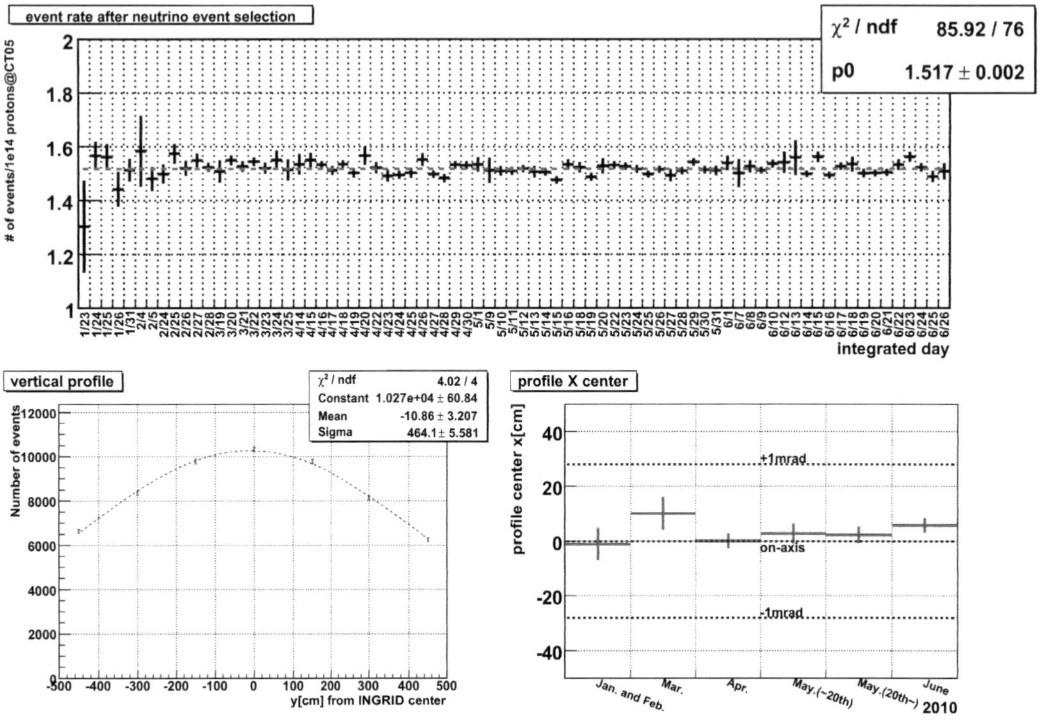

Figure 3. The T2K neutrino beam quantities measured by the INGRID detector: (Top) the neutrino event rate normalized by the number of protons on target as a function of the time, (Bottom left) the neutrino beam profile in the vertical view, and (Bottom right) the reconstructed neutrino beam direction (the profile center) in each run period in the horizontal view.

As an example, two new detector technologies are developed and realized in T2K ND280. One is a new semiconductor photon sensor, called MPPC (Multi-Pixel-Photon-Counters) manufactured by Hamamatsu Photonics. The other is called Micromegas (Micro MEsh GAseous Structure) which is based on the technology of Printed Circuit Board (PCB). The Micromegas is used in the gas amplification part of the large TPC detectors in T2K-ND280. Below, we briefly introduce MPPC in T2K.

2.4.1. MPPC The T2K ND280 uses 60,000 MPPCs in total. MPPC is the new photon-detector, and T2K is the first experiment to use the large amount of MPPCs. The T2K collaboration develops the custom designed MPPC with a collaboration of Hamamatsu Photonics. We show the T2K MPPC specification in Table 1. Advantages of MPPC are (1) very compact, (2) operating at the not-high voltage (~70V), (3) tolerance to the magnetic field, (4) high gain ($\sim 10^5$ or higher), (5) good photon-counting capability and (6) high photon detection efficiency. A few disadvantages are the high dark count rate and the temperature dependence of the performance. In the accelerator neutrino experiments, the high dark count rate is not a serious problem because of the synchronization of the timing with the beam. In T2K, we measured a basic performance of all MPPCs before installation. The result of measurements is presented in [18].

Journal of Physics: Conference Series **308** (2011) 012004 doi:10.1088/1742-6596/308/1/012004

Table 1. Specifications of T2K-MPPC (S10362-13-050C).

Item	Spec.
Active area	1.3×1.3 mm^2
Pixel size	50×50 μm^2
Number of pixels	667
Operation voltage	~ 70 V
Photon Detection Efficiency	> 15 %
Dark count	~ 1 Mcps (> 0.5 photo-electrons)

3. Future neutrino oscillation experiments with near neutrino detectors

The future of the neutrino oscillation experiments are fruitfully discussed in this workshop and the reports will be available in the workshop proceedings. In a general consensus, when we discover $\nu_\mu \rightarrow \nu_e$ appearance, we will start a new experiment to probe CP violation in neutrinos. Hereafter, we discuss a scenario of the future experiment based on T2K with the J-PARC facility. If T2K discovers $\nu_\mu \rightarrow \nu_e$ appearance, we will propose to build the more gigantic neutrino detector at the far site together with the upgrade of the J-PARC neutrino beam. There are two options of detectors discussed in Japan: one is the megaton size water Cherenkov detector, called Hyper-K, and the other is the 100 kton size liquid Argon TPC.

3.1. A near detector for a future experiment with Hyper-K

In the future experiment with Hyper-K, we will measure the asymmetry of neutrino oscillations between neutrinos and anti-neutrinos. In the experiment, the better understanding of the anti-neutrino beam and the cross sections of anti-neutrinos are more important because they are not well understood yet. In the experiment, the near detector should have a good performance to study anti-neutrino interactions. A feature of anti-neutrino QE interaction, $\overline{\nu}_\mu + p \rightarrow \mu^+ + n$, is the exiting of one positive muon with a neutron in the final state. As an example, in SciBooNE, the anti-neutrino QE interaction is identified by selecting the event with no activities around the interaction vertex except for the muon. Since the T2K ND280 has a capability of the sign selection of charged partciles, the anti-neutrino event could be effectively identified by selecting the positive muon without an extra vertex activity. It is one of the advantageous feature of T2K ND280.

3.2. A near detector for a future experiment with a big liquid Argon TPC

In the experiment with the large liquid Ar TPC, we will study the CP violation by measuring the oscillation probability at the first and the second oscillation maxima. Especially, the oscillation probability at the second peak contains a large effect of the CP phase interference. In the experiment, the energy resolution and the response function must be well understood and be tested to probe CP violation. So, the energy resolution and response functions should be tested by the near detector. In addition to the response function, the effects of the feed-down background events from the high energy region to the low energy around the second oscillation maximum should be studied. The feed-down background could be caused by the imperfect detector performance and the uncertainty of the neutrino interactions. The near detector must address these issues and demonstrate the high performance of the liquid Ar TPC.

3.3. Requirements of the near detectors for the future experiments

In order to consider requirements of the future near detectors, first we look over the performance of the current near detectors introduced in Section 2.3. Some results are provided by the SciBooNE experiment, in which the fine-segmented scintillator tracker called SciBar [19] plays an important role. We show the tracking performance of SciBar in Figure 4 for CC-QE interactions simulated by NEUT [20] with the SciBooNE neutrino beam (~ 1 GeV). The SciBar detector has

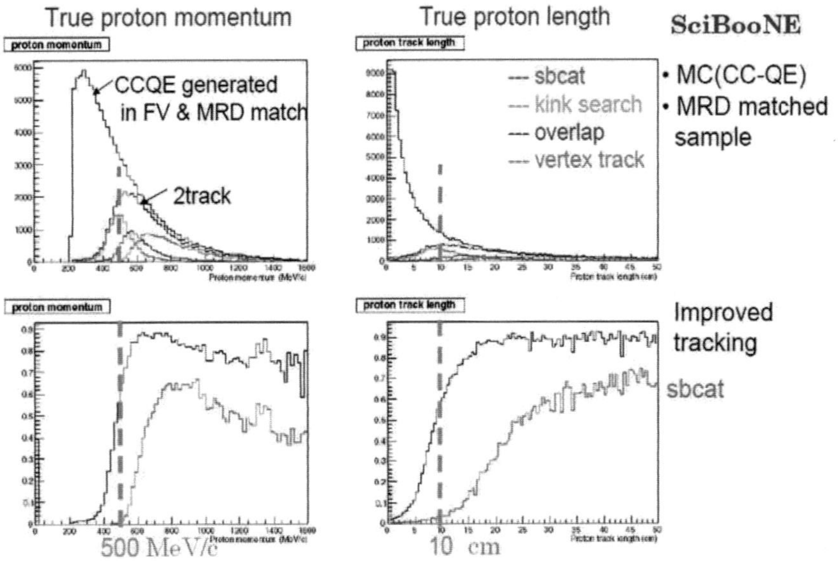

Figure 4. The tracking performance of the SciBar detector in the SciBooNE experiment. The left-top plot shows the proton momentum distribution in CC-QE interactions predicted by NEUT for the generated events and the reconstructed events with two tracks. The left-bottom plot is the detection efficiency of the proton track as a function of the proton momentum with the SciBooNE improved (black line) and the standard (red line) tracking algorithms . The right top plot is the range of the proton in CC-QE for the generated events and the reconstructed events. The right bottom plot is the detection efficiency of the proton track as a function of the proton range with the improved (black line) and the standard (red line) tracking algorithms.

a scintillator segmentation of 2.5×1.3 cm^2 in cross section and 300 cm in length. The scintillators are alternately arranged vertically and horizontally, which results in event reconstruction in two views with the effective resolution of 2.5×2.6 cm^2 in cross section. Because a tracking algorithm requires two or more hits in one view, the naive threshold for the track length is 5 cm. We see the 5 cm minimum threshold in the right-bottom plot of Figure 4. We also find that a effective threshold is around 10 cm and many protons in CC-QE interactions are distributed below the threshold.

In T2K-ND280, FGD (Fine-Grained-Detector) has a finer segmentation of 1×1 cm^2 than that of SciBar. T2K-ND280 will improve the tracking performance by a factor of two or three than that of SciBooNE by considering the finer segmentation. However, even with the better performance (~ 5 cm effective threshold), still we will only reconstruct ~ 50 % of CC-QE protons for ~ 1 GeV neutrinos as seen in Figure 4.

In future neutrino experiments, if we want to reconstruct the protons in CC-QE interaction with much higher efficiency, it requires a few mm segmentation resulting in ~ 1 cm tracking capability. Today, one of the promising detector technology with 1 cm tracking capability is a liquid Ar TPC. The other option will be the scintillating fiber detector although we have to develop the economical way to readout the image of fibers by developing a new photon sensor. It might be possible to develop the fine segmented photon-imaging sensor based on the MPPC technology with some R&D.

3.4. A possible upgrade of the T2K near detector

In a future experiment of T2K with the J-PARC facility, the T2K near detector will play an essential role to understand the property of the T2K neutrino beam including an anti-neutrino beam, to improve knowledge of neutrino cross sections and to develop the future technology for neutrino detectors. It is natural to consider a possible upgrade of the T2K near detector for the future experiments [1].

As discussed in Section 3.3, a possible upgraded detector is a liquid Ar TPC with much finer resolutions, which could improve the measurements of neutrino interactions. In addition, the development of liquid Ar TPC technology could help R&D for the big liquid Ar TPC at the far site. It is exciting to consider the possibility to install the compact liquid Ar TPC inside of the ND280 magnet. We could also propose the extension of T2K with an anti-neutrino beam, which is primarily a rehearsal of the CP-violation neutrino experiment discussed in Section 3.1. With the experiment with anti-neutrinos, we will study the property of anti-neutrino beam in detail and study the anti-neutrino cross sections precisely. For the measurements of anti-neutrino interactions, the detection of a neutron in the final state could be important to tag anti-neutrino interactions by suppressing the background of neutrino interactions which have a non-negligible contribution with a conventional neutrino beam. For this neutron tagging, a superior performance of liquid Ar TPC may have an advantage.

A new finest grained detector with scintillating fibers could be another option of the new T2K near detector. The detector itself consists of scintillating fibers of a few mm diameter, and the image from the fibers must be captured with a high-resolution large-area photon camera. The camera, a fine-pixelated photon detector with a large photo-coverage, is quite challenging and is an interesting device for the future R&D. Today, the fine-pixelated photon detector could be realized based on the MPPC technology with fine segmentation. The scintillating fiber neutrino detector with a new large-area fine-resolution photon imaging sensor could be an attractive option as the new T2K near detector. Such a photon imaging sensor have a huge possibility to be used in many applications in different fields.

In addition to the high resolution detectors discussed above, a miniature of the far detector at the near site is also useful to develop and test the detector technology used in the far detector. In the case of Hyper-K for the far detector, a small-size water Cherenkov detector with the same photon-sensors and the same electronics as those for Hyper-K is utilized as a testing ground of the detector technologies. We could also measure the neutrino cross sections with water by using the small Cherenkov detector.

In order to realize the above possibilities and to propose a new realistic experiment to probe CP symmetry in neutrino oscillation, we need to start R&D for the future immediately.

4. Summary and Outlook

In this paper, we look over the essential functions of near neutrino detectors together with the physics results achieved with the detectors. There is no question that we need a near detector in the long baseline accelerator neutrino oscillation experiments and that the development of the near detector is strongly coupled with the realization of a gigantic far detector. Since the

[1] Since the upgrade of T2K-ND280 was never discussed in the T2K collaboration, this is simply an idea of the author.

far detector in the future experiments could be expensive and could require a long construction time, an upgrade of the near detectors is useful to develop the key detector technology for a far detector and will produce many physics results before the new far detector realized. We start considering a new near detector or the upgrade of the current one for future neutrino experiments.

Acknowledgments

The author acknowledges the K2K, SciBooNE and T2K collaborations for the results presented in this paper. We acknowledge supports by MEXT and JSPS with the Grant-in-Aid for Young Scientists S 20674004, Scientific Research on Priority Areas "New Developments of Flavor Physics", the global COE program "The Next Generation of Physics, Spun from Universality and Emergence".

References

[1] Y. Fukuda *et al.* [Super-Kamiokande Collaboration], Phys. Rev. Lett. **81**, 1562 (1998)
[2] M. H. Ahn *et al.* [K2K Collaboration], Phys. Rev. D **74**, 072003 (2006)
[3] P. Adamson *et al.* [MINOS Collaboration], Phys. Rev. Lett. **101**, 131802 (2008)
[4] N. Agafonova *et al.* [OPERA Collaboration], Phys. Lett. B **691**, 138 (2010)
[5] Y. Itow *et al.* [The T2K Collaboration], arXiv:hep-ex/0106019
[6] M. G. Catanesi *et al.* [HARP Collaboration], Nucl. Phys. B **732**, 1 (2006)
[7] A. Bravar [NA61/SHINE Collaboration], AIP Conf. Proc. **1222**, 431 (2010)
[8] S. Nakayama *et al.* [K2K Collaboration], Phys. Lett. B **619**, 255 (2005)
[9] A. Rodriguez *et al.* [K2K Collaboration], Phys. Rev. D **78**, 032003 (2008)
[10] M. Hasegawa *et al.* [K2K Collaboration], Phys. Rev. Lett. **95**, 252301 (2005)
[11] S. Mine *et al.* [K2K Collaboration], Phys. Rev. D **77**, 032003 (2008)
[12] C. Mariani [K2K Collaboration], AIP Conf. Proc. **1189** (2009) 339
[13] R. Gran *et al.* [K2K Collaboration], Phys. Rev. D **74**, 052002 (2006)
[14] K. Hiraide *et al.* [SciBooNE Collaboration] Phys. Rev. D **78**, 112004 (2008)
[15] Y. Kurimoto *et al.* [SciBooNE Collaboration], Phys. Rev. D **81**, 033004 (2010)
[16] Y. Kurimoto *et al.* [SciBooNE Collaboration], Phys. Rev. D **81**, 111102 (2010)
[17] T. Nakaya [K2K, Super-K, SciBooNE and T2K Collaborations], PoS **NUFACT08** (2008) 010
[18] M. Yokoyama *et al.*, Nucl. Instrum. Meth. A **622**, 567 (2010) [arXiv:1007.2712 [physics.ins-det]].
[19] K. Nitta *et al.*, Nucl. Instrum. Meth. A **535**, 147 (2004)
[20] Y. Hayato, Nucl. Phys. Proc. Suppl. **112**, 171 (2002)

The WArP Experiment

R. Acciarri[c], M. Antonello[b], B. Baibussinov[f], P. Benetti[a], F. Calaprice[e], E. Calligarich[a], M. Cambiaghi[a], N. Canci[c], C. Cao[e], F. Carbonara[d], F. Cavanna[c], S. Centro[f], M.B. Ceolin[f], A. Chavarria[e], A. Cocco[d], F. Di Pompeo[bc], G. Fiorillo[d], C. Galbiati[e], L. Grandi[bc], B. Loer[e], G. Mangano[d], A. Menegolli[a], G. Meng[f], C. Montanari[a*], O. Palamara[b], L. Pandola[b], F. Pietropaolo[f], G.L. Raselli[a], M. Roncadelli[a], M. Rossella[a], C. Rubbia[b], R. Saldanha[e], E. Segreto[b], A. Szelc[b], S. Ventura[f], C. Vignoli[a].

[a] Università degli Studi di Pavia e INFN, Pavia, Italy

[b] INFN - Laboratori Nazionali del Gran Sasso, Assergi, Italy

[c] Università degli Studi dell'Aquila e INFN, L'Aquila, Italy

[d] Università degli Studi di Napoli e INFN, Napoli, Italy

[e] Princeton University Department of Physics, Princeton, New Jersey, USA

[f] Università degli Studi di Padova e INFN, Padova, Italy

[*] Corresponding author, e-mail: claudio.montanari@pv.infn.it

Abstract. Cryogenic noble liquid detectors are presently considered one of the best options for WIMP Dark Matter searches, especially when extensions to multi ton scale sensitive masses are foreseen. The WArP experiment is the first one that exploits the unique characteristics of liquid Argon to make a highly sensitive search for WIMP Dark Matter candidates. In 2008, a double phase detector has been assembled in the Gran Sasso National Laboratory with 140 kg sensitive mass and a discovery potential in the range of 5 x 10^{-45} cm^2 in the spin-independent WIMP-nucleon cross-section. In addition to standard neutrons and gamma-rays passive shields, WArP implements an 8 ton liquid Argon active shield with 4π coverage. The detector was commissioned and put into operation during the first half of 2009 for a first technical run. This first run lasted about three months and then it was stopped for some detector repairs and modifications in the summer of 2009. A second run was started at the beginning of 2010. Detector design, construction and assembly are described, together with the results of the technical run and the very first results of the 2010 run.

1. Introduction

Increasing efforts are being made to unveil the nature of Dark Matter, especially since it's existence was definitively certified by astrophysical and cosmological observations [1]. Several possibilities for Dark Matter constituency are seriously considered although a central role is presently played by direct WIMP searches due to the potential link with the SUSY hypothesis.

Journal of Physics: Conference Series **308** (2011) 012005 doi:10.1088/1742-6596/308/1/012005

In the last ten years several technologies have grown up from the laboratory scale to full size experiments running or to be run in several underground locations [2]. Despite the increasing efforts made by the present experiments, the positive signal claimed by the DAMA Collaboration in terms of seasonal variations [3] still stands without an independent confirmation or clear disproof. In this scenario there are already many proposals for upgrades of the presently adopted technologies to sensitive masses of the order of few tons with increasing complexity in terms of shielding and background control in order to fully explore the predicted regions of the minimal SUSY hypothesis. Noble liquid detectors play a central role in this future scenario due to their potentiality for the instrumentation of very large sensitive masses. WArP is the very first experiment that exploits the characteristics of Liquid Argon (LAr) for direct Dark Matter searches.

The WArP technology was established in several years of R&D studies. They were started in CERN at the beginning of the nineties with pioneering works with liquid Xenon in the framework of the general R&D activities of the Icarus experiment [4]. When the concept of double phase chamber as a tool for direct Dark Matter searches was put forward [5], around the mid of the nineties, we opted for argon as our primary choice for a number of reasons [6].

As a liquid scintillator, argon has a very good light yield: about 20 eV are required, for a relativistic electron, to produce a scintillation photon, just slightly lower than the one of liquid xenon [7]. Scintillation light is emitted at 128 nm wavelength, in the Vacuum Ultraviolet Region, and specific methods have to be implemented to wave shift and efficiently collect it. Contrary to liquid xenon the fast (singlet) and the slow (triplet) components of the scintillation light are very widely separated: the fast component is emitted with a decay time of about 7 ns while the slow component has a decay time of about 1400 ns [8]. This separation allows for very efficient identification of nuclear recoils against gamma and beta backgrounds by means of pulse shape discrimination with methods similar to the ones commonly in use with specific organic scintillators. Liquid argon has a very high electron mobility, it is available in large quantities at low cost and can be easily purified from electronegative impurities allowing for very long drift distances, of the order of several meters, for free electrons as amply shown by the results obtained by the Icarus Collaboration [9]. On the other side, argon is contaminated by ^{39}Ar, a β active isotope with 269 years half-life and end point at 565 keV; ^{39}Ar is produced by interaction of cosmic rays in the upper atmosphere and is giving a β activity in commercially available liquid argon of about 1 Bq / kg [10]. Argon nuclei have no spin, so that spin-independent only interactions are allowed. The relatively low atomic mass of argon results in a lower integrated cross-section for elastic scattering off nuclei, which is proportional to A^2 due to the coherence effect, when compared to other materials like xenon and germanium. On the other hand the recoil spectrum is more energetic because of the nuclear form factor, so that at a recoil energy threshold of about 30 keV, the integrated rates in argon, xenon and germanium are almost the same but with recoil energies in argon that extend well above the ones from the other targets.

After the pioneering works done in CERN, the R&D activities were also carried on in the laboratories of the Collaboration groups: in Pavia, Padova, Naples, LNGS and later in Princeton. Various technical issues have been investigated in details: electrons extraction from liquid to gas, secondary light production and it's proportionality to the extracted charge, light wave-shifting, methods for light collection and their optimization, photomultipliers (PMTs) behavior at cryogenic temperatures, materials behavior and their radioactivity, effects of argon pollutants, procurement of low activity argon either by isotopic separation of commercial argon or by extraction of "aged" argon from underground sources [11]. A key role has been played by a 2.3 liters prototype built at the beginning of years 2000. This device has been widely used (and it is continuously used) mainly to qualify the technical solutions to be implemented in the 100 liters detector. In 2004 it was taken to LNGS to extend the R&D activity. It has been run underground in several conditions. With full passive shields it was able to provide a very convincing limit on spin-independent WIMP-nucleon cross-section [12], even though the internal backgrounds were intentionally not kept at optimal levels, thus proving the goodness of the reached technological advances.

Design and construction of the 100 liters started in 2004 soon after the approval of the experiment proposal by INFN. In 2006 also NSF decided to approve the WArP proposal and provided the funds necessary to complete the detector instrumentation.

2. Design and construction of the 100 liters detector

The basic scheme of the 100 liters detector is shown in figure 1. Starting from the center, the detector consists in:

a) the inner detector: a double phase chamber with 100 liters of liquid volume surmounted by a gas volume where the readout PMTs are placed;
b) the active shield completely surrounding the inner detector, with a minimum thickness of 60 cm, for a total liquid argon volume of about 5600 liters, and readout by a set of PMTs placed on the external surface;
c) an internal passive shield made of polyethylene with 10 cm thickness, to filter neutrons coming from the main cryostat walls;
d) the main cryostat, made of stainless steel, with double walls, superinsulated under vacuum, with an internal volume of 30 m^3 of which 23 m^3 are to be filled with LAr;
e) an external passive shield made of 10 cm thick lead, encapsulated in stainless steel boxes, plus a 70 cm thick layer of polyethylene;
f) an external, anti-seismic, sustaining structure made of carbon steel;
g) circuits for the cryostat cooling and for the purification and recirculation of the liquid argon.

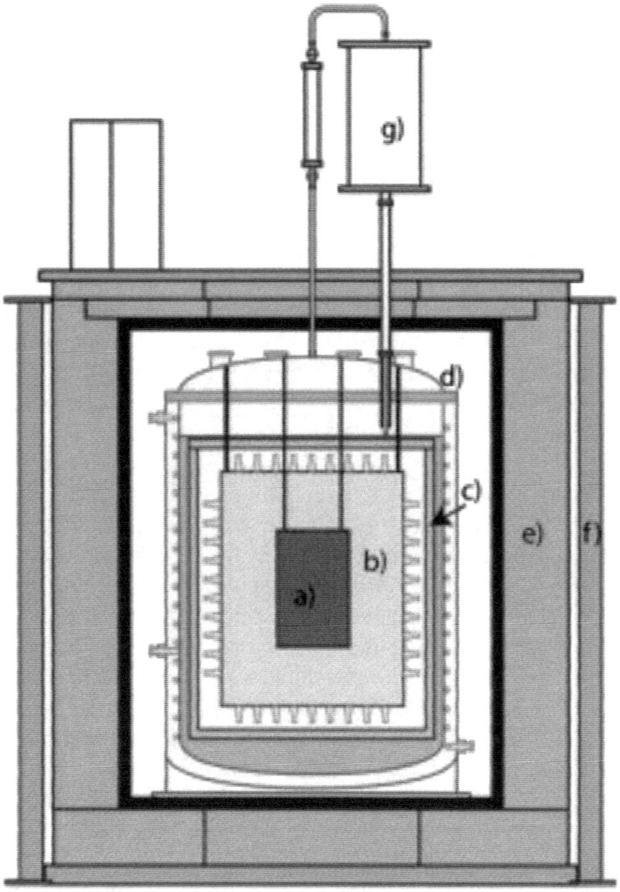

Figure 1

All the internal surfaces of the inner detector and the active shield are covered with a reflector / wave-shifting layer made of a highly reflective plastic foil onto which a highly efficient wavelength shifter (TetraPhenylButadiene – TPB) is deposited. For optimal uniformity and efficiency, the wavelength shifter deposition is made by thermal evaporation under vacuum, with a specially designed evaporation system. The inner detector and the active shield are optically separated, but the argon volume is the same so that no independent filling and purification systems are required.

The inner detector can be subdivided into three regions: the drift region, in liquid phase; the extraction and multiplication region, in gas phase; the light detection region, where the PMTs are placed, in the gas phase. The drift region is delimited by a cathode (an Oxygen Free copper disk with about 50 cm diameter) to be supplied from -57 kV to -75 kV to achieve a drift field between 500 v/cm and 1 kV/cm, a stainless steel grid with inner diameter of about 50 cm, placed 60 cm above the cathode, to be put at -15 kV, and a set of circular stripes (copper stripes printed on a Kapton^{TM} substrate) placed every 2 cm between the cathode and the grid, acting as field shaping electrodes (race-tracks) and ensuring in the drift region a uniform field that drives ionization electrons through the grid to the liquid surface. Race-tracks are arranged in a truncated conical shape to prevent ionization electrons from being trapped between the race tracks themselves. Supply voltage is distributed from the cathode to each race-track and to the grid, by means of a resistive divider chain. Above the first grid a second and a third grids are placed, in the gas phase. The field between the first and the second grid is arranged in such a way to provide complete extraction from the liquid of the ionization electrons and to accelerate them in the gas phase in order to produce a suitable secondary, proportional, light signal. Electrons are then decelerated above the second grid and collected on the third one. Voltage is supplied to the second and third grid by means of dedicated high voltage channels. A set of 31, 3" PMTs plus 6, 2" PMTs is placed 3 cm above the third grid. All the PMTs, including the ones of the active shield, are specifically designed to work at LAr temperature, they have glass windows and peak sensitivity (in the range of 20%) at 430 nm wavelength, well matched with the peak emission of the wavelength shifter. The photocathode coverage of the inner detector is about 10%. All PMTs have their divider chains mounted directly on them; divider chains are made of circuits printed on a Kapton^{TM} substrate with discrete components (resistors and capacitors) selected for operation at LAr temperature. All PMTs are negatively biased; the anode signals are readout at ground. PMTs of the inner detector are coated with TPB in a polystyrene matrix to ensure good localization of the secondary signal in the horizontal plane. The upper part of the inner detector is contained into a vacuum superinsulated dewar placed upside down. This dewar is acting as a diving bell with a set of 12 small holes placed in correspondence of the desired liquid level (5 mm above the first grid). A set of heating resistances is placed just below the liquid surface; they continuously boil off the liquid and the produced gas in excess is evacuated through the holes, thus ensuring that the liquid level is precisely positioned. A copper sheet completely surrounds the inner detector. This sheet is grounded, ensuring that the drift fields are confined in the central volume.

The active shield volume is approximately of the shape of a rotational ellipsoid with a vertical axis of 220 cm and an horizontal axis of 180 cm for a total volume (excluding the volume occupied by the inner detector) of 5600 liters (corresponding to 7800 kg of LAr). The active shield mechanical structure is made of Oxygen Free copper pipes and plates and sustains the photomultipliers and the wave-shifting / reflector layers. Both the active shield and the inner detector are suspended to the main cryostat upper flange with adjustable mechanics that allow positioning and alignments to 0.5 mm precision. A total of 300 PMts (264 of 3" diameter plus 36 of 2" diameter) are placed on the external surface of the active shield for a total photocathode coverage of about 7% of the active shield surface. Two different types of electronics are used to readout the PMTs. The one for the active shield was manufactured by CAEN on a custom design made by the Princeton group and is essentially based on signal digitization and multiplexing after charge integration. The one for the inner detector is based on fast digitization (1 GHz) of the non-integrated signals by means of high performance digitizers manufactured by Aquiris.

Construction of the 100 liters device started in 2005. All construction materials have been subject to strict selection for low radioactive contamination. Samples of all the employed materials have been measured in the LNGS Low Activity Facilities prior to their acceptance for the final construction. After the selection, the main source of unvetoed background for the WArP experiment, resulting from materials measurements and from Montecarlo data, is coming from the photomultipliers and is accounted for less than 1 event / year of exposure.

The detector mechanical components were first delivered in Pavia where they have been completely assembled in order to control and qualify the assembly procedures and the mechanical tolerances. The detector was then dismounted and the components were sent to a specialized industry for cleaning before the final delivery to LNGS. PMTs were subject to very severe testing procedure: each PMT have been tested by the manufacturer in liquid nitrogen for stability and at room temperature for compliance to the specifications (quantum efficiency, dark counts, gain). The phototubes were then sent to Napoli where they undergo an additional set of tests (operational stability and intrinsic noise) in liquid nitrogen. The PMTs were then sent to LNGS for pre-assembly operations installation.

Figure 2

Final detector assembly started in July 2007, when the underground installation area, in LNGS Hall B, became available. The external shield basement, the main cryostat and the anti-seismic supporting structure were first put in place. The assembly clean and dehumidified room was then built and put in operation. Assembly of the internal passive shield followed. In parallel several pre-assembly operations took place in the external laboratories above ground, they consisted in: pre-assembly of the PMTs with Kapton™ bases, cutting and pre-assembly of the reflector foils, thermal evaporation of the wavelength shifter on the reflector foils and quality control of evaporated samples. All the pre-assembled components were taken in the clean room underground for the final installation. The final detector assembly started in June 2008 from the inner detector and was completed in December of the same year. All the PMTs were tested in dark conditions before that the detector was finally moved in the main cryostat around the mid of December. Vacuum pumping started soon after. The detector plant was completed in January 2009 with the installation of the external cooling and purification systems. In parallel, the external shield lead walls were installed together with a first part of the lateral polyethylene external shield.

3. Detector Commissioning

The first phase of the Detector Commissioning was started on Dec. 22nd, 2009 with the vacuum pumping of the cryostat internal volume, combined with a number of cycles of inert gas flushing (N_2 at room temperature) for the reduction of the residual H_2O content of the internal materials. This phase lasted about five months (from Dec. 23rd to May 5th, 2009).

Initially, leak search, localization and repairs required about one month and successively vacuum pumping lasted up until the design residual pressure was reached as measured by the pressure probes located in various points of the cryostat. During this period external cabling of the PMT's and of the slow-control sensors (LAr levels, Temperature, Pressure) has been completed. A full test of the PMT signals and HV mapping has also been then performed and validated.

In this phase of the commissioning the composition of the residual gas has been continuously monitored by mass spectrometry. Air components (from residual leaks) and water (from material outgassing) have been identified and monitored during their progressive decrease (e.g. H_2O outgassing rate was, in the last period of the vacuum pumping $\approx 6 \times 10^{-5}$ mbar lt/sec). The vacuum pumping period was assumed as completed when the residual pressure reached the design value in the range $2 - 6 \times 10^{-6}$ mbar (end of March, 2009).

The second phase of the commissioning procedure (detector Cooling and LAr Filling) was started on May 5th, the first day of resuming into normal operation at LNGS underground – after the suspension of the activities due to the earthquake (during the month of April, vacuum pumping indeed continued without interruptions).

The *Cooling phase* lasted less than 3 days – via low-quality LAr forced recirculation in the piping system inserted in the vacuum jacket of the cryostat, in thermal contact with the internal wall. This produced a smooth cooling down of the materials inside the cryostat to less than 200 K (in the innermost element of the detector), thus stopping the water outgassing (as seen by the drop of the H_2O partial pressure). The ultimate residual pressure just below 1×10^{-6} mbar was then reached.

The following step with the "LAr filling" lasted four more days. Good-quality LAr (equivalent to 5.5 grade) at a speed of about 300 lt/hr was conveyed inside the cryostat through a dedicated vacuum insulated filling line from a movable LAr storage. The line included a large two-stages filter (Oxygen reactant for O_2 removal and molecular sieves for H_2O trapping).

During the first day of the filling phase all the inner components (internal Polyetilene shield and detector) were cooled down to LAr temperature and successively the LAr level inside the cryostat increased to the detector operation design level (full immersion of the active veto detector).

On May 13th, the WArP-100 detector Technical Commissioning was successfully completed.

A technical run of the WArP 100lt detector started with the activation of the various components of the detector and of the cryogenic system (May-July, 2009):

- GAr recirculation (liquefaction and purification) activated, allowing direct measurement of the actual total heat load of the system (estimated at a level of ≤ 1 kW, in agreement with design value).
- GAr/LAr precise level configuration for the two-phase 100lt inner detector successfully established (gas Argon pocket generated and maintained by resistive heaters inside the "bell-shaped" immersed cryostat).
- PMT activation with voltage bias at nominal value with slow HV ramp and individual test.
- DAQ & Trigger for 100lt detector, after initial debugging phase, in run with progressive tuning and optimization of the performance.
- Off-line event reconstruction successfully tested and used for first data analysis.

- HV system for the Inner Detector (connection from external power supply to inner detector cathode): while ramping up the HV a major discharge occurred inside the cryostat along the connection line. This unforeseen event prevented to establish the EF in the LAr volume of the 100 lt detector, necessary for operation in two-phase configuration.

From the analysis of the events collected in the inner detector with no drift field, in particular from the reconstruction of the ^{39}Ar beta decay spectrum, we obtained a light yield for betas and gammas of about 1.6 photoelectrons/keV. A systematic measurement of the light yield from reflector / waveshifter samples co-produced with the elements of the 100 liters detector showed that the light yield could be affected up to about a factor 35% by the quality of the wavelength shifter that was delivered in several batches.

The technical run was stopped at the end of July 2009, the main cryostat was emptied and the detector was brought back to room temperature in pure argon atmosphere. The cryostat was then opened and the detector was brought back in the assembly clean room. Two major interventions were then performed on the detector: the substitution of the high voltage cable and feed-through with a new version, the substitution of the reflective / waveshifting layers of the inner detector with new ones produced with a high quality batch of wavelength shifter. The detector was reinstalled in the main cryostat at the beginning of February 2010; vacuum pumping started immediately after. It took about one month to reach the vacuum level obtained in the first run ($\approx 10^{-5}$ mbar at room temperature). The commissioning procedure followed the same steps of the first run. The detector filling was completed on March 16, 2010.

The high voltage for the drift field was tested at -45 kV just after the filling: applied voltage and current were verified to be stable. Photomoltupliers of the inner detector were then slowly brought to operation at the nominal supply voltage and gain, with no drift field. The light yield for beta and gammas, measured from the beta decay spectrum of ^{39}Ar, resulted in about 1.6 photoelectrons/keV, approximately the same value measured in the previous run. The run continued with the test of the outer detector (active shield) behavior and of the inner detector performance with applied drift field. Fast purification of argon in the liquid phase was started. Results of these tests are still under analysis and discussion. The second run was finally ended in the second half of June 2010.

References

[1] D. N. Spergel et al., (WMAP Collaboration), Astrophys. J. Suppl. **170** (2007) 377:
M. Tegmark et al., (SDSS Collaboration), Phys. Rev. **D74** (2006) 123507;
D. N. Schramm and M. S. Turner, Rev. Mod. Phys. **70** (1998) 303;
R. H. Cyburt, B. D. Fields, and K. A. Olive, Phys. Lett. **B567** (2003) 227.

[2] For a review see for example: N.J. Spooner, "*Direct Dark Matter Searches*", J. Phys. Soc. Jap. **76** (2007) 111016, also as arXiv:0705.3345v1[astro-ph].

[3] R. Bernabei et al., "*First results from DAMA/LIBRA and the combined results with DAMA/NaI*" Eur. Phys. J. **C 56** (2008) 333-355.

[4] P. Benetti, et al., NIM **A327** (1993) 203-206;
P. Benetti, et al., NIM **A329** (1993) 361-364;
F. Arneodo, et al., NIM **A449** (2000) 147-157.

[5] D. Cline, et al., Astropart. Phys., **12** (2000) 373-377.

[6] C. Rubbia, et al., Letter of Intent, "*A programme to search for WIMP particles in liquid argon at the LNGS*", University of Pavia, July 1999.

[7] See for example T.Doke, "*Fundamental Properties of Liquid Argon, Krypton and Xenon as Radiation Detector Media*", Portug. Phys., **vol.12** (1981) 9-48.

[8] See for example: A. Hitachi et al. Phys. Rev. B, **27** (1983) 5279 and references therein

[9] See for example: S. Amerio, et al., NIM **A527** (2004) 329-410.

[10] H. H. Loosli, Earth and Planetary Science Letters **63** (1983) 51;
P. Benetti, et al., NIM **A574** (2007) 83-88.

[11] For a list of the papers of the WArP Collaboration see the site: http://warp.lngs.infn.it
[12] P. Benetti, et al., Astropart. Phys., **28** (2008) 495-507.

ArDM: a ton-scale LAr detector for direct Dark Matter searches

A. Marchionni[1], C. Amsler[2], A. Badertscher[1], V. Boccone[2], A. Bueno[3], M. C. Carmona-Benitez[3], J. Coleman[4], W. Creus[2], A. Curioni[1], M. Daniel[5], E. J. Dawe[6], U. Degunda[1], A. Gendotti[1], L. Epprecht[1], S. Horikawa[1], L. Kaufmann[1], L. Knecht[1], M. Laffranchi[1], C. Lazzaro[1], P. K. Lightfoot[6], D. Lussi[1], J. Lozano[3], K. Mavrokoridis[4], A. Melgarejo[3], P. Mijakowski[7], G. Natterer[1], S. Navas-Concha[3], P. Otyugova[2], M. de Prado[5], P. Przewlocki[7], C. Regenfus[2], F. Resnati[1], M. Robinson[6], J. Rochet[2], L. Romero[5], E. Rondio[7], A. Rubbia[1], L. Scotto-Lavina[2], N. J. C. Spooner[6], T. Strauss[1], C. Touramanis[4], J. Ulbricht[1], and T. Viant[1]

[1] ETH Zurich, Institute for Particle Physics, CH-8093 Zürich, Switzerland
[2] Physik-Institut, University of Zürich, Winterthurerstrasse190, CH-8057 Zürich, Switzerland
[3] University of Granada, Dpto. de Fsica Terica y del Cosmos & C.A.F.P.E, Campus Fuente Nueva, 18071 Granada, Spain
[4] University of Liverpool, Liverpool L69 3BX, United Kingdom
[5] CIEMAT, Div. de Fisica de Particulas, Avda. Complutense, 22, E-28040, Madrid, Spain
[6] University of Sheffield, Department of Physics and Astronomy, Hicks Building, Hounsfield Road, Sheffield, S3 7RH, UK
[7] The Andrzej Soltan Institute for Nuclear Studies, Hoza 69, 00-681 Warsaw, Poland

E-mail: alberto.marchionni@cern.ch

Abstract. The Argon Dark Matter (ArDM-1t) experiment is a ton-scale liquid argon (LAr) double-phase time projection chamber designed for direct Dark Matter searches. Such a device allows to explore the low energy frontier in LAr with a charge imaging detector. The ionization charge is extracted from the liquid into the gas phase and there amplified by the use of a Large Electron Multiplier in order to reduce the detection threshold. Direct detection of the ionization charge with fine spatial granularity, combined with a measurement of the amplitude and time evolution of the associated primary scintillation light, provide powerful tools for the identification of WIMP interactions against the background due to electrons, photons and possibly neutrons if scattering more than once. A one ton LAr detector is presently installed on surface at CERN to fully test all functionalities and it will be soon moved to an underground location. We will emphasize here the lessons learned from such a device for the design of a large LAr TPC for neutrino oscillation, proton decay and astrophysical neutrinos searches.

1. Introduction

Astronomical observations give strong evidence for the existence of non-luminous and non-baryonic matter, presumably composed of a new type of elementary particles. The leading candidate is a cold thermal relic gas of Weakly Interacting Massive Particles (WIMPs) [1], which feel the gravitational interaction, but are otherwise interacting more weakly than standard

Table 1. Main properties of LAr: temperature at 1 atm and corresponding density, attenuation length for 50 keV γ, electron mobility at the boiling point, average energies needed for the production of an electron-ion pair [3] or of one scintillation photon [4] (measured for 1 MeV electrons), LAr scintillation wavelength, fast and slow scintillation lifetimes.

T at 1 atm (K)	ρ (g/cm^3)	λ_{att} 50 keV γ (g/cm^2)	$\mu_{electron}$ (cm^2/Vs)	W$_{ion}$ (eV)	W$_\gamma$ (eV)	λ_{scint} (nm)	τ_{fast} (ns)	τ_{slow} (μ_s)
87.2	1.396	1.4	500	23.6	25	128	7	1.6

weak interactions. Direct detection could be achieved by observing the energy deposited when WIMPs elastically scatter from an ordinary target material nucleus, requiring the measurement of recoils of target nuclei with kinetic energy in the range of 5 - 100 keV.

The Argon Dark Matter (ArDM-1t) experiment [2] is using a ton-scale liquid argon (LAr) target for direct WIMP searches, operated as a double-phase time projection chamber with imaging and calorimetric capabilities.

Liquid argon represents a promising target with a favorable form factor and it is only sensitive to spin-independent interactions. It is easily commercially available, since argon constitutes 1% of air, and it provides adequate self-shielding against the background of low energy photons and electrons. In addition liquid argon is very attractive as detector material, since it has excellent scintillation properties and, given its relatively high electron mobility, is suitable for drifting electrons. Its main properties are summarized in table 1 and will be discussed more deeply in the following.

Ionizing radiation leads to the formation of excited molecular states in either singlet or triplet states, which decay radiatively with fast and slow lifetimes, respectively [5]; the lifetimes are largely different, τ_{fast} = 7 ns and τ_{slow} = 1.6 μs, allowing to easily discriminate between the two components. The relative abundance of singlet and triplet states depends on the ionization density, and so on the type of ionizing radiation [5]. In addition, the phenomenon of recombination, effectively transforming ionization into scintillation [6], also depends on the ionization density [7, 8]. Based on these facts, both the time evolution of the argon scintillation light [9, 10] and the relative fraction of scintillation to ionization charge [11] are, in LAr, powerful discriminants of nuclear recoil signals against γ and electron backgrounds.

Thanks to the high electron mobility of LAr, large LAr volumes can be operated as Time Projection Chambers, providing both imaging and energy measurements from the detection of the ionization charge. Relatively high electric fields can be used over extended regions, resulting in very high electric potential values, which are possible in LAr due to its very high dielectric strength. High drift fields could possibly help in the detection of the highly quenched ionization charge from nuclear recoils, as discussed in section 5.

A WIMP interaction leading, i.e., to a 30 keV nuclear recoil is expected to produce about 300 VUV photons, together with a few tens of free electrons, both values depending on the electric field strength and still not fully known. This is a very small value for the released charge, which would not be detectable without amplification. A solution is offered by the use of a detector operated in double-phase (liquid-gas) [12], where the ionization charge produced in the liquid is extracted into the gas phase by a sufficiently high electric field at the liquid/gas interface, and there amplified in the pure Ar gas [2, 13] or observed through proportional scintillation [14].

ArDM [2] is a one ton argon detector, operated in double-phase, with independent ionization and scintillation readout. The goal is to demonstrate the feasibility of an argon-based ton-scale experiment to efficiently detect WIMP induced nuclear recoils with sufficient background discrimination. Characteristic features of this experiment are the direct charge readout with good spatial resolution, following amplification in the gas phase, and the possibility to reach high drift fields. The ionization electrons, after drifting in the LAr volume, are extracted into the gas phase and there amplified, directly extrapolating from the more delicate GEM detectors, by thick macroscopic GEMs, also named Large Electron Multipliers (LEMs).

Large liquid argon (LAr) detectors, up to 100 kton scale, are presently being considered for proton decay searches and neutrino astrophysics as well as far detectors for the next generation of long baseline neutrino oscillation experiments, aiming at neutrino mass hierarchy determination and CP violation searches in the leptonic sector [13]. The operation of a single LAr module of 100 kton as a charge imaging LAr TPC operated in double phase was first suggested by [13], in order to compensate the charge loss due to a very long drift with the subsequent amplification of the ionization charge in the pure argon vapor phase. ArDM represents in this respect an important step to assess the feasibility of generation of high drift electric fields and of the amplification and readout of the ionization charge in a LAr TPC operated in double phase.

An overview of the ArDM detector is presented in section 1; more detailed descriptions of the cryogenics, scintillation light readout, high voltage generation and ionization charge detection are given in sections 3, 4, 5 and 6, respectively. The status of the detector, and particularly its novel features, are summarized in section 7.

2. Overview of the detector and prospects

Figure 1 shows a conceptual layout of the ArDM apparatus [2]. The sensitive liquid argon volume is defined by the drift volume, delimited by a cathode at the bottom and by a set of field shapers on the sides. It has a diameter of 80 cm and a maximum drift length of 120 cm.

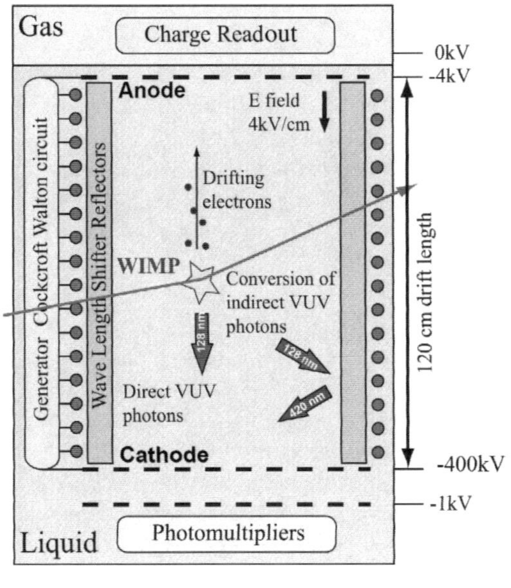

Figure 1. Conceptual layout of ArDM.

Figure 2. View of the ArDM detector from the bottom

Charged particles lead to production of ionization charge and to excitation of argon atoms, yielding scintillation light emitted isotropicaly along the ionization path at a wavelength of 128 nm. This is converted to blue light (420 nm) by a wavelength shifter (WLS) deposited on foils attached to the field shapers [15]. The shifted and reflected light is detected by an array of photomultipliers (PMTs) immersed in the liquid at the bottom of the detector, protected from the high electric potential of the cathode by a shielding grid. Direct LAr scintillation light impinging on the PMTs is also detected, since PMTs are coated with WLS.

A 400 kV Cockroft-Walton voltage multiplier [16], immersed in LAr, has some of its stages directly connected to the field shapers, and the last stage connected to the cathode, providing a field, possibly exceeding 3 kV/cm, to drift the electrons to the top surface of the liquid towards the gas phase. Two extraction grids positioned across the liquid-vapour interface generate a $\approx$ 3 kV/cm field to extract the electrons to the gas phase, where they are driven inside the holes of a Large Electron Multiplier which provides charge multiplication [17]. The amplified charge is finally collected by a two-dimensional projective anode with fine spatial resolution [18].

Figure 2 shows a photograph of the detector, displaying, from the bottom to the top, the cryogenic photomultipliers with the shielding grid on top, the cathode, the set of field shapers and, hanging from the top, the Cockroft-Walton generator. In addition it shows the wavelength shifting foils, while illuminated with a UV lamp.

The ^{39}Ar, a beta-emitter with a lifetime of 269 years and a Q-value of 565 keV, naturally present in the atmospheric argon [19, 20] and causing a background rate of about 1 kHz in ArDM, will eventually be the limiting factor in the sensitivity of the experiment. It may be possible to replace the atmospheric argon contaminated with ^{39}Ar with argon extracted from underground wells, shown to be depleted from ^{39}Ar by more than an order of magnitude [21].

The ArDM detector is presently installed on surface at CERN to fully test all functionalities and it will be later moved to an underground location. ArDM could possibly investigate WIMP cross sections down to 10^{-45} cm^2, depending on background rejection. Within a canonical WIMP halo mode, a WIMP with a mass of 100 GeV and a cross-section of 10^{-44} cm^2 would yield about 1 recoil event per day per ton of argon above a recoil energy threshold of 30 keVr.

3. Cryogenics and safety

The detector is housed in a vacuum tight stainless steel dewar vessel, with an inner diameter of 100 cm and a height of about 2 m, containing approximately 1.4 m^3 of LAr, along with a layer of saturated argon vapor at the top. Before filling with LAr, the detector vessel is evacuated to a high vacuum, below 10^{-6} mbar, to minimize the contamination of the liquid from remaining impurities and outgassing. The purity of the liquid argon is maintained by a custom-made purification cartridge, based on CuO powder, separated from the main dewar to allow the insertion of a radiation shield against neutrons (see figure 3). A cryogenic pump allows recirculation of the LAr in the detector vessel through the purification cartridge.

The detector vessel and the purification cartridge are fully surrounded by a jacket of LAr, named argon bath, which is used for the initial cooling of the detector and for maintaining stable cryogenic conditions. The surrounding LAr jacket is enclosed by an outer shell of stainless steel, where in between it is maintained a vacuum better than 10^{-4} mbar for thermal insulation. Presently the LAr bath is open to the atmosphere so that the temperature is maintained at 87 K, corresponding to a saturated argon vapor pressure of 1 bar. This implies that the pressure of the saturated gas on top of the liquid in the detector vessel is also kept at atmospheric pressure.

For the future we want also to operate the LAr bath as a closed system and keep the temperature stable by the use of cryocoolers. This allows more flexibility in the choice of the temperature, possibly cooling below a saturated argon vapor pressure of 1 bar, with no dependance on variations of the atmospheric pressure. It also removes the need for a continual refilling of the LAr bath, impracticable for a continuous operation in an underground environment. The needed cooling power, between 300 and 400 W at LAr temperature, has been estimated by measuring the LAr consumption when operating in the open LAr bath configuration. It is dominated by the heat input through the top flange, which is not equipped with an insulating vacuum. We are in the process of installing two $\sim$ 250 W crycoolers, with

Figure 3. Schematics of the detector vessel (on the right), surrounded by the radiation shield, and of the purification column (on the left).

Figure 4. View of the electronic equipment racks of ArDM, including the PLC system and vacuum and cryogenic controls.

enough flexibility to allow operation in a subcooled state.

As the detector contains such a large amount of cryogenic liquid in a closed volume, it is necessary to protect it with the aid of pressure relief devices. Since pressure relief valves do not meet the tight leak rate requirements needed to maintain the purity of the LAr in the detector vessel, the system has been equipped with two burst discs, with a burst pressure of 0.7 barg. Pressures and temperatures of the system, together with the vacuum levels for thermal insulation, are continuously monitored through a Programmable Logic Controller (PLC) system. The PLC controls all the vacuum and cryogenic processes: operation of the vacuum pumps, recirculation of the LAr in the dector vessel, refilling of the LAr bath and in the future the operation of the cryocoolers. A view of the ArDM electronic equipment racks, also housing the PLC system and all vacuum and cryogenic controls, is shown in figure 4.

4. Light readout with cryogenic PMTs and wavelength shifting techniques

The emission spectrum of LAr scintillation is peaked at 128 nm, which would require the use of VUV sensitive PMTs (e.g. MgF_2 windowed), not commercially available with large area photocathodes. An alternative and still efficient solution is the use of reflectors coated with a wavelength shifter (WLS) coupled with standard bialkali photomultiplier tubes. In ArDM [15] the reflectors consist of Tetratex (TTX) foils, 254 μm thick, coated by evaporation with 1 mg/cm^2 of TetraPhenyl Butadiene (TPB), which presents a fluorescence decay time of 1.68 ns and an emission spectrum between 400 and 480 nm. A reflection coefficient close to 97% has been measured for evaporated TTX samples.

The shifted and reflected light is collected by 14 PMTs (8" Hamamatsu R5912-02MOD with bialkali photocathodes and Pt-underlay, see figure 5) located at the bottom of the cryostat and completely immersed in LAr. The PMTs were also coated by evaporation with a thin layer (0.05 mg/cm^2) of TPB for direct scintillation light detection. Each PMT is soldered with its leads onto a 3mm thick printed circuit board (see figure 6), providing the voltage divider for cathode and dynodes, using only passive electronic components. A 5 mm thick stainless steel plate with hexagonal holes serves as a holder for the PMTs. The PMTs are firmly held in place by simply attaching the printed boards to the support plate.

The PMTs were calibrated by using short light pulses from a 400 nm blue LED, placed in the

Figure 5. A photograph of the assembly of the 14 8" PMTs on a stainless steel holder plate.

Figure 6. View of a cryogenic voltage divider.

vapor region above the liquid argon surface. The PMT gains were set to 2.5×10^7 by individually adjusting the bias voltages. A typical pulse shape for scintillation in LAr induced by electrons, gammas and cosmic muons, selected by requiring a fast/slow component ≤ 0.4, is shown in figure 7. A fit to the signal [22] shows the necessity of an intermediate component, providing lifetimes of $\tau_1 = 17.8 \pm 1.0$ ns, $\tau_2 = 1556 \pm 10$ ns, $\tau_3 = 326 \pm 10$ ns, and corresponding contributions of 21, 74 and 5%, for the fast, slow and intermediate components, respectively. No attempt was made to extract the fast component LAr lifetime by correcting for the time response of the PMTs and of the data acquisition system.

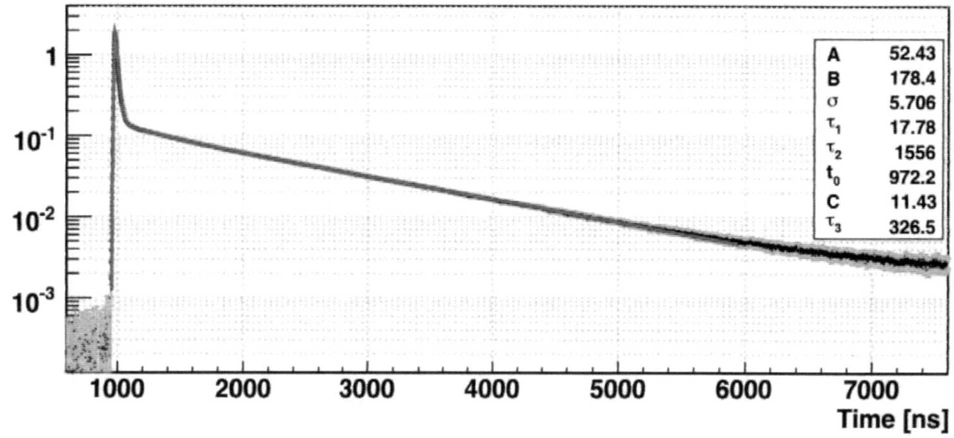

Figure 7. Typical LAr scintillation signal induced by minimum ionizing particles.

A preliminary set of only seven PMTs were tested in LAr with ArDM on surface at CERN in 2009. By using external radioactive sources it was measured a light yield at zero electric field between 0.3-0.5 phe/keVee depending on the position within the detector volume [23]. A new recent test performed in 2010 with a completed detector with 14 PMTs is still being analyzed.

During these tests of ArDM we found very convenient to monitor the LAr purity by measuring the τ_2 lifetime. The τ_2 lifetime is quite sensitive to traces of impurities, down to 0.1 ppm, allowing to easily monitor a deterioration of the purity. In particular we use this method to quickly detect possibly contaminated LAr batches at the delivery time. This is particularly important when filling large detectors, which require the delivery of many LAr batches, and the addition of only one bad batch could seriously deteriorate the LAr purity of the whole detector. In ArDM we measured the τ_2 lifetime over about 600 h with no significant deterioration, allowing to set a limit of ≤ 0.1 ppm for the N_2 and O_2 concentrations [22, 24].

5. Development of High Voltage systems

The development of high voltage systems is crucial for LAr TPCs, which need a stable and uniform drift field over extended LAr volumes. In case of ArDM the maximum drift length is 120 cm, but drift lengths up to 20 m have been envisaged for future large LAr TPCs [13]. There are clear advantages in going to higher drift electric fields:

- shorter charge collection times, in order to reduce the attenuation of the drift electrons due to their attachment to electronegative impurities present in LAr, e.g. oxygen. In fact the ratio of collected to released ionization charge is well described by $Q/Q_0 = e^{-t/\tau}$, where t is the drift time and τ the electron lifetime, inversely proportional to the concentration of impurities according to $\tau[\mu s] = 300/\rho_{O_2}[ppb]$. In LAr the drift velocity of electrons increases less than linearly for electric fields above 100 V/cm, still there is considerable gain at higher fields. It reaches 1.55 mm/μs at 0.5 kV/cm for a temperature T=89 K, and increases by 30% by doubling the electric field intensity from 0.5 to 1 kV/cm, and again by 30% from 1 to 2 kV/cm [25].
- Attachment cross sections of electrons to O_2 impurities in LAr have been measured in [26] to decrease as a function of the electric field . Anyhow in [27] it was found only a weak dependance of the attachment cross-section on electric fields in the range 0.05-1 kV/cm.
- Recombination of electrons and positive ions from the initial ionization charge decreases with increasing electric fields, particularly for heavily ionizing particles. Measurements in LAr have been reported by [28–34]. In reference [34] recombination factors for relativistic particles with unit charge are fitted with a Birks law dependance as a function of the ratio of the particle stopping power dE/dx to the applied drift electric field. This is purely a phenomenological expression and, as cautioned in [34], it is not valid in the high recombination region, that is in either low electric fields or high ionization density. The dependance of recombination on the electric field is more pronounced in LAr than LXe [31]. In LXe it was also found an almost complete field independence of the ionization yield for nuclear recoils [35] , a striking result not completely understood. The possibility to go to high drift electric fields in ArDM, up to 3 kV/cm, will be important to experimentally assess recombination effects as a function of electric field for nuclear recoils.

ArDM has a cylindrical drift volume of 80 cm diameter and 120 cm height, delimited by a cathode grid at the bottom and by 30 ring-shaped electrodes, called field shapers, spaced by 40 mm. The solution adopted in ArDM for the high voltage generation is to have a 210 stages Cockcroft-Walton voltage multiplier directly immersed in LAr [16] (see figure 2), driven by a 50 Hz AC charging voltage with a maximum $V_{pp} = 2.5$ kV. Each capacitor in the Cockroft-Walton scheme is made out of four (two) 82 nF metallized film polypropylene capacitors in parallel for stages 1 to 170 (171 to 210) and each diode is realized by three high voltage avalanche diodes in series. The redundancy built in the system prevents from a critical loss of functionality of the circuit. The voltage multiplier can be operated at a low frequency charging voltage, well outside of the relevant bandwidth of the charge and light readout electronics, since a negligible current is drawn by the voltage multiplier: in fact the DC voltages generated by the circuit are directly connected to the cathode and to the field shapers. Given the large number of Cockroft-Walton stages, it is possible to choose appropriately the connections to the field shapers in order to achieve a good linearity in the voltage distribution.

Voltage multipliers with a large number of stages are affected by the shunt capacitance due to the diode capacitance and the stray capacitance between two adjacent stages, causing the output voltage to increase less than linearly with the stage number, even with no current load. In case of ArDM this non-linearity has been measured, resulting in a voltage on the last stage of 72% of the maximum theoretical attainable voltage. This results in a cathode voltage of 378 kV, when driving the circuit at V_{pp} = 2.5 kV, and a drift electric field exceeding 3 kV/cm. This solution allows to reach interestingly high values for the drift electric field, in order to reduce the charge attenuation due to the presence of electronegative impurities in LAr and to limit the phenomenon of recombination. Externally generated high voltages would need the use of cryogenic feedthroughs, presently tested only up to 150 kV [36].

In a preliminary test in ArDM, the Cockroft-Walton voltage multiplier has been successfully operated up to 70 kV.

6. Charge collection and amplification

The detection of a few tens of ionization electrons is one of the most challenging aspects of the ArDM detector. The sensitive volume of the detector is defined by the drift volume, already described in section 5. The ionization electrons are drifted towards the liquid surface on top, possibly with a high drift electric field value to reduce attenuation due to electronegative impurities in LAr, are extracted into the gas phase by means of two extraction grids precisely positioned across the liquid, and then amplified in the pure Ar gas by thick macroscopic GEMs called Large Electron Multipliers.

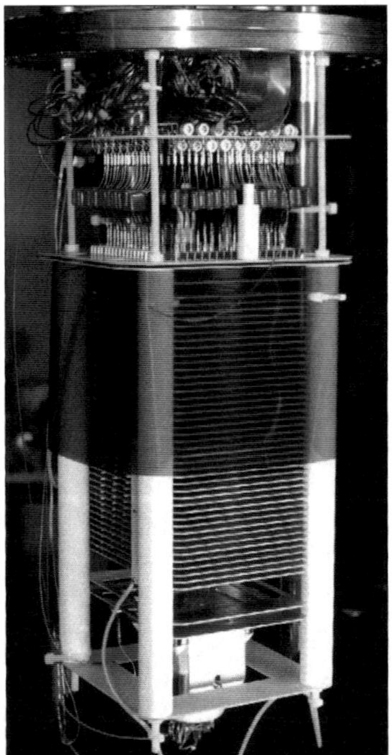

Figure 8. Picture of the LAr LEM-TPC test setup with an active area of 10 × 10 cm^2 and a drift length of 21 cm.

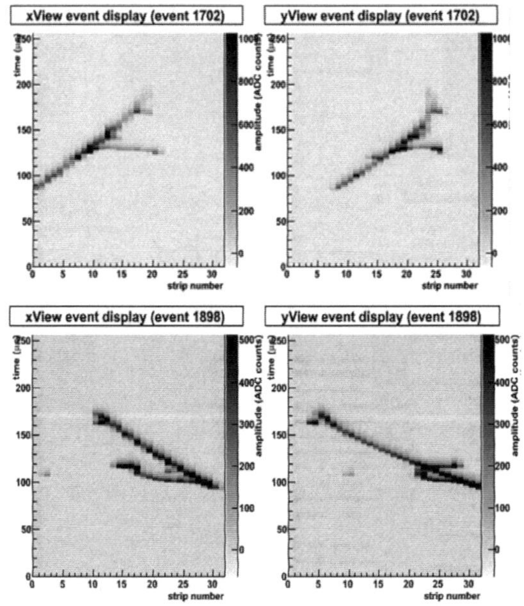

Figure 9. LEM-TPC displays of two muon events (top and bottom pictures) crossing the LAr sensitive volume with visible delta rays: left) drift time vs x-channel number; right) drift time vs y-channel number.

In order to get charge amplification in pure argon gas, without any quencher, it is necessary to confine the electron avalanche, which would be otherwise spread by the argon scintillation photons produced in the high electric field regions by proportional scintillation. The geometry of the GEM is particularly suited to be used in pure Ar gas, because it generates the high electric fields, necessary for charge amplification, only inside the holes, effectively confining the electron avalanche. GEMs have been successfully tested in pure Ar gas at normal pressure and temperature, obtaining gains of the order of 1000 [37]. They have also been tested in pure Ar in double phase conditions using triple-stage GEMs, reaching gains of the order of 5000 [38], in order to compensate for the essentially inverse exponential dependance of the first Townsend coefficient with gas density [39]. Large Electron Multipliers (or Thick GEMs) [40, 41] with mm size holes in a mm thick printed circuit board, are a natural evolution of GEMs on a more rigid structure better suited for a cryogenic environment.

In order to develop a LEM for the ArDM detector, we started a series of tests on LEMs of different thickness and manufacturing procedure and with different schemes for the charge readout. We constructed a 3 l active volume double-phase LAr TPC with a LEM readout system, as shown schematically in figure 8, which we call LAr LEM-TPC. Details on the construction and operation of the device are presented in [17, 18, 42, 43]. In order to get a full 3-dimensional reconstruction of an ionizing event, it is necessary, in addition to the drift time information, to induce signals on at least two independent spatial views, typically realized by two sets of orthogonal strips in case of a bi-dimensional readout. We found important to decouple the mechanism of charge amplification in the LEM from the charge readout, so we chose to drift the amplified charge to a two-dimensional projective readout anode, providing two independent views with 3 mm spatial resolution [18]. With a single 1 mm thick LEM, coupled to a two-dimensional projective anode, we achieved an effective gain of 27 on the released ionization charge, without any degradation in charge collection resolution. Examples of recorded events are shown in figure 9, displaying muon events crossing the sensitive volume, accompanied by delta rays. A signal to noise ratio larger than 200 has been obtained for minimum ionizing particles, thanks to charge amplification, resulting in images of very high quality.

For the charge detection of WIMP initiated nuclear recoils in ArDM, corresponding to an ionization charge down to a few tens of electrons, charge gains of about 500 are necessary, since a signal to noise ratio of 10 for 1 fC input charge is typical for our charge preamplifiers. Such gains should be achievable in setups with two 1 mm thick LEMs operated in cascade, as it will be soon verified in an experimental test.

Figure 10 is a view of the present charge readout assembly in ArDM, showing, from bottom to top, the upper field-shaper, the two extraction grids precisely kept at 10 mm distance by a fiberglass ring spacer, and the charge readout electrode, positioned 10 mm above the upper grid. Also visible is one of three capacitive level meters, directly mounted on the extraction grids, to monitor the liquid level between the two grids. The whole assembly is hanging from a mechanical support capable of vertical and tilting movements, adjustable from outside the dewar, which allows a precise positioning and leveling of the extraction grids across the liquid-vapour interface.

In a first test of drift and extraction of the ionization charge, the Cockroft-Walton voltage multiplier has been operated at 70 kV, corresponding to an electric drift field of ~ 0.5 kV/cm. For this test a temporary electrode with 32 pads readout, with no charge amplification, has been used for the collection of the extracted charge, as shown in figure 11. The same pad structure is present on the lower and upper faces of the charge readout plane. Charge signals are readout from the upper pads, capacitatively coupled to the lower pads, which are kept at high voltage. By triggering on muons crossing the LAr sensitive volume by means of external scintillator plates, both scintillation light and ionization charge have been detected. In addition proportional scintillation light from the extraction of the ionization charge into the gas phase, due to the high electric field region between the two extraction grids, has been observed.

Figure 10. View of the charge readout assembly.

Figure 11. Picture of the temporary charge readout electrode, subdivided into 32 pads.

7. Conclusions

ArDM, a one ton liquid argon double-phase TPC, has been assembled at CERN and it will soon be moved to an underground location to be operated as Dark Matter detector. It relies on the detection of both primary scintillation light and ionization charge with good spatial resolution for the identification of nuclear recoils initiated by WIMP interactions.

Novel features of ArDM are the possible use of high electric fields, generated by a Cockroft-Walton voltage multiplier directly immersed in LAr, and the direct readout of the ionization charge, extracted from the liquid to the gas phase, and there amplified by the use of Large Electron Multipliers. The light detection system, recently completed with 14 PMTs, has been experimentally shown to be able to detect low energy deposits as those expected from nuclear recoils. In a first test, the Cockroft-Walton multiplier has been operated stably at 70 kV, resulting in an electric drift field of ~ 0.5 kV/cm. The drifted charge has been extracted into the gas phase by means of two extraction grids, precisely positioned across the liquid level. Proportional scintillation light from the extraction of the ionization charge into the gas phase, due to the high electric field region between the two extraction grids, has been observed. A temporary electrode with pads readout, with no charge amplification, has been used for the collection of the extracted charge.

Successful charge extraction from the liquid argon phase and multiplication in the gas phase have been routinely achieved in a smaller test setup by using Large Electron Multipliers. An effective gain of 27 on the released ionization charge has been reached with a single 1 mm thick LEM of 10×10 cm^2. The charge is finally collected by a two-dimensional projective anode with 3 mm spatial resolution. Gains of about 500, as necessary for the charge detection of WIMP initiated nuclear recoils, should be possible in a setup with two 1 mm thick LEMs operated in cascade, and it will be soon verified in an experimental test. We do not foresee any fundamental problem in the construction of a ~ 1 m^2 LEM, as necessary for ArDM, and we are expecting soon the delivery of a ~ 0.3 m^2 system for a different application.

ArDM will explore the low energy frontier in LAr with a charge imaging device, but at the same time it represents an important step to test some of the features of the design of very large LAr TPCs [44].

References

[1] Steigman G and Turner M S 1985 *Nucl. Phys.* **B253** 375

[2] Rubbia A 2006 *J. Phys. Conf. Ser.* **39** 129–132 (*Preprint* hep-ph/0510320)
[3] Miyajima M *et al.* 1974 *Phys. Rev.* **A9** 1438–1443
[4] Doke T, Masuda K and Shibamura E 1990 *Nucl. Instrum. Meth.* **A291** 617 – 620
[5] Hitachi A *et al.* 1983 *Phys. Rev.* **B27** 5279–5285
[6] Kubota S *et al.* 1978 *Phys. Rev. B* **17** 2762–2765
[7] Hitachi A *et al.* 1987 *Phys. Rev. A* **35** 3956–3958
[8] Doke T *et al.* 1988 *Nucl. Instrum. Meth.* **A269** 291 – 296
[9] Boulay M G and Hime A 2006 *Astropart. Phys.* **25** 179–182
[10] Lippincott W H *et al.* 2008 *Phys. Rev.* **C78** 035801 (*Preprint* 0801.1531)
[11] Benetti P *et al.* 1993 *Nucl. Instrum. Meth.* **A327** 203 – 206
[12] Dolgoshein B *et al.* 1973 *Sov. J. Particles Nucl.* **4** 70–77
[13] Rubbia A 2004 (*Preprint* hep-ph/0402110)
[14] Benetti P *et al.* 2008 *Astropart. Phys.* **28** 495–507 (*Preprint* astro-ph/0701286)
[15] Boccone V *et al.* (ArDM Coll.) 2009 *JINST* **4** P06001 (*Preprint* 0904.0246)
[16] Horikawa S *et al.* 2010 Proceedings of GLA2010 (*Preprint* 1009.4908)
[17] Badertscher A *et al.* 2008 (*Preprint* 0811.3384)
[18] Badertscher A *et al.* 2010 (*Preprint* 1012.0483)
[19] Loosli H 1983 *Earth and Planetary Science Letters* **63** 51 – 62
[20] Benetti P *et al.* (WARP) 2007 *Nucl. Instrum. Meth.* **A574** 83–88 (*Preprint* astro-ph/0603131)
[21] Acosta-Kane D *et al.* 2008 *Nucl. Instrum. Meth.* **A587** 46 – 51
[22] Boccone V 2010 *Development of the Light Readout for the ArDM Dark Matter Search* Ph.D. thesis University of Zurich
[23] Amsler C *et al.* (The ArDM) 2010 *JINST* **5** P11003 (*Preprint* 1009.3641)
[24] Acciarri R *et al.* 2009 *Nucl. Instrum. Meth.* **A607** 169 – 172
[25] Walkowiak W 2000 *Nucl. Instrum. Meth.* **A449** 288–294
[26] Bakale G, Sowada U and Schmidt W F 1976 *The Journal of Physical Chemistry* **80** 2556–2559
[27] Amoruso S *et al.* 2004 *Nucl. Instrum. Meth.* **A516** 68–79
[28] Gruhn C R and Edmiston M D 1978 *Phys. Rev. Lett.* **40** 407–409
[29] Scalettar R T, Doe P J, Mahler H J and Chen H H 1982 *Phys. Rev. A* **25** 2419–2422
[30] Shibamura E *et al.* 1987 *Nucl. Instrum. Meth.* **A260** 437 – 442
[31] Thomas J and Imel D A 1987 *Phys. Rev. A* **36** 614–616
[32] Aprile E, Ku W H M, Park J and Schwartz H 1987 *Nucl. Instrum. Meth.* **A261** 519 – 526 ISSN 0168-9002
[33] Cennini P *et al.* 1994 *Nucl. Instrum. Meth.* **A345** 230–243
[34] Amoruso S *et al.* (ICARUS) 2004 *Nucl. Instrum. Meth.* **A523** 275–286
[35] Aprile E *et al.* 2006 *Phys. Rev. Lett.* **97** 081302
[36] Amerio S *et al.* (ICARUS) 2004 *Nucl. Instrum. Meth.* **A527** 329–410
[37] Bressan A *et al.* 1999 *Nucl. Instrum. Meth.* **A423** 119–124
[38] Bondar A *et al.* 2006 *Nucl. Instrum. Meth.* **A556** 273–280 (*Preprint* physics/0510266)
[39] Aoyama T 1985 *Nucl. Instrum. Meth.* **A234** 125 – 131
[40] Otyugova P 2008 *Development of a Large Electron Multiplier (LEM) Based Charge Readout System for the ArDM Experiment* Ph.D. thesis ETH Zurich, Diss. ETH No. 17704
[41] Bondar A *et al.* 2008 *JINST* **3** P07001 (*Preprint* 0805.2018)
[42] Badertscher A *et al.* 2010 *Nucl. Instrum. Meth.* **A617** 188–192 (*Preprint* 0907.2944)
[43] Badertscher A *et al.* 2010 Proceedings of GLA2010 (*Preprint* 1010.2482)
[44] Marchionni A 2009 (*Preprint* 0912.4417)

1st International Workshop towards the Giant Liquid Argon Charge Imaging Experiment IOP Publishing
Journal of Physics: Conference Series **308** (2011) 012007 doi:10.1088/1742-6596/308/1/012007

From ArgoNeuT to MicroBooNE

Bonnie T. Fleming
Yale University, New Haven, CT
E-mail: bonnie.fleming@yale.edu

Abstract. The final unknowns in the neutrino oscillation mixing matrix, including the search for CP Violation in the neutrino sector, offer extraordinary physics opportunities, which make a compelling case for developing the best possible detectors. Liquid Argon Time Projection Chambers (LArTPCs) combine fine-grained tracking with total absorption calorimetry to achieve sensitivities well beyond conventional neutrino detectors. The ArgoNeuT and MicroBooNE projects serve as steps in a development path towards massive kiloton scale LArTPCs for long baseline electron neutrino appearance physics. These experiments combine development goals with exploration of low energy neutrino interaction physics. Progress on ArgoNeuT and MicroBooNE en route to massive LArTPCs for the LBNE program are presented here.

1. Introduction

The recent exciting new results in neutrino physics have both answered many puzzling questions and created many more. Most notably, the first conclusive evidence that neutrinos oscillate and have mass occurred only about 10 years ago. The consequences of this seminal discovery include the possibility that neutrinos and their anti-particles could behave differently suggesting CP Violation in the neutrino sector. As well, neutrinos may have other unexpected properties or unexpected partner particles. Results from the MiniBooNE [2] experiment have ruled out the possibility of two-neutrino oscillations at high Δm^2. However, they also contain a puzzling new excess below the oscillation region, < 475 MeV. A 129 event excess ($\sim 3\sigma$) in this region suggests possible new physics [3, 4, 5, 6, 7, 8, 9, 10, 11]. No matter what the explanation, this surprising low energy excess must be understood.

These surprises offer extraordinary physics opportunities, best addressed with precision detection techniques. Liquid argon Time Projection Chambers (LArTPCs) [1] combine fine-grained tracking with total absorption calorimetry to achieve sensitivities well beyond conventional neutrino detectors. For long baseline neutrino physics, the combination of electron efficiency and neutral current pion (NCπ^0) background rejection make the detectors ~ 6 times [12] more sensitive than Water Cerenkov detectors.

Recent progress on the growing US LArTPC R&D effort has been built on the pioneering efforts of the ICARUS collaboration's LArTPC program. The US program combines test stand measurements with data taking in LArTPCs of increasing sizes towards a final goal of 20 kton modules for the LAr20 detector within LBNE [13]. An overview of the test stands and experiments underway in the US is shown in Figure 1. TPCs scaled in size from test stands to LBNE are shown in Figure 2. Both ArgoNeuT and MicroBooNE are key steps in this path towards massive LArTPCs for CP violation searches.

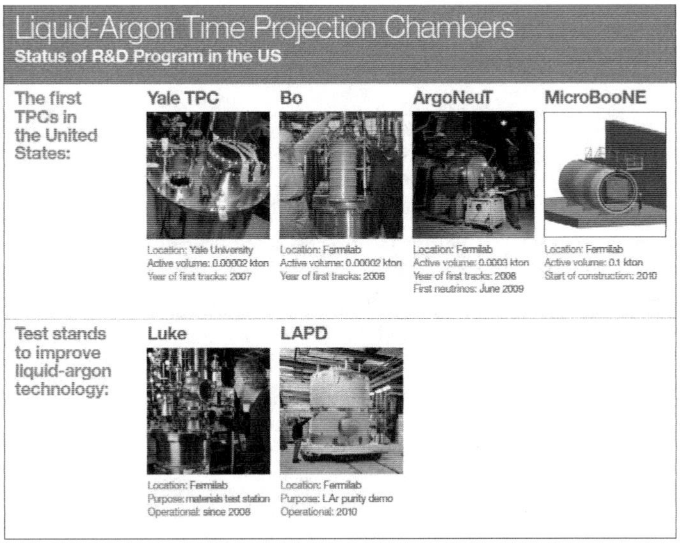

Figure 1. Present program for LArTPC development in the United States

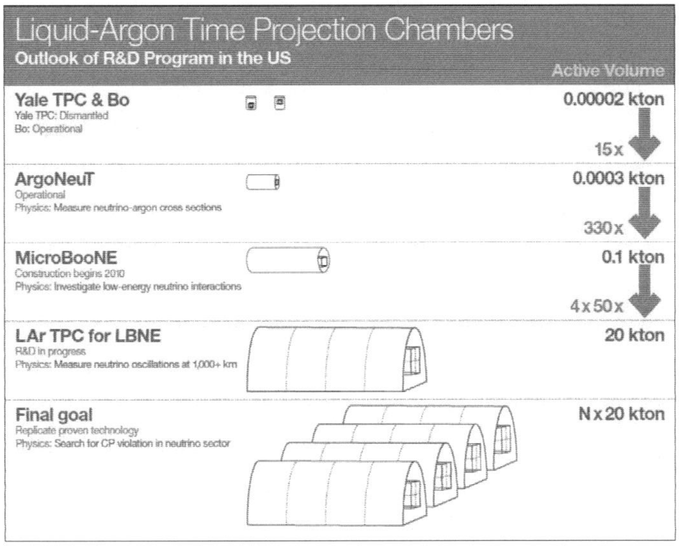

Figure 2. Scaling of LArTPC prototypes and experiments up to LAr20 and beyond

2. ArgoNeuT

The ArgoNeuT project [14] is a test beam experiment which has run under MOU agreement with Fermilab through February 2010. The cryostat, holds ~500 liters of LAr total, and ~175 liters within the 0.5m x 0.5m x 1m TPC. The LArTPC sat just upstream of the MINOS near detector and was exposed to the on-axis NuMI neutrino beam, using the MINOS near detector as a muon catcher. ArgoNeuT was commissioned above ground in 2008 and moved below ground in May of 2009. ArgoNeuT filled and collected neutrino interactions during a first month of commissioning with the NuMI neutrino beam in May of 2009. Commissioning continued throughout the 2009 summer shutdown after which ArgoNeuT collected on the order of ~10k neutrino and anti-neutrino interactions in its 5 month run exposed to the NuMI anti-neutrino beam. ArgoNeuT was removed in late February of 2010 and will undergo upgrades before a proposed ArgoNeuT phase 2 run.

The ArgoNeuT test has allowed for a first ever large sample of low energy neutrino interactions in an LArTPC collected over a relatively long and stable run. Over the 5 month run, excluding a 2 week down time due to an off-the-shelf cryocooler failure, the test operated at appoximately 99% uptime. The detector ran in a "shiftless" mode with emergency monitoring via a pager.

The ArgoNeuT data is being analyzed towards measurements of neutrino cross sections on argon. The 10k neutrino events collected have spurred analysis of and the development of a full MC simulation and automated reconstruction package both for ArgoNeuT and for LArTPC experiements beyond. This analysis framework package, dubbed LArSoft, is described in Section 4.

3. MicroBooNE

MicroBooNE is the next step beyond ArgoNeuT in the US based plan, in developing a fully instrumented LArTPC for neutrino physics. It is a 150 ton LArTPC experiment to be designed, constructed, and installed by 2013, with beam delivery of 6.5E20 protons on target (pot) over 2-3 years of data taking beyond this in the Booster Neutrino Beam at Fermilab.

MicroBooNE has a combination of physics goals and development goals. The physics goals encompass measurements of low energy neutrino interaction phenomena on Argon. The development goals align the MicroBooNE design with the path to larger detectors so that as much as possible can be learned about design and engineering of very large LArTPCs.

MicroBooNE received Stage 1 approval from the Fermilab director in 2008 and has recently been approved for CD-1 by the Department of Energy towards data taking to begin in 2013.

3.1. MicroBooNE physics

MicroBooNE's physics goals include addressing the low energy excess observed by the MiniBooNE experiment and performing a unique set of low energy neutrino cross section measurements. For the MiniBooNE low energy excess, the e/γ separation capability of LArTPCs, unavailable in all previous low energy accelerator neutrino experiments, nearly eliminates muon neutrino induced backgrounds and increases electron efficiency resulting in a $3-5\sigma$ significant signal in a detector – the MicroBooNE detector – almost an order of magnitude smaller than MiniBooNE. The Figures below show the expected signal over background in MicroBooNE given the two different interpretations of the low energy excess from the MiniBooNE experiment, electrons or photons.

In addition, rare events such as coherent pion production, kaon production for proton decay studies, and ν_e CCQE interactions can be studied using MicroBooNE's high efficiency detector. Some of these interaction channels have never before been measured at these energies. All of these measurements are described in more detail in the MicroBooNE proposal, proposal addendum, and the Conceptual Design Report [15].

3.2. MicroBooNE development

The MicroBooNE detector is an 89 ton active volume LArTPC enclosed in a cylindrical cryostat $\sim$4m in diameter and $\sim$ 12 m long. The inner TPC is 2.6m x 2.6m in cross section and 10.4m long in the beam direction. Ionization electrons from charged particles are drifted via a 500V/cm electric field, at 1.6 mm/μs to wire chamber planes on the beam right side of the detector, where they are readout. A PMT array behind the TPCs wire chamber planes records prompt scintillation light used for background rejection, and to determine T_0 for non-beam events. The conceptual design of the PMT and TPC array are shown in Figure 5.

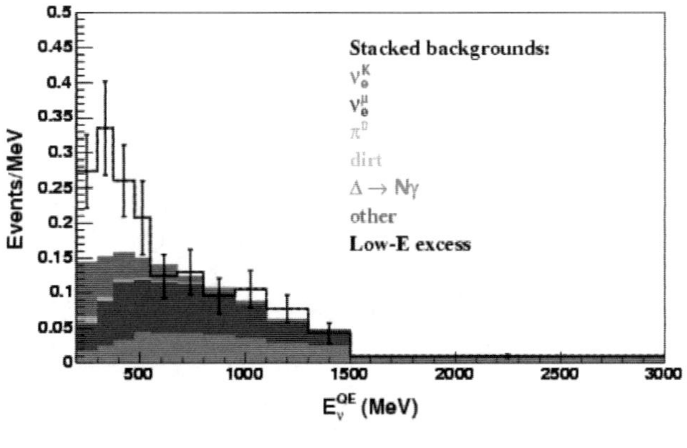

Figure 3. Expected signal over background from the MicroBooNE experiment if the MiniBooNE low energy excess is electrons

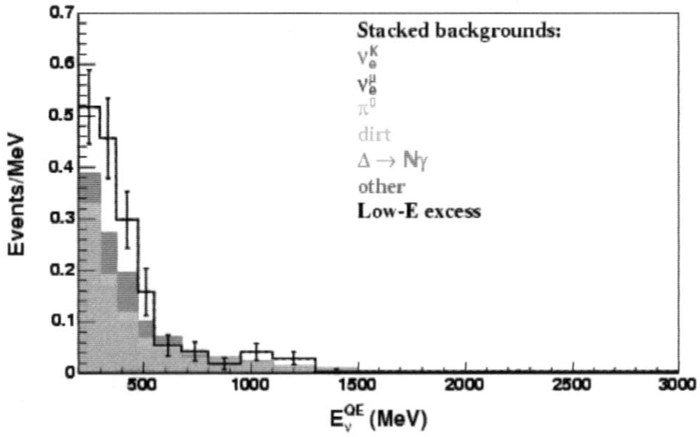

Figure 4. Expected signal over background from the MicroBooNE experiment if the MiniBooNE low energy excess is photons

In addition to MicroBooNE's physics goals, the experiment has development goals which include achieving purity in a fully instrumented non-evacuable vessel, development of cold low-noise electronics, and development of design and construction techniques to be employed in larger detectors. These development goals are summarized in other proceedings from this workshop.

In addition to these hardware development goals, MicroBooNE will provide a large sample of fully contained neutrino interactions which motivates development of a fully automated simulation and reconstruction package. The package under development called LArSoft, or Liquid Argon Software, is designed to be adaptable for other LArTPC experiments. Presently, it is actively being used by ArgoNeuT, MicroBooNE, and the LBNE efforts. Progress on LArSoft is described in the section below.

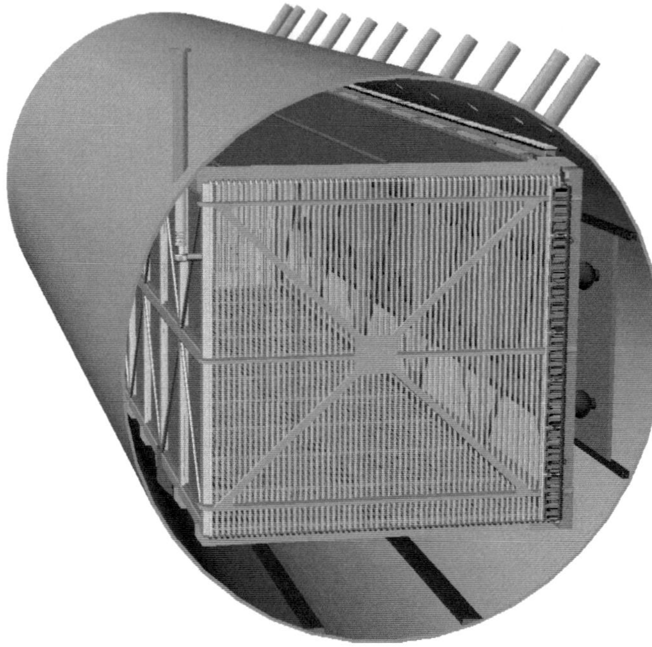

Figure 5. Conceptual design of time projection chamber in MicroBooNE cryostat

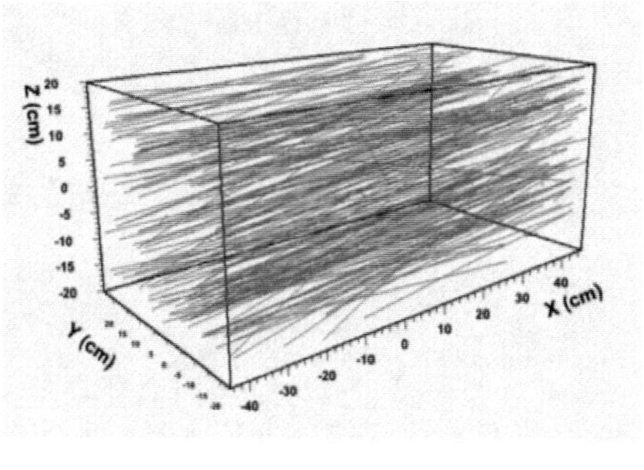

Figure 6. 3D reconstruction of a large sample of through going muons in the ArgoNeuT detector

4. LArSoft

The LArSoft simulation and reconstruction package is under active development by experiments including ArgoNeuT, MicroBooNE, and LBNE. Data and Monte Carlo events are reconstructed in 2D and 3D images. Figure 6 shows a large sample of through going muons in ArgoNeuT reconstructed in 3D. The reconstruction chain includes hit finding, line finding, vertex finding, density clustering, 3D image reconstruction, and calorimetry. Figure 7 shows the hough transform based line finding and huff vertex finding applied to a candidate neutrino interaction from the ArgoNeuT data set. Figure 8 shows the density based (DBSCAN [16]) clustering algorithm applied to a candidate neutral current pion event from ArgoNeuT. Rapid development progress is underway with a full reconstruction chain for data and Monte Carlo anticipated soon.

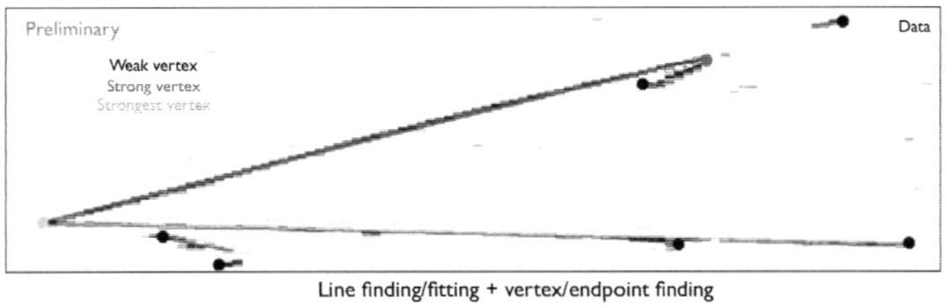

Figure 7. Line finding and vertex finding algorithms applied to a candidate neutrino interaction in the ArgoNeuT Detector

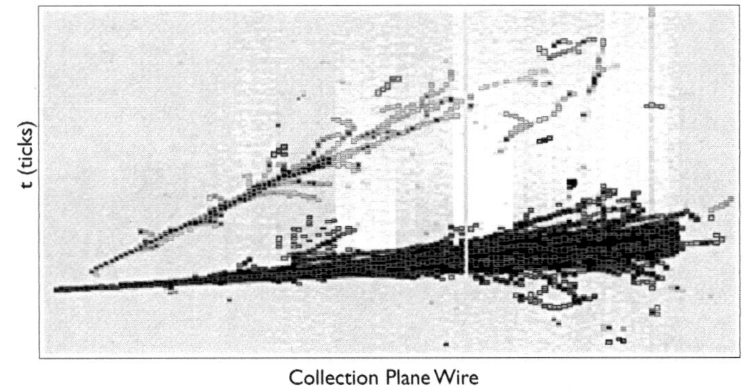

Figure 8. Density based clustering algorithm, DBSCAN, applied to a candidate neutrino interaction in the ArgoNeuT Detector

5. Conclusions

A development program is underway in the US towards construction of very massive LArTPC detectors for the LBNE experiment. ArgoNeuT and MicroBooNE are key experiments in this development path. Both have physics and development goals being pursed which have been described here.

References

[1] C. Rubbia, "The Liquid-argon time projection chamber: a new concept for Neutrino Detector", CERN-EP/77-08 (1977).

[2] A. A. Aguilar-Arevalo *et al.* [The MiniBooNE Collaboration],"A Search for Electron Neutrino Appearance at the Delta m**2 1 eV**2 Scale," Phys. Rev. Lett. **98**, 231801 (2007) [arXiv:0704.1500 [hep-ex]].

[3] M. Maltoni and T. Schwetz,"Sterile neutrino oscillations after first MiniBooNE results,"arXiv:0705.0107 [hep-ph].

[4] S. Goswami and W. Rodejohann, "MiniBooNE Results and Neutrino Schemes with 2 sterile Neutrinos: Possible Mass Orderings and Observables related to Neutrino Masses," arXiv:0706.1462 [hep-ph].

[5] H. Pas, S. Pakvasa and T. J. Weiler,"Sterile - active neutrino oscillations and shortcuts in the extra dimension," Phys. Rev. D **72**, 095017 (2005) [arXiv:hep-ph/0504096].

[6] C. Giunti and M. Laveder, "ν_e Disappearance in MiniBooNE," arXiv:0707.4593 [hep-ph].

[7] T. Katori, A. Kostelecky and R. Tayloe,"Global three-parameter model for neutrino oscillations using Lorentz violation,"Phys. Rev. D **74**, 105009 (2006) [arXiv:hep-ph/0606154].

[8] A. de Gouvea and Y. Grossman, "A three-flavor, Lorentz-violating solution to the LSND anomaly," Phys. Rev. D **74**, 093008 (2006) [arXiv:hep-ph/0602237].

[9] X. Q. Li, Y. Liu and Z. T. Wei, "Neutrino decay as a possible interpretation to the MiniBooNE observation with unparticle scenario," arXiv:0707.2285 [hep-ph].

[10] A. E. Nelson and J. Walsh, "Short Baseline Neutrino Oscillations and a New Light Gauge Boson," arXiv:0711.1363 [hep-ph].

[11] J. A. Harvey, C. T. Hill and R. J. Hill, "Anomaly mediated neutrino-photon interactions at finite baryon density," arXiv:0708.1281 [hep-ph].

[12] See, for example, talk by Mark Dierckxsens, BNL's uDiG workshop, 2008.

[13] www-lbne.fnal.gov

[14] The Argon Neutrino Test, Fermilab Test Beam Experiment, T962. http://t962.fnal.gov

[15] Website for the MicroBooNE Experiment: www-microboone.fnal.gov

[16] M. Ester, H.-P. Kriegel, J. Sander, X. Xu, "A density-based algorithm for discovering clusters in large spatial databases with noise", Proceedings of Second International Conference on Knowledge Discovery and Data Mining, 1996.

Journal of Physics: Conference Series **308** (2011) 012008 doi:10.1088/1742-6596/308/1/012008

A tagged low-momentum kaon test-beam exposure with a 250L LAr TPC (J-PARC T32)

T. Maruyama, on behalf of J-PARC T32 Collaboration[1]

High Energy Accelerator Research Organization (KEK), Tsukuba, Ibaraki 305-0801, Japan

E-mail: `takasumi.maruyama@kek.jp`

Abstract. At the beginning of 2010, we presented at the J-PARC PAC an R&D program towards large (100 kton scale) liquid argon TPCs, suitable to investigate, in conjunction with the J-PARC neutrino beam, the possibility of CP violation in the neutrino sector and to search for nucleon decay. As a first step we proposed a test experiment to identify and measure charged kaons, including their decays, in liquid argon. The detector, a 250L LAr TPC, is exposed to charged kaons, in a momentum range of 540 – 800 MeV/c, in the K1.1BR beamline of the J-PARC slow extraction facility. This is especially important to estimate efficiency and background for nucleon decay searches in the charged kaon mode ($p \rightarrow \bar{\nu}K^+$, etc.), where the kaon momentum is expected to be in the few hundred MeV/c range. A prototype setup has been exposed in the K1.1BR beamline in the fall of 2010. This paper describes the capabilities of the beamline, the construction and setting up of the detector prototype, along with some preliminary results.

1. Introduction

Liquid argon Time Projection Chambers (TPC) have been proposed and studied by the ICARUS Collaboration for decades (see Ref. [1] and references therein). Their efforts culminated with the underground operation of a 600 ton-scale detector at LNGS. The LAr TPC provides similar track position resolution as bubble chambers, with the advantage of being continuously sensitive devices with real time readout. In its original design, the LAr TPC suffers from scalability problems to the relevant mass scale for next generation experiments. The main challenges for the extrapolation are (a) very small charge signals (no amplification in liquid) (b) charge attenuation and diffusion along the drift path (c) large wire chambers in cryogenic environment and (d) wire electric capacitance and resistance.

Since several years, we have investigated the feasibility of a new giant next-generation and multi-purpose neutrino observatory based on the LAr LEM TPC concept with a total mass in the range of 100 kton [2], devoted to particle and astroparticle physics [3, 4, 5, 6, 7].

In 2008, ETHZ and KEK groups have begun to collaborate towards the realization of a very large 100 kton-scale detector [8]. In synergy, a coherent R&D programme is being developed under the program of IPNS and KEK Detector Technology Project and under the CERN Recognized Experiment (RE18) and RD51 Collaboration [9].

[1] ETHZ-KEK-Iwate-Waseda.

At the 9th J-PARC PAC meeting of January 2010, we presented an R&D path towards a proposal to search for CP violation in the leptonic sector and for proton decay, using a 100 kton scale LAr TPC on the Okinoshima island [10], Japan, positioned along the J-PARC neutrino beam on a baseline of 658 km (J-PARC PAC Proposal P32 [11]). The performance of liquid argon TPCs has been mainly studied with cosmic rays and on neutrino beams, however a test using a well-defined charged particle beam has not been performed so far. As part of our R&D path, we proposed to expose a 250L LAr TPC to a charged particle test beam of a few hundred MeV/c momentum, enriched with kaons, at the J-PARC slow extraction facility (test experiment T32). In fact one of the most important advantages for using a liquid argon TPC detector for proton decay searches is the direct identification of the kaon track for proton decays in the kaon mode. Our proposal was readily accepted, further attracted the interest of Iwate and Waseda University groups, and we performed a first measurement campaign with a preliminary setup in October 2010 at J-PARC.

In the two body decay $p \rightarrow \bar{\nu}K^+$, the charged kaon is emitted with a monochromatic momentum of 340 MeV/c in case of a proton at rest, but then smeared by the nucleon Fermi motion. Such kaons are not directely visible in water Cherenkov detectors due to the high momentum threshold ($\sim$ 600 MeV/c) of kaons for Cherenkov radiation in water. A positive identification of the kaon track is essential to minimize the background from atmospheric neutrino interactions. Kaons are identifiable in a LAr detector by measuring the local energy loss along the tracks as a function of the residual range, and by their decay topology [6].

Fig. 1 shows a MC simulation of kaon decays in a $40 \times 40 \times 80$ cm^3 LAr TPC, along with local dE/dx as a function of residual range and a dE/dx based likelihood distribution for kaons, pions and muons. The likelihood shows a good separation of kaons from pions and muons. In addition the topology of the decay and possibly the reconstruction of the decay products provide additional constraints on the identification of the primary particle. In a 100 kt LAr detector there are in total $\sim$ 200000 atmospheric neutrino events/year, a fraction of which are single track events, constituting a background to the $p \rightarrow \bar{\nu}K^+$ search; this test will allow to experimentally determine the efficiency of kaon identification and the rejection factor of pions, muons and electrons.

We plan to build a 250L LAr TPC operated in double phase, where the ionization charge, extracted from the liquid, is amplified in the gas phase by means of a Large Electron Multiplier and readout by a two-dimensional projective anode, as described in [12] and references therein. Due to time limitations between January and October 2010, we first built a 250L prototype detector, operated in liquid phase and with coarse readout sampling. This detector has been exposed to charged particle beams, in a momentum range of 200–800 MeV/c, in the K1.1BR beamline of the J-PARC slow extraction facility. The beamline is equipped with an electro-static separator providing a K/π ratio of $\sim$ 1. Kaon identification by the LAr detector is cross checked against the identification of the incident particles provided by the beam monitor equipment in the beamline.

The beamline in the slow extraction facility, equipped with an electro-static separator, is described in section 2. An overview of the experimental setup, including a discussion of the beamline instrumentation, is given in section 3. The components and operation of the cryogenic system are described in detail in section 4, while the construction and assembly of the prototype detector are detailed in section 5. Some preliminary results are shown in section 6.

2. K1.1BR beamline

The J-PARC slow extraction facility (hadron hall) and the location of the K1.1BR beamline are shown Fig. 2. The K1.1BR beamline has been designed by the TREK Collaboration [13].

The basic concept of the K1.1BR beamline is a low momentum (0.8 GeV/c) separated K$^+$ beam, by means of a single electro-static separator. The layout of the K1.1BR line, about 20 m

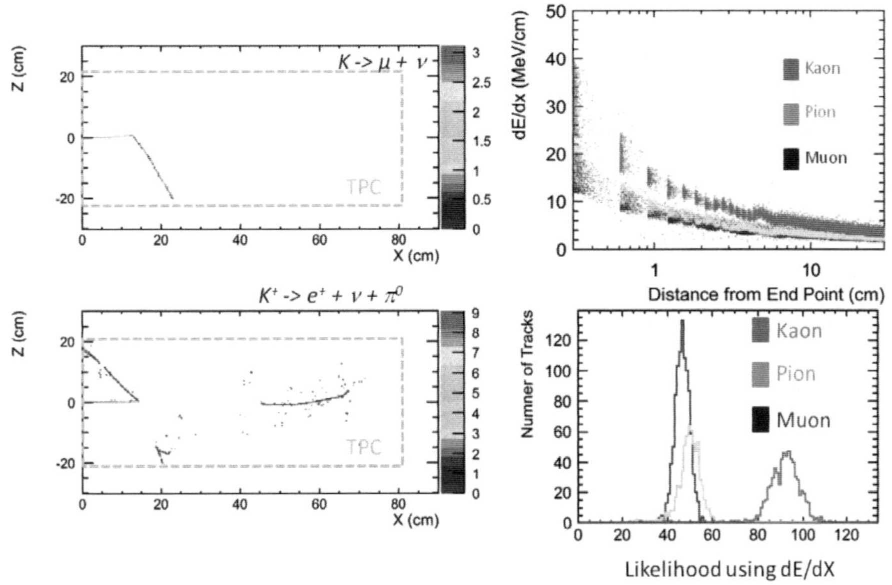

Figure 1. Expected performance of the 250L prototype. Left plots; examples of kaon decays. Top-right; local dE/dx as a function of residual range, bottom-right; dE/dx based likelihood distribution for kaon identification

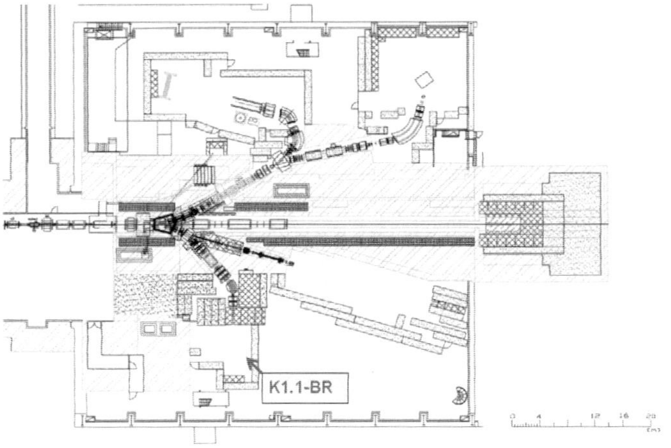

Figure 2. J-PARC hadron hall and the location of the K1.1BR beamline

long, is given in Fig. 3.

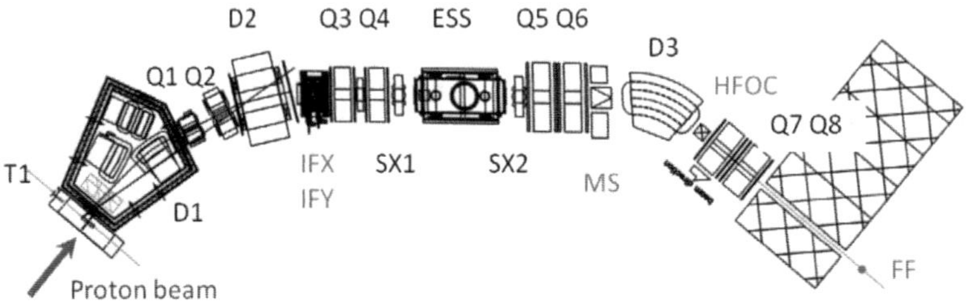

Figure 3. Magnet layout of the K1.1BR beamline

The secondary beam takes off after the T1 production target before the first bending magnet D1. A quadrupole doublet Q1-Q2 focuses the beam at a vertical focus IFY, downstream of a second bending magnet D2. An horizontal acceptance slit is positioned at the location of the IFY vertical focus. A further quadrupole doublet Q3-Q4 makes the beam vertically parallel in the ESS electro-static separator and contains the beam horizontally. The ESS is a 2.0 m long electro-static separator, generating an electric field of 50 kV/cm over an 11 cm vertical gap. The following quadrupoles Q5 and Q6 generate a vertical focus at the mass slit (MS) position and an horizontal focus at HFOC, after the D3 bending magnet. Additionally two sextupole magnets (SX1, SX2), before and after ESS, are used to correct for higher order aberrations. The vertical slit at MS removes most of the pions, and the horizontal slit at HFOC removes 'cloud' pions (from the decay of neutral kaons close to the production target, constituting a widespread source) and slit-scattered pions. A pion rejection factor of $\sim$ 500 is expected, resulting in a K/π ratio of $\sim$ 1.

The final quadrupole doublet Q7-Q8 provides the final focus, with horizontal and vertical beam rms widths of 0.8 cm and 0.5 cm, respectively. The momentum spread Δp/p is $\pm$3 %. The absolute number of expected kaons is 2×10^6 for 5.5×10^{13} 30 GeV protons on target.

It is possible to tune the K1.1BR beamline down to 200 MeV/c momentum, but with a kaon content sharply falling below 800 MeV/c. In order to get an acceptable kaon flux at <800 MeV/c momentum, we use a degrader at the end of the beamline, where particles lose energy by ionization. We have been using a combination of lead glass blocks and a lead brick of 12.5 cm and 2.5 cm thickness, respectively.

3. T32 setup

3.1. Overview

Fig. 4 shows the overview of the T32 setup. Beamline instrumentation, over a length of $\sim$ 5m, is used for the definition of beam particles and for their identification. The 250L TPC chamber is situated in the downstream part of the beamline. Just upstream of the chamber it is possible to insert a degrader to decrease the momentum of the beam particles.

3.2. Beam instrumentation

Beam instrumentation has been provided by the TREK Collaboration. As shown in Fig. 4, there is a beam hodoscope, a beam defining counter (BDC), two MWPCs, two time of flight counters (TOF), a gas Cherenkov counter and a Fitch-type differential Cherenkov. In addition we have placed an additional beam defining counter (BDC_T32) just upstream of the 250L LAr TPC and downstream of the degrader position.

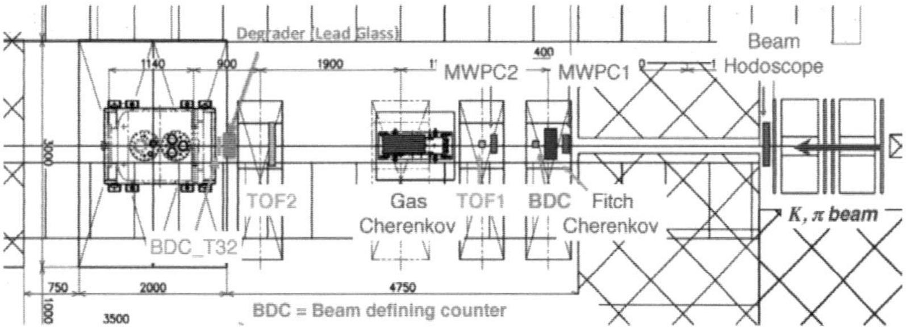

Figure 4. K1.1BR beamline showing the beam monitor equipment and the location of the 250L cryostat.

In the Fitch Cherenkov counter kaons and pions go through a 40 mm-thick acrylic radiator (see Fig. 5–Left), where they emit Cherenkov light at definite angles depending on their velocities. The Cherenkov light from pions is reflected by a mirror around the radiator and detected by an inner PMT ring (π–ring). The light from kaons is reflected by a backward parabolic mirror and detected by an outer PMT ring (K-ring). Each ring is composed of 14 PMTs with a Winston cone at the entrance.

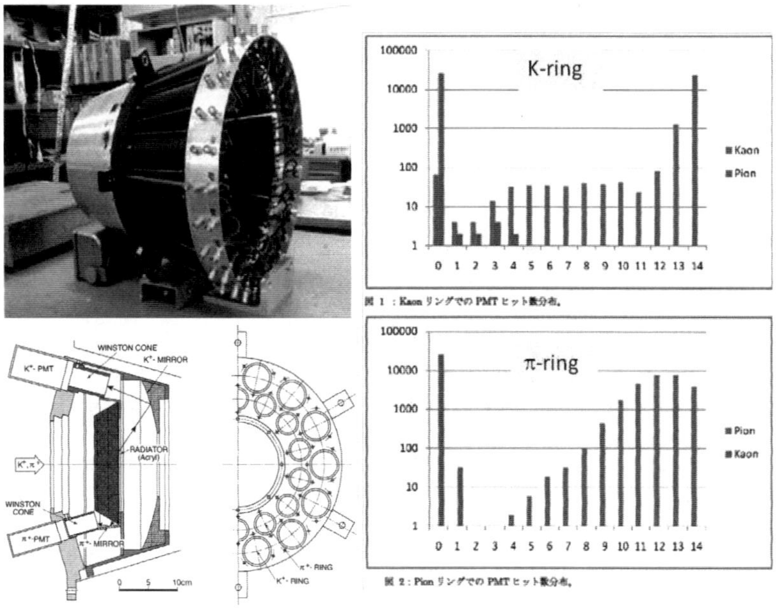

Figure 5. Configuration (Left) and simulated performance (Right) of the Fitch Cherenkov counter.

The PMT hit multiplicity in each ring of the Fitch counter is used to trigger online on pions or kaons. Fig. 5–Right shows the simulated performance. The kaon selection efficiency is $(99.5 + 0.5 - 0.9)\%$, while rejection factor of pion is close to 100%.

The gas Cherenkov counter is used in the online trigger to reject electrons and positrons in the beam.

The TOF counters are at a distance of 3.5 m. With a measured time resolution of $\sim$ 200 ps, they provide an excellent separation of kaons from pions and muon at 800 MeV/c, where the time of flight difference is $\sim$ 2 ns.

4. The cryogenic system

4.1. Cryostat

The cryostat of the detector has been borrowed from the MEG experiment (search for $\mu^+ \rightarrow e^+\gamma$ conversion), and used to house a prototype LXe calorimeter exposed to gamma and electron test beams [14]. Fig. 6 shows a picture and a 3D drawing of the cryostat, originally built by JECC TORISHA Co., ltd.

Figure 6. Picture and 3D drawing of the 250L cryostat

It consists of two concentric vessels, an inner and an outer one. The inner vessel, of about 75 cm diameter and 100 cm length, is filled with LAr. The maximum size of a parallelepiped which could fit inside the inner vessel is $50 \times 50 \times 100$ cm^3 (250L). The volume between the two vessels is evacuable and additionally it is thermally insulated with super-insulating material (20 layers of aluminized mylar film). The inner vessel is supported by four cylinders made of GFRP (Glass Fiber Reinforced Polymer) to reduce the heat inflow. The resulting heat load is only $\sim$30W at LAr temperature, as measured by monitoring the evaporation of LAr.

There are 4 flanges to access the inner volume. One of the two top flanges is devoted to cryogenics, and it houses the cryoocoler input, the LAr and LN_2 inlets, and the safety devices. On the other top flange, there are signals and high voltage feedthroughs for the detector and the slow control instrumentation. The front flange is used to insert the TPC detector into the vessel. A 9 cm diameter beam window is present at the center of the front flange, and it is equivalent to a material thickness of 0.13 X_0. The back flange is connected to a vacuum pump for vacuum insulation.

4.2. Cooling and purification system

A schematic of the cryogenic setup and purification system is shown in Fig. 7–Left.

A Gifford-MacMahon(GM) cryocooler and a LN_2 coil inside the inner vessel provide the necessary cooling. They are mounted on one of the top flanges, as shown in Fig. 7–Middle. The cryocooler cold head has $\sim$ 100 mm diameter, with a length of 193 mm inside the inner vessel.

The available cooling power at LAr temperature is 160 W. In stable operating conditions, after filling and in absence of gas recirculation, a heater on the cold head is used to prevent the argon from freezing, since the cryocooler power is larger than the heat load. The LN_2 coil is located on the same top flange housing the cryocooler. The coil is made of 5 turns of 1 mm thick 3/8" stainless steel pipe, for a total length of $\sim$ 400 cm. The coil provides a cooling power of more than 600 W for a LN_2 flow of 20 L/hour.

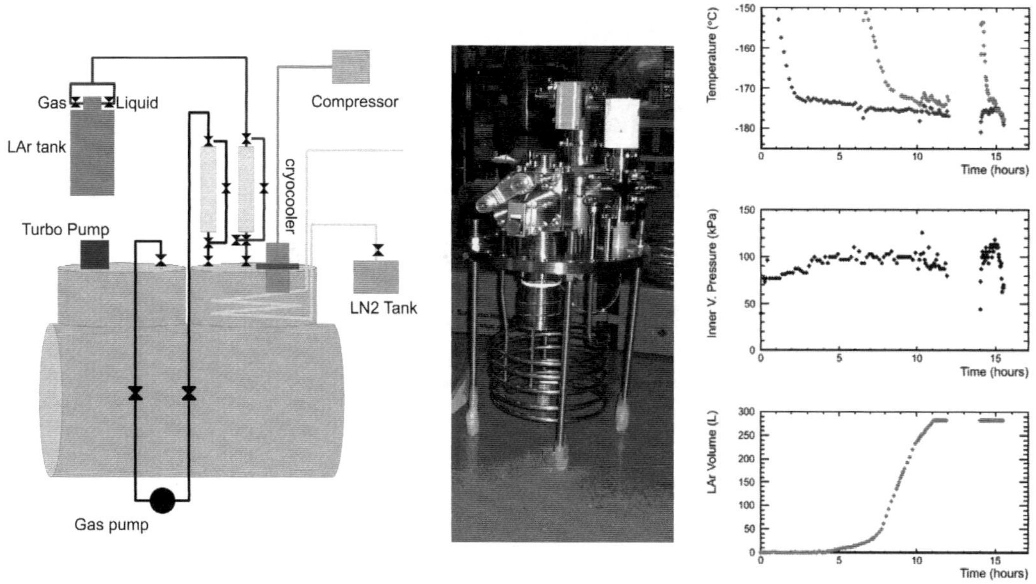

Figure 7. Left: schematic of the cryogenic setup and purification system. Middle: a picture of the GM cryocooler (the copper cold head is visible at the bottom) and of the LN_2 coil. Right: monitoring of the system while filling with LAr. Top plot shows the temperature of the cryocooler head (blue) and of the detector anode (red). Middle plot shows the inner vessel temperature, bottom plot the LAr volume.

In order to achieve a good argon purity, commercial LAr during the initial filling is passed through a purification cartridge. Additionally a gas recirculation and purification system has been implemented as shown in Fig. 7–Left. We use a custom-made purification cartridge [15] for the liquid argon filling phase, with a diameter of 60 mm and a length of 600 mm, containing activated copper-coated alumina granules [16] to remove O_2 and type 4A molecular sieve material to remove H_2O, as previously used in [17]. The two components are kept in separate volumes in the cartridge, with sintered metal disks at the inlet and outlet of the cartridge for the containment of the granules. The cartridge contains 1125 g and 363 g of copper-coated granules and molecular sieve material, respectively, corresponding to a volume ratio of 2:1.

In the recirculation system, argon gas from the inner vessel is forced by a metal bellows pump to circulate through a gas purification filter and then inserted back into the inner vessel. The piping is made of 3/8" stainless steel tubes, using VCR connections and welds to minimize leaks. The gas purification filter, a SAES [18] type MC3000, provides at the output <0.1 ppb of oxygen and water, at a maximum gas flow rate of about 80 slpm.

4.3. Monitoring, control and safety

The monitoring and control of the cryogenic system is managed by a PLC from Keyence [20]. The detailed setup is shown in Fig. 8.

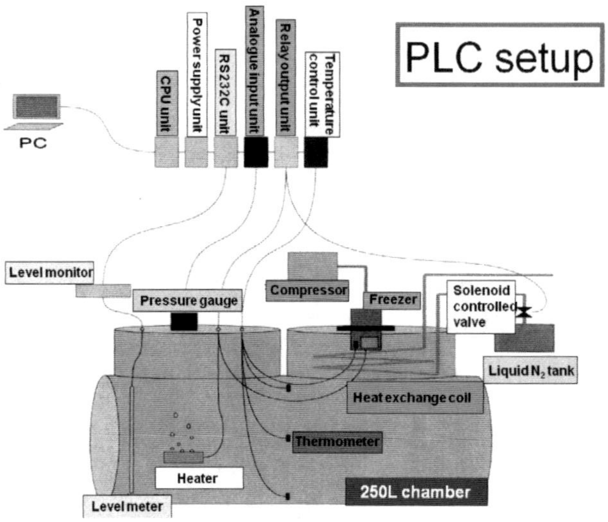

Figure 8. Setup of the PLC monitoring and control.

The following items are monitored and controlled by the PLC:

- temperatures using PT100 sensors;
- pressure in the inner vessel. If the pressure exceeds a set value, LN_2 is let to flow in the cooling coil;
- heater on the cryocooler cold head. This is controlled by the PLC in order to keep the temperature in the vessel at a given set value.

As the inner vessel contains a large amount of cryogenic liquid in a closed volume, it is necessary to protect it with the aid of pressure relief devices. The safety of the system is assured by a purge line equipped with a rupture disk, a safety valve, and an electro-valve. The safety valve opens at 0.15 MPa (Gauge), while the burst pressure of the rupture disk is 0.2 MPa (Gauge). The electro-valve is opened in case of electric power outage. Additionally two oxygen monitors are installed in the K1.1BR beamline area for personnel protection.

4.4. 250L Filling Procedure

The process of filling the inner vessel with LAr requires particular attention not to exceed the maximum allowable pressure in the vessel and not to compromise the purity of the argon because of outgassing during the transition phase to LAr temperature. We describe here in detail each step performed:

- Evacuation of the inner vessel using a molecular turbo pump (Edwards, 300L/s capacity) and a getter pump (SAES, 40 L/s capacity), directly mounted on the top flange for $\sim$ 1 week. We achieved a vacuum level of 1×10^{-4} Pa, with an outgasssing rate of $\sim$ 1 Pa/hour.
- Filling of the inner vessel with argon gas, passing through the purification cartridge, up to 80–100 kPa(Gauge).
- Initial cooling of the vessel using the GM cryocooler and the LN_2 heat exchanger coil.

- Gas recirculation is turned on at the same time, with a typical gas flow of 70 L/min.
- As temperature and pressure inside the vessel decrease, we insert argon gas and we keep the pressure in the inner vessel at ∼ 80 kPa(Gauge).
- After ∼5 hours, the vessel is cooled down to LAr temperature, so that LAr starts accumulating at the bottom of the vessel. The typical liquefaction speed, limited by the cooling power available in the cryocooler and in the LN_2 coil, is 10 L/hour.
- At this point, we start filling the vessel directly with LAr, going through the purification cartridge. It takes ∼3 hours to cool down the purification cartridge to LAr temperature. The filling speed, limited by the impedance of the purification cartridge, is ∼ 50 L/hour.
- Filling is stopped once the readout anode is immersed in LAr.
- We recirculate the argon gas, evaporated from the liquid inside the vessel, in order to improve or maintain the initial purity.

Figure 7–Right shows the monitoring values while filling the inner vessel with LAr. The top plot shows the temperature of the cryocooler head (blue) and of the detector anode (red), the middle plot shows the inner vessel temperature and the bottom plot the volume of LAr from the level meter measurement. Note that the filling was suspended at ∼12 hours and resumed at ∼14 hours; the LAr volume measurement saturates at 280 L because of a level meter limitation.

5. The LAr TPC

5.1. TPC detector

For the first test campaign in October 2010, we built a prototype detector with coarse readout sampling, operated in liquid argon phase. A set of PMTs, at the bottom of the TPC, is used to get a trigger for cosmic ray muons from the argon scintillation light. Table 1 summarizes the main detector parameters.

Total LAr volume	∼ 300 L
Field cage dimension	42 cm × 42 cm × 78 cm
Fiducial volume	40 cm × 40 cm × 76 cm
Fiducial mass	170 kg
Electric Field	0.2–0.3 kV/cm
Maximum cathode voltage	12 kV
Readout method	charge collection in liquid phase
Readout pitch	1.0 cm
Number of readout channels	76

Table 1. Parameters of the 250L prototype detector exposed at the J-PARC K1.1BR beam in October 2010.

The field cage determines the fiducial volume of the detector. Its dimensions are $42 \times 42 \times 78$ cm^3, for a total LAr fiducial mass of 170 kg. It is made with a printed circuit board (PCB) technique, where the field shapers are 8 mm wide gold-plated copper strips, with 10 mm pitch, on a 1.6 mm thick PCB (see top pictures in Fig. 9). The copper strips are connected to a voltage divider, made of a chain of 10 MΩ high voltage resistors. The cathode plane (see Fig. 9), located at the bottom of the field cage, needs to be transparent to the argon scintillation light generated in the fiducial volume, detected by the PMTs at the bottom. The cathode is built as a grid, with 100 μm stainless steel wires, spaced by 5 mm. Electric field simulations of this configuration show a good uniformity of the drift electric field, with distortions of ∼ 10% at 1 cm distance from the field cage walls.

Journal of Physics: Conference Series **308** (2011) 012008 doi:10.1088/1742-6596/308/1/012008

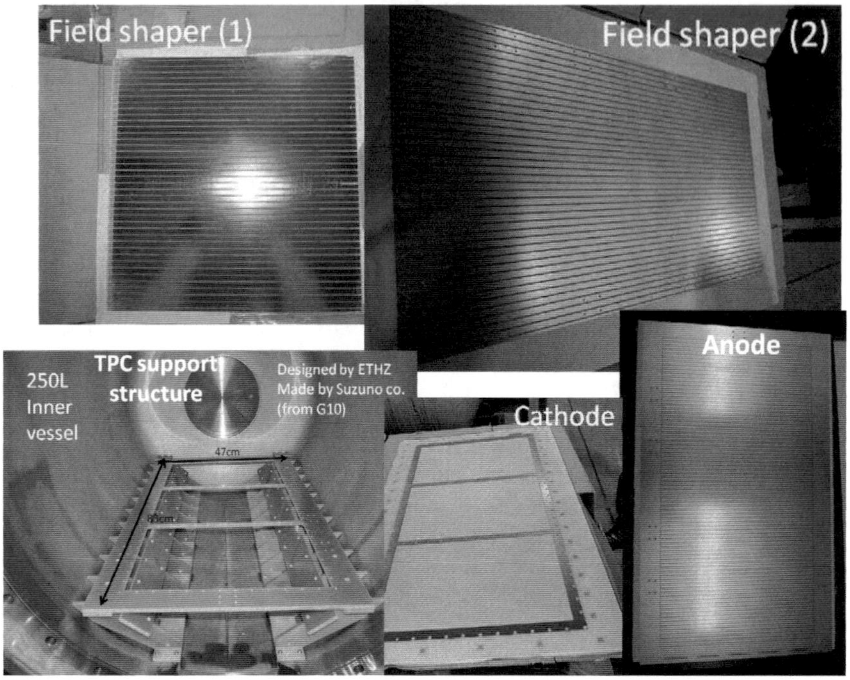

Figure 9. Detector components and support structure

Ionization electrons are drifted upwards to a collecting anode, immersed in LAr, segmented in 1 cm wide strips for a total of 76 readout channels (see Fig. 9). Fig. 10 shows the assembled detector inside the inner vessel.

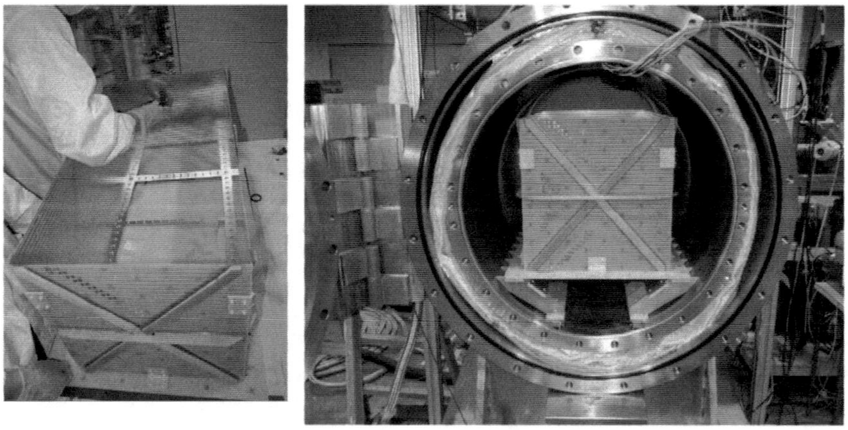

Figure 10. TPC detector during assembly (left) and after insertion inside the inner vessel on the support structure at the bottom (right).

Two cryogenic PMTs are located below the cathode, as shown in Fig. 11. They are Hamamatsu type R6041-02ASYM MOD, operated at $\sim$ 800 V with a typical gain of 1.0×10^6.

The PMTs are supported by stainless holders with protection grids on top. The grids are made of a stainless steel crossed wire pattern of 100 μm width and 10 mm pitch, obtained from etching, and they are kept at ground voltage to protect the PMTs from the high voltage values of the nearby cathode plane.

Figure 11. PMTs with stainless steel holders at the bottom of the vessel.

Figure 12. TPB coated PMT, illuminated by UV light.

In order to shift the 128 nm scintillation light from LAr to a sensitive region of the PMTs,we use tetraphenyl-butadiene (TPB) as wavelength shifter, which has an emission spectrum between 400 and 480 nm. The PMTs are coated with TPB, embedded in a polymer matrix. We used a mixture of TPB (0.10 g), Paraloid (1.01 g) and Toluene (19.99 g). Fig. 12 shows a TPB coated PMT illuminated by UV light.

5.2. Trigger, electronics and DAQ

Different triggers have been used during our data taking period, combined with different settings of the K1.1BR beamline, as summarized here:

- a cosmic ray trigger, by triggering on the coincidence of the two PMTs at the bottom of the 250L TPC, which detect the LAr scintillation light produced in the fiducial volume;
- a basic beam trigger, formed by the coincidence of the Beam Defining Counter BDC, the two time of flight counters TOF1 and TOF2 and the BDC_T32 counter placed just upstream of the 250L chamber (see Fig. 4);
- a positron trigger, from the coincidence of the beam trigger with the signal from the gas Cherenkov;
- a kaon trigger, from the coincidence of the beam trigger with the signal from the Fitch Cherenkov.

The collected charge on the segmented anode of the 250L LAr chamber is readout by the SY2791 system [19] from CAEN, developed in collaboration with ETHZ. Charge sensitive preamplifiers, with a sensitivity of ~ 12 mV/fC, are directly mounted on the CAEN modules. The ouput signals are digitized by 12 bit ADCs with 2.5 MHz sampling frequency.

In order to synchronize the event number between the LAr TPC detector and the beamline instrumentation we use a Master Trigger Module (MTM). The MTM module delivers a spill and a trigger number via VME (see Fig. 13 right) to the LAr TPC electronics and to the beamline instrumentation. Fig. 13 shows the configuration of the system. Data from the LAr TPC and the beamline instrumentation are merged at a later stage by using the MTM information.

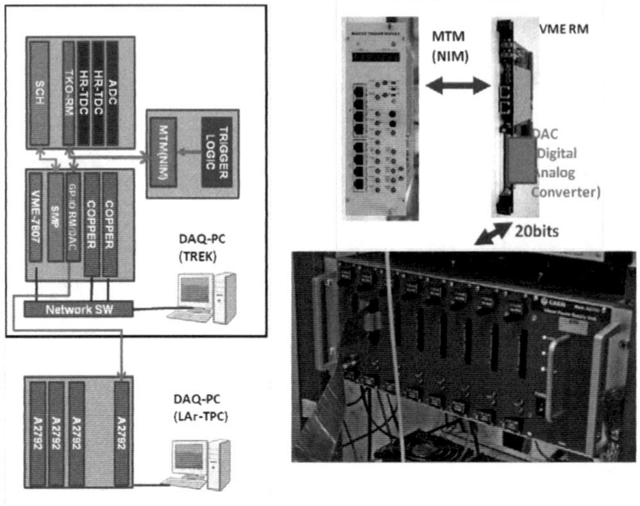

Figure 13. Event tag system between beamline information and LAr TPC. The SY2791 system from CAEN is shown at the bottom on the right

6. Preliminary results and future plans

After filling in October 2010, the setup has been stably operating for a week, corresponding to the allocated time at the J-PARC K1.1BR hadron hall. The detector has been exposed to charged particle beams in the 200 – 800 MeV/c momentum range.

Some images of collected events are shown in Fig. 14. The event displays show drift time vs. channel number, where channel '0' is the upstream channel and the grey scale is proportional to the signal amplitude. A digital filter has been applied to the signals to remove the coherent noise common to all channels.

The top left picture shows an event with three overlapping beam particles crossing the detector, interpreted, from top to bottom, as a pion, a positron and a proton. The event was acquired by triggering on positrons, with the beam momentum set to 800 MeV/c and with no degraders inserted. The top right picture shows a candidate for a pion decay at rest, obtained by triggering on beam particles, with positron rejection, in a 200 MeV/c beam. The events at the bottom in Fig. 14 have been taken with a kaon trigger at 800 MeV/c momentum. A lead glass block and a lead brick were inserted just before the 250 L chamber to degrade the momentum of beam particles entering the chamber. Bottom left picture is a candidates for $K^+ \to \mu^+\nu_\mu$ decay at rest. The bottom right picture can be interpreted as a $K^+_{\mu 3}$ decay in flight.

The analysis of the data and comparison to a detailed MC simulation of the setup and beam properties, is presently ongoing. By the observation of cosmic muons, we measured an electron lifetime in LAr better than $300\mu s$, corresponding to an upper limit of $\sim$ 1 ppb of electronegative impurities.

At the same time we have been proceeding with the construction of a double phase LAr TPC, equipped with a Large Electron Multiplier and a two-dimensional projective anode with 3 mm readout sampling, for a total of 512 channels. All components have been manufactured, and we are presently assembling them. A test with cosmic rays has been planned at CERN. If successful, we will consider and proposed another installation in the K1.1BR beamline.

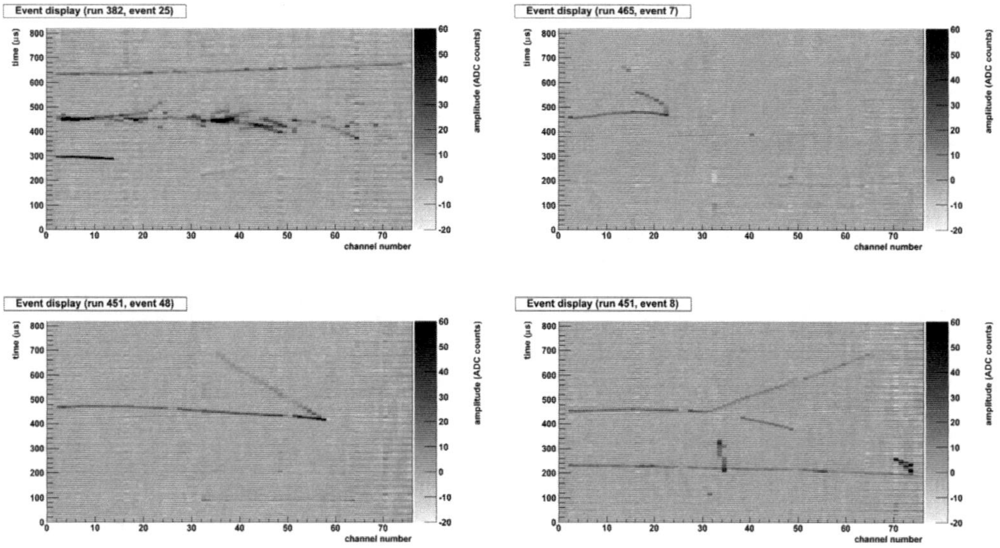

Figure 14. Displays of collected events, after removal of coherent noise. Drift time is shown vs channel number, grey scale is proportional to signal amplitude.

7. Acknowledgments

We would like to warmly thank the TREK collaboration, and in particular Jun Imazato, for information on the K1.1BR beamline and for the use of the beamline instrumentation. We also thank the MEG collaboration for lending us the cryostat for an extended period of time. We especially thank Satoshi Mihara for his valuable information on the cryostat. We greatly appreciate the help received by the J-PARC Hadron Group for the installation of our setup. We also thank the RD51 Collaboration for useful discussions and help of the CERN TS/DEM group. We acknowledge the financial and technical support of our funding agencies and in particular KEK IPNS, KEK Detector Technology Project, the Swiss National Science Foundation and ETH Zurich.

References

[1] S. Amerio *et al.* [ICARUS Collaboration], *Design, construction and tests of the ICARUS T600 detector*, Nucl. Instrum. Meth. A **527**, 329 (2004).

[2] A. Rubbia, *Experiments for CP-violation: A giant liquid argon scintillation, Cerenkov and charge imaging experiment?*, [arXiv:hep-ph/0402110].

[3] I. Gil Botella and A. Rubbia, *Decoupling supernova and neutrino oscillation physics with LAr TPC detectors*, JCAP **0408** (2004) 001 [arXiv:hep-ph/0404151].

[4] A. G. Cocco, A. Ereditato, G. Fiorillo, G. Mangano and V. Pettorino, *Supernova relic neutrinos in liquid argon detectors*, JCAP **0412** (2004) 002 [arXiv:hep-ph/0408031].

[5] A. Meregaglia and A. Rubbia, *Neutrino oscillation physics at an upgraded CNGS with large next generation liquid argon TPC detectors*, JHEP **0611**, 032 (2006) [arXiv:hep-ph/0609106].

[6] A. Bueno, Z. Dai, Y. Ge, M. Laffranchi, A.J. Melgarejo, A. Meregaglia, S. Navas, A. Rubbia *Nucleon decay searches with large liquid argon TPC detectors at shallow depths: Atmospheric neutrinos and cosmogenic backgrounds*, JHEP **0704**, 041 (2007) [arXiv:hep-ph/0701101].

[7] A. Rubbia, *Underground Neutrino Detectors for Particle and Astroparticle Science: the Giant Liquid Argon Charge Imaging ExpeRiment (GLACIER)*, J. Phys. Conf. Ser. **171**, 012020 (2009) [arXiv:0908.1286 [hep-ph]].

Journal of Physics: Conference Series **308** (2011) 012008 doi:10.1088/1742-6596/308/1/012008

[8] A. Badertscher *et al.*, *A Possible Future Long Baseline Neutrino and Nucleon Decay Experiment with a 100 kton Liquid Argon TPC at Okinoshima using the J-PARC Neutrino Facility*, [arXiv:0804.2111 [hep-ph]].
[9] RD51 Collaboration, `http://rd51-public.web.cern.ch/`
[10] M. Yoshioka *et al.*, *Okinoshima site study*, to appear in the Proc. of the 1st International Workshop towards the Giant Liquid Argon Charge Imaging Experiment (GLA2010).
[11] A. Badertscher *et al.*, P32: Towards a Long Baseline Neutrino and Nucleon Decay Experiment with a next-generation 100 kton Liquid Argon TPC detector at Okinoshima and an intensity upgraded J-PARC Neutrino beam, presented at the 9th Program Advisory Committee (PAC) for Nuclear and Particle Physics Experiments at the J-PARC 50GeV Proton Synchrotron (J-PARC PAC), see `http://j-parc.jp/NuclPart/pac_1001/pdf/KEK_J-PARC-PAC2009-10.pdf`
[12] A. Badertscher *et al.*, *First operation of a double phase LAr Large Electron Multiplier Time Projection Chamber with a two-dimensional projective readout anode*, Nuclear Inst. and Methods in Physics Research, A 641 (2011), pp. 48-57 [arXiv:1012.0483 [physics.ins-det]].
[13] TREK homepage, http://trek.kek.jp/
[14] S. Mihara [MEG Collaboration], "R&D work on a liquid-xenon photon detector for MEG experiment at PSI," Nucl. Instrum. Meth. A **518**, 45 (2004).
[15] M. Tanaka, *Recent Results from Liquid Argon R&D at KEK*, to appear in the Proc. of the 1st International Workshop towards the Giant Liquid Argon Charge Imaging Experiment (GLA2010).
[16] Engelhard Corporation, MSDS for Cu-0226 S 14 x 28 (2002).
[17] A. Curioni *et al.*, Nucl. Instrum. Meth. A **605**, 306-311 (2009).
[18] `http://www.saespuregas.com`
[19] `http://www.caen.it/nuclear/syproduct.php?mod=SY2791`
[20] `http://www.keyence.co.jp/`

MEG liquid xenon detector

Satoshi Mihara
KEK IPNS, 1-1 Oho, Tsukuba, Ibaraki 305-0801, Japan
E-mail: satoshi.mihara@kek.jp

Abstract. We constructed a gamma-ray detector using 900 L of liquid xenon to identify a 52.8-MeV gamma ray from the lepton-flavor violating muon decay, $\mu \to e\gamma$. The detector uses only scintillation light originating from liquid xenon to reconstruct the gamma-ray energy, interaction position, and arrival time. A new photomultiplier that can be operated in liquid xenon has been newly developed for this detector. Details of the detector is described in this paper along with the present performance of the detector.

1. Introduction

Liquid xenon is believed to be an excellent material for gamma-ray detection because of its suitable features such as high density, efficiency and decay time of scintillation light emission and high transparency for the light. To achieve large acceptance with good uniformity, it is beneficial for the liquid to be stably maintained at pressures around the atmospheric pressure in a cryostat equipped with a refrigerator with moderate cooling power. The ionization charge induced in liquid xenon is also commonly used for gamma-ray and/or charged particle measurement, although the detector response becomes slow for such measurements because of the limited electron drift speed in the liquid. A hybrid method that combines scintillation light and ionization charge measurements has been studied to improve the performance of a detector. Extensive studies have been conducted to enable the use of such new detectors in many applications [1, 2].

2. MEG experiment

The mu-e-gamma (MEG) experiment aims at detecting a lepton-flavor violating $\mu^+ \to e^+\gamma$ decay with higher sensitivity than that obtained in the previous experiment, in which the limit was set at 1.2×10^{-11} at 90% confidence level (CL)[3]. The ultimate aim of the MEG experiment is to achieve a sensitivity of 10^{-13} for the $\mathrm{Br}(\mu^+ \to e^+\gamma)$ [4].

The signal of $\mu^+ \to e^+\gamma$ is a simultaneous emission of a positron and a gamma ray, each with an energy equivalent to half of the muon mass, 52.8 MeV, at an opening angle of 180°. The gamma ray is detected by a liquid xenon photon detector, which is described in detail in this paper. A positron detector is composed of a drift chamber tracking system [5], time measuring counter system [6], and super-conducting magnet with a gradient magnetic field [7].

There are two major background sources in this experiment. One is the radiative muon decay, $\mu^+ \to e^+\nu_e\bar{\nu_\mu}\gamma$, in which two neutrinos carry tiny amounts of energy. The other is an accidental overlap of an electron emission from the Michel decay, $\mu^+ \to e^+\nu_e\bar{\nu_\mu}$, and a gamma-ray emission from a radiative muon decay or positron annihilation in flight.

The setup of the experiment is shown in Figure 1. Muon beam is provided at the PiE5 beam line of Paul Scherrer Institute (PSI), where the world's most intense surface muon beam is available. The PSI proton accelerator generates a 580-MeV 2-mA proton beam that creates pions and muons as daughter particles of the pions. The repetition cycle of the proton is 19.75 ns, which is smeared by the pion life time (26 ns), resulting in the generation of a quasi-continuous muon beam that is used in the experiment. The muon beam is transported to the experimental area and filtered by an electrostatic Wien filter to remove positron contamination. The beam is focused and stopped on a muon stopping target located at the center of the spectrometer magnet.

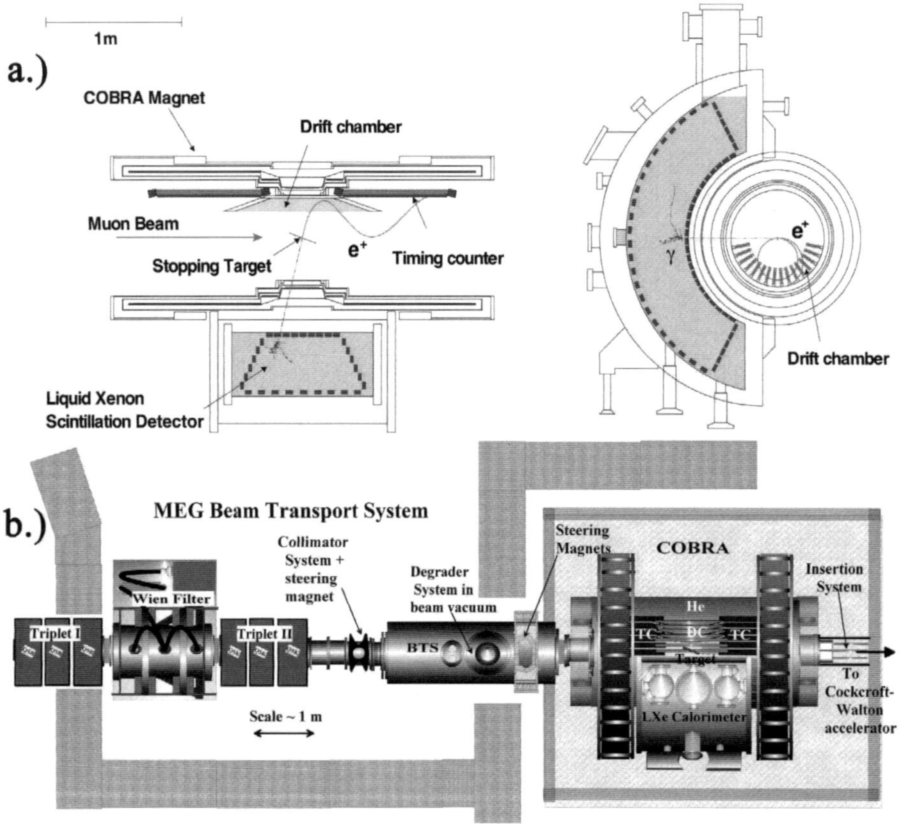

Figure 1. Schematic of MEG experiment. a) Details of positron spectrometer and photon detector. b) MEG beam transport system.

The target is surrounded by a positron tracking system composed of 16 drift chamber modules; this system measures the positron momentum and predicts the hit position on the timing counters located at both ends of the magnet. The liquid xenon detector is located at the right of the beam line beside the spectrometer magnet and measures the gamma ray originating from the target. The detector covers 10% of the acceptance with good uniformity.

The construction of the experimental setup was completed in 2007 followed by a short engineering run. In 2008 and 2009, physics data were acquired using the completed setup.

Journal of Physics: Conference Series **308** (2011) 012009 doi:10.1088/1742-6596/308/1/012009

The result obtained using the 2008 data has been recently published; in this result the limit on Br($\mu^+ \rightarrow e^+\gamma$) was set at 2.8×10^{-11} (90% CL)[8]. Currently, the analysis of 2009 data is in progress, and further acquisition of physics data is continuing in 2010 and is expected to continue in future years.

3. MEG liquid xenon photon detector

A schematic view of the MEG xenon photon detector is shown in Figure 2. The detector measures gamma rays from the muon stopping target by using liquid xenon as the scintillation material. We do not use ionization charge information but use only scintillation light for the measurement. Consequently, the detector can be quite simple, and the detector can provide a reasonably fast response, which is essential for searching a rare decay mode. A total of 900 L of liquid xenon is used in the detector. The active volume is viewed from all sides by photomultipliers operational in the liquid. The gamma-ray energy is evaluated by summing the charge integration of all photomultiplier signals, position of gamma-ray first interaction positions in the liquid is evaluated by weighting the photomultiplier positions on the inner (gamma-ray entrance) side with the observed charge, and gamma-ray arrival time is estimated by using the time information of the photomultiplier signals. Waveforms of all photomultipliers are recorded using waveform digitizers that are newly developed for the experiment [9]. Details of the detector components are described in this section.

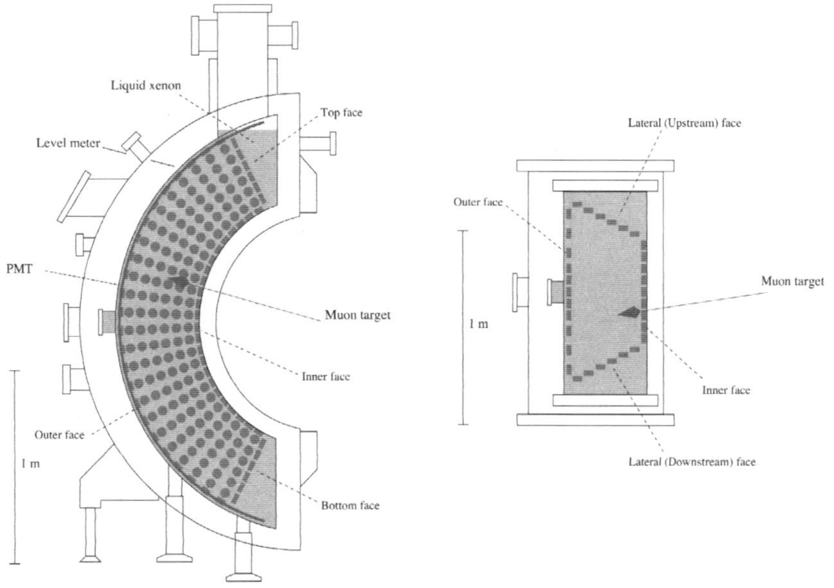

Figure 2. Schematic views of MEG liquid xenon photon detector. Side (left) and top (right) views.

3.1. Prototype R&D

The detector has been constructed after carrying out numerous R&D activities using two prototypes with different sizes. The first prototype was constructed to demonstrate the principle of the detector, that is, the measurement of gamma rays using only liquid xenon scintillation light by photomultipliers submersed in the liquid. This prototype has an active volume of 2.3 L viewed by 32 photomultipliers. The detector response was investigated by using gamma rays below 2 MeV from radioactive sources, and this response was compared with a simulation incorporating energy loss of gamma rays in liquid xenon and the scintillation light emission and transmission in the liquid [10].

The second prototype was constructed to demonstrate the detector performance for higher energy gamma rays around 50 MeV and to gain the operation experience required to conduct the $\mu^+ \to e^+\gamma$ search experiment, which runs for several months once the physics data acquisition is started. The second prototype has an active volume of 68.6 L surrounded by 228 photomultipliers. The detector performance was investigated by using gamma beams up to 40 MeV generated in inverse-Compton scatterings and also using electron beams up to 60 MeV [11]. A sophisticated technique to calibrate photomultipliers in liquid xenon has been developed in this prototype [15] together with a purification method to remove impurities such as water, oxygen, and nitrogen that can affect scintillation light emission and transmission [16, 17]. A method to evaluate the light absorption, separately from elastic scattering of light, has also been developed [13].

3.2. Detector construction

The MEG liquid xenon photon detector consists of the following components:

(i) Photomultiplier to detect scintillation light of liquid xenon
(ii) Cryostat and refrigerator to maintain liquid xenon in a stable state
(iii) Xenon purifier to remove impurities possibly contained in xenon
(iv) Xenon storage to store xenon during detector maintenance

The photomultiplier (Hamamatsu R9869) has been developed for use in the MEG liquid xenon photon detector. This photomultiplier uses a metal-channel dynode structure to achieve a reasonable electron multiplication gain ($>10^6$) inside a short package (height: 3.2 cm) and a high tolerance against the magnetic field as large as 100 G, as shown in Figure 3. The package is made of stainless steel, which allows the photomultiplier body to be pressure resistant up to 0.3 MPa. The photocathode was carefully designed to minimize the reduction in the quantum efficiency in the liquid xenon temperature range (165–167 K). Electronic parts used in the bleeder circuit have been carefully selected to minimize the influence of outgassing impurities from these parts. The diagram of the bleeder circuit is shown in Figure 4. Heat dissipation from the bleeder circuit is minimized by optimizing the resistor chain (16 MΩ) with zener diode protection in the last two dynodes, which keeps the voltage constant under a high counting rate environment. In the MEG liquid xenon detector, 846 photomultipliers are used. They are first installed to separate holders and then assembled in the detector cryostat. There are five different types of holders: top, bottom, lateral, outer, and inner. The top and bottom face holders are identical, and they hold 54 photomultipliers each. The lateral face holders have a trapezoidal shape and hold 6 photomultipliers each. The outer face hodlers have a simple rectangular shape and hold 9 photomultipliers each, except at the central part where the photomultiplier density is maintained higher than that at the other parts for calibration purposes. The inner face holders are fabricated in such a manner that the amount of material used is reduced and the liquid xenon penetration is minimized to avoid gamma-ray interaction inside the holder. The holder body is made of PEEK and the other parts are primarily made of plastic. The photomultiplier cables are molded in expoxy glue, and the remaining space around the bleeder circuit of the photomultipliers is filled with glass beads. Each inner face holder contains 9 photomultipliers, and a minimum possible distance is maintained between neighboring photomultipliers.

For photomultiplier calibration, blue LEDs with different light attenuations are installed on the photomultiplier holders so that the photomultipliers can be illuminated uniformly with different light intensities. Five gold-plated tungsten wires of 100-μm diameter are permanently suspended in the active volume and fixed on the photomultiplier holders; five ^{241}Am alpha

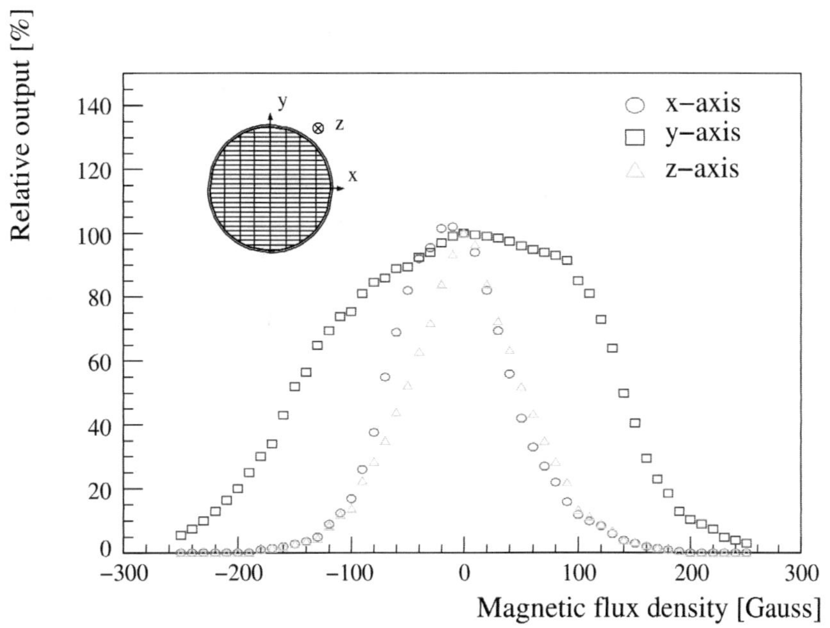

Figure 3. Tolerance of photomultiplier (Hamamatsu R9869) used in MEG liquid xenon detector against magnetic field.

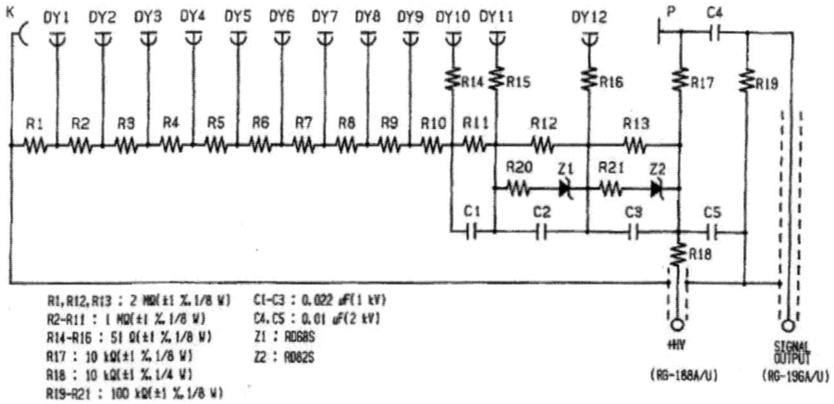

Figure 4. Bleeder circuit diagram of photomultiplier (Hamamatsu R9869).

sources are deposited on each of these wires [15]. Figure 5 (left) shows a micropicture of the source, and Figure 5 (right) shows a photograph of a suspended wire in the second prototype. In the MEG liquid xenon detector, wires are suspended in a similar manner. Alpha particles with fixed energy are constantly emitted from the sources and thereby a constant amount of scintilaltion light is emitted regularly. Because of the difference in the mechanisms of scintillation light emission, the photomultiplier waveforms for the gamma-ray and alpha events are different, and thus, the alpha source signal can be easily separated from the gamma-ray signals.

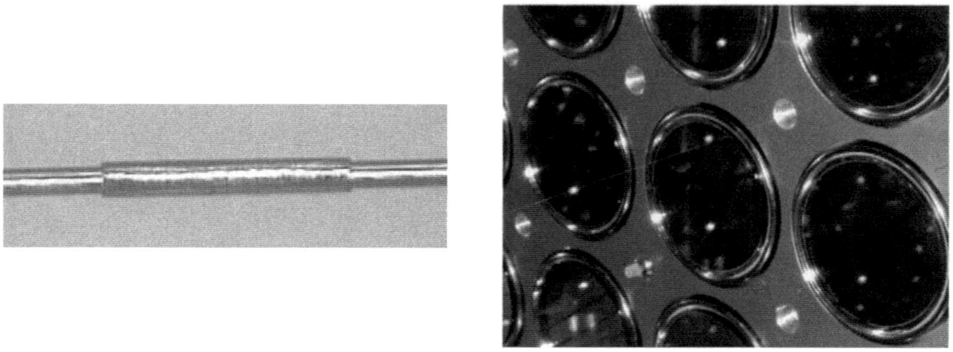

Figure 5. ^{241}Am alpha source deposited on 100-μm gold-plated tungsten wire (left) and source wire fixed on photomultiplier holder in second prototype (right).

The detector is C-shaped; this shape enables the detector to cover the acceptance uniformly. Therefore, the cryostat, which contains liquid xenon is also C-shaped, as shown in Figure 2. The cryostat is composed of an inner vessel that contains liquid xenon and an outer vessel for vacuum heat insulation. The wall thickness of the gamma-ray entrance side in the outer vessel is kept small (400 μm, stainless steel) to reduce gamma-ray interaction with the wall. The thickness of the wall of the inner vessel is also restricted to 400 μm and is made of the same stainless steel material; this wall is backed by a honeycomb panel sandwiched between two carbon fiber sheets. The panel was specially developed for the MEG liquid xenon detector. The honeycomb itself is made of aluminum with a cell thickness of 0.0254 mm, size of 4.67 mm, and height of 19 mm. Each carbon fiber sheet is composed of 8 layers of prepreg and has a thickness of 0.13 mm. These layers are glued together to form a sheet using two-component epoxy adhesive, which is known to be used at low temperatures. The specifications of the panel is summarized in Table 1. The panel is fixed to the inner vessel with screws at the edges of the panel, as shown in Figure 6 (left). The most difficult part of the panel design is to devise a method to transit the thickness of the panel from the inner area where the aluminum honeycomb exists (thick part) to the edges of the panel where only carbon fiber sheets are stacked (thin part), as illustrated in Figure 7. The total force that is applied perpendicular to the panel by the xenon pressure is as large as

Figure 6. Honeycomb panel fixed on inner vessel (left) and reinforcement parts installed at panel edge (right).

80 kN. This would cause a huge shearing force at the transition area, which is almost at the limit of the carbon fiber strength. Therefore, we added some reinforcement parts, as shown in Figure 6 (right), to the transition area to induce a change in the direction of the force applied at the edge. The honeycomb backing enables us to operate the detector up to pressure of 0.2 MPa when the thermal insulation layer is evacuated. The cryostat is equipped with a rapture disk to blow the xenon gas when the pressure reaches 0.19 MPa. The outlet of the rupture disk is connected to a check valve to save the xenon gas when the pressure decreases below 0.16 MPa.

Table 1. Specifications of honeycomb panel used in MEG liquid xenon detector.

Honeycomb material	Aluminum 5052
Hexagonal cell thickness	0.0254 mm
Hexagonal cell size	4.76 mm
Hexagonal cell height	19 mm
Carbon fiber sheet	T300 (Toray)
Carbon fiber sheet thickness	1.5 mm
Prepreg layers per sheet	0.13 mm × 8
Glue	EA 9361 (Hysol)

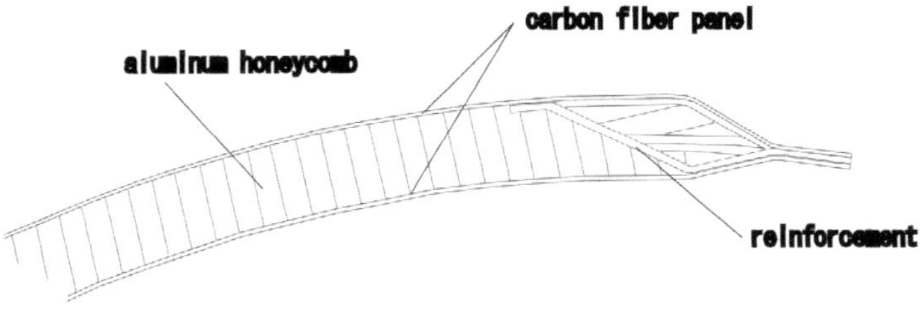

Figure 7. Cross section of transition area of carbon-fiber honeycomb panel.

A pulse-tube refrigerator is mounted on the cryostat to enable the recondensation of xenon. The refrigerator has been developed by our group [1] and is optimized for cooling liquid xenon to around 165 K. It has a cooling power of 200 W at 165 K without any electrical noise nor mechanical vibration thanks to the cooling principle of the pulse-tube refrigerator [12]. The refrigerator itself does not require maintenance, enabling us to preserve the purity of liquid xenon once it is purified. Liquid nitrogen cooling pipes are used to enable the refrigerator to achieve efficient cooling. One spiral pipe is installed around the refrigerator mounted on a central chimney, another pipe is welded on to the outer surface of the wall of the inner vessel, and one pipe each is welded on to the outer surfaces of the upstream and downstream covers of the inner vessel. The pipes on the outer surfaces of the wall and covers (Figure 8) are used particularly for pre-cooling the vessel before it is filled with liquid xenon.

[1] Currently, the refrigerator of this type is produced by Iwatani Co. Ltd.

Figure 8. Cooling pipes welded on the outer surfaces of the wall (left) and covers (right).

Impurity removal from liquid xenon is an important issue that should be addressed to achieve the best performance of the detector [13]. Scintillation light is known to be emitted in liquid xenon according to the following mechanism [18]:

$$\mathrm{Xe}^* + \mathrm{Xe} \to \mathrm{Xe}_2^* \to 2\mathrm{Xe} + h\nu$$

or

$$\begin{aligned} \mathrm{Xe}^+ + \mathrm{Xe} &\to \mathrm{Xe}_2^+ \\ \mathrm{Xe}_2^+ + \mathrm{e} &\to \mathrm{Xe} + \mathrm{Xe}^{**} \\ \mathrm{Xe}^{**} &\to \mathrm{Xe}^* + \text{heat} \\ \mathrm{Xe}^* + \mathrm{Xe} &\to \mathrm{Xe}_2^* \to 2\mathrm{Xe} + h\nu \end{aligned}$$

Because a bound Xe_2 ground state does not exist, xenon molecules themselves do not absorb light. However, electronegative impurities such as oxygen and nitrogen, if present in the liquid, can affect light emission by disturbing the recombination process. Oxygen contamination can also affect the transmission of the scintillation light because the light absorption cross section of oxygen overlaps the spectrum of the scintillation light originating from xenon. Water contamination affects scintillation light transmission to a larger extent than oxygen because of the significant overlap of the absorption cross section of water with the spectrum [13]. In order to remove these impurity molecules, three types of vacuum pumps and gas and liquid purifiers are included in the experimental system.

Evacuation of the cryostat is carried out before pre-cooling, first by using a turbo-molecular pump with a pumping speed of 400 L/h down to a presssure of 10^{-2} Pa, and then by operating a cryopump with a pumping speed of 2500 L/s for water removal along with the turbo-molecular pump operation. Because the cryostat contains photomultipliers that cannot be heated above 50 °C, use of the cryopump is very beneficial for removing water molecules from the cryostat. After the vacuum level reaches 10^{-3} Pa, a getter pump starts operation to achieve vacuum levels below 10^{-4} Pa.

Liquid xenon is filled in the inner vessel after evacuation and pre-cooling. Although the vessel is evacuated sufficiently good from the view point of remaining gas ratio to the amount of used xenon, impurity molecules that have been captured on the surface of material in the cryostat dissolve into the liquid after liquid filling. These impurities are removed by using the xenon purification system. The purification system has two purifiers, as mentioned previously. One is a metal-heated getter purifier that can remove almost all types of gas molecules except rare gas molecules. Xenon extracted from the cryostat is evaporated, transferred to the purifier, and then returned to the cryostat to be liquefied again. A diaphragm pump is used to circulate xenon without introducing any additional impurities. In this purification method, the speed of gas circulation is limited by the cooling power, resulting in a purification speed of 0.06 L/h, which is not sufficient to purify 900 L of liquid xenon. The other purifier is a copper-based purifier that is used to remove impurities in the liquid phase. A faster purification speed can be achieved by using this purifiers because of the lower amount of heat inflow during xenon circulation. Liquid xenon is transferred to a purifier cryostat located beside the detector through a drain at the bottom of the detector; the purified xenon is returned to the detector cryostat through two purifier cartridges. One of the cartridges is the main purifier filled with pure copper that removes impurities by oxidization. The other cartridge is filled with molecular sieves with a pore size of 13A to enable the removal of water molecules. A centrifugal pump that is designed for use with liquid xenon is employed to circulate the liquid; this enables us to achieve a purification speed of 70 L/h.

The components required for detector operation have already been described above, but there is another component that is required when the experiment is being performed. This is the xenon storage system that is used to store xenon during detector maintenance [14]. Xenon is stable at 5.8 MPa with a density of 1.1 g/cm^3 at 17 °C (critical point). In the MEG experiment, eight high-pressure tanks are used to store xenon, each of which can contain up to 360 kg of xenon corresponding to 120 L of liquid xenon. Recovery of xenon is carried out in the tanks by cooling the tanks with liquid nitrogen, which freezes xenon in the tanks. The vapor pressure of solid xenon is almost zero; thus, we can recover xenon from the detector with minimum loss. This high-pressure tank storage system is simple and safe, but it takes more than one week to transfer xenon from the detector to the tanks. Such a long transfer period is not acceptable in case of an emergency such as the breakdown of the insulation vacuum during operation. To solve this problem, a cryogenic tank with a volume of 1100 L is connected to the system so that xenon can be transferred from the detector using the centrifugal pump of the purification system. The cyogenic tank is equipped with a pulse-tube refrigerator as well as a liquid nitrogen cooling pipe for pressure control. The incoming heat while maintaining xenon at 165 K is estimated to be less than 20 W, which is sufficiently small to be cooled by using only the refrigerator. In addition, the cryogenic tank is designed such that it can tolerate pressure of up to 0.6 MPa; thus, xenon can be stored safely without the supply of cooling power for 100 h. Figure 9 shows a schematic view of the purification and storage system. In this system, xenon can be recovered to the cryogenic tank within two days. Xenon is transferred from the cryogenic tank to the detector by maintaining a pressure difference between them. The xenon pressure in the cryogenic tank can be increased using a heater equipped at the bottom of the tank.

3.3. Detector operation

Before starting detector operation, the detector cryostat is evacuated sufficiently, as previously described. Simultaneously, xenon is liquefied in the cryogenic tank using the pulse-tube refrigerator and by liquid nitrogen cooling. Liquid xenon is subsequently transferred from the cryogenic tank to the detector, and the operation is started. After the detector cryostat is filled with liquid xenon, the gas pressure is controlled using a heater and by liquid nitrogen cooling with continuous operation of the pulse-tube refrigerator. The heater reduces the refrigerator cooling power whereas liquid nitrogen cooling is used to increase the cooling power. The detector status is continuously monitored and controlled by a system developed for the MEG experiment. This control system uses one microcontroller in a single node that can handle multichannel inputs

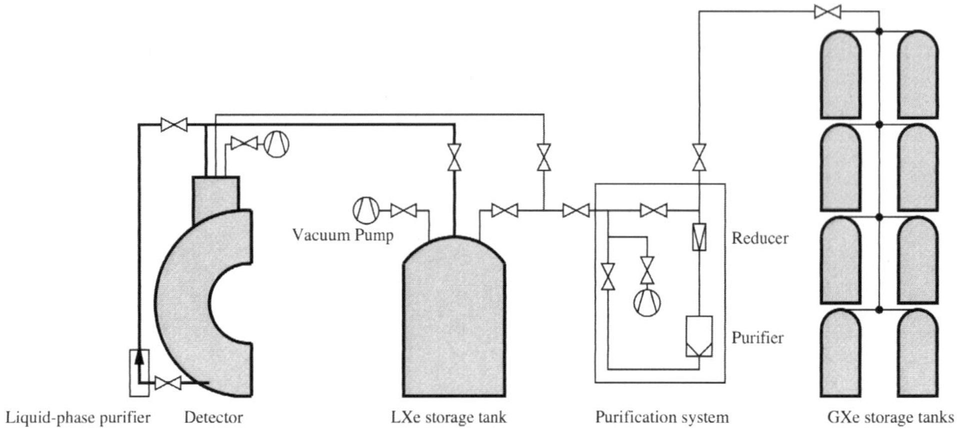

Figure 9. Xenon purification and storage system.

and outputs. All nodes are connected to the Ethernet and are integrated to the data acquisition system. The system periodically records the detector status on data files so that the online detector status can be inspected easily while performing data analysis.

The calibration of the detector is carried out in several steps. Among these steps, photomultiplier calibration is the most basic and important step. The behavior of the photomultipliers has to be carefully measured and calibrated under realistic operating conditions. The gain of the photomultipliers is determined by flashing blue LEDs in liquid xenon. The quantum efficiencies of the photomultipliers are determined using scintillation light induced by alpha sources installed in the active volume [15]. The measurement result of the alpha source signal is compared with a simulation result that incorporates information regarding energy loss, scintillation light emission, and scintillation light transmission in the liquid [15]. The determination of the quantum efficiencies is carried out using an iteration process because there are at least two undefined parameters in the system, i.e., the absorption length of the scintillation light in the liquid and the quantum efficiencies. These parametes are determined step by step and more precise results are obtained as the iteration process continues. In this process, photomultiplier calibration data in cold gas xenon is used to determine the initial parameter of quantum efficiencies because scintillation light transmission in gas is not affected by the Rayleigh scattering which in liquid has to be carefully treated. Furthermore, an Am-Be radioactive source has recently been used for more precise determination of the quantum efficiencies of the photomultipliers.

The detector response to higher energy gamma rays is investigated using gamma rays of 17.6 MeV obtained from a reaction of p(Li, Be)γ, which is resonant at E_p = 440 keV with a width of approximately 15 keV, and also from π^0 decays produced via a charge exchange process $\pi^- + p \rightarrow \pi^0 + n$. For using the gamma rays obtained from the p(Li, Be)γ reaction, a Cockcroft-Walton (CW) proton accelerator is installed at the rear end of the experimental area. The accelerator can provide 1–100-μA 300–900-keV protons to a LiF target that can be positioned at the center of the detector by a linear motion system. The specifications of the CW accelerator are summarized in Table 2. Fig. 10 shows the detector response to 17.6-MeV gamma rays. This calibration is repeated three times a week to monitor the change of the detector status and offline corrections. The detector performance to the signal gamma ray (52.8 MeV) is evaluated by using 54.9-MeV gamma rays originating from the π^0 decays and by carrying out an extrapolation based on the Monte Carlo simulation from 54.9 to 52.8 MeV.

Table 2. Specifications of MEG CW accelerator.

	Nominal	Measured at PSI
Terminal energy range	300–900 keV	200–1100 keV
Energy ripple	<500 V_{rms}	<50 V_{rms}
Angular divergence	<(5×5) $mrad^2$	∼(4×4) $mrad^2$
Spot size at 3 m	<(3×3) cm^2	<(1×1) cm^2
Energy setting reproducibility	0.1%	OK
Energy stability FWHM	0.1%	OK
Range of current	1–100 μA	0.1–135 μA
Current stability	3%	OK
Current reproducibility	10%	OK
Start-up time	<20 m	<15 m

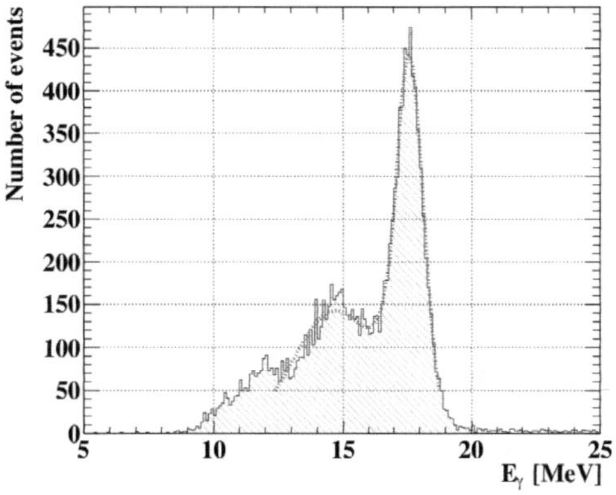

Figure 10. Detector response to 17.6-MeV gamma rays obtained from p(Li, Be)γ

The production of π^0 is realized through a charge exchange (CEX) process $\pi^- + p \rightarrow \pi^0 + n$. For executing this calibration, the beam line setting is modified to provide a π^- beam. A liquid hydrogen target with a pressure control using liquid helium replaces the muon stopping target. For identifying the 54.9-MeV gamma rays originating from π^0 decays, an NaI counter composed of 9 crystals, each of which has a dimension of $62.5 \times 62.5 \times 305$ mm^3, and equipped with an avalanche photodiode readout is installed on the opposite side of the magnet to the liquid xenon detector. The counter is mounted on a three-dimensionally movable stage to detect $\pi^0 \rightarrow \gamma\gamma$ decays with the opening angle between two gamma rays close to 180° over the acceptance of the liquid xenon detector. Two plastic scintillation counters with lead converters are placed in front of the NaI counter for carrying out time calibration. The setup of the CEX calibration is schematically shown in Figure 11.

Figure 12 shows a distribution of π^0 data measured using the liquid xenon detector and the NaI counter. The opening angle between two gamma rays originating from π^0 decays is selected to be approximately 180°, thus providing gamma ray energy close to both edges of energy

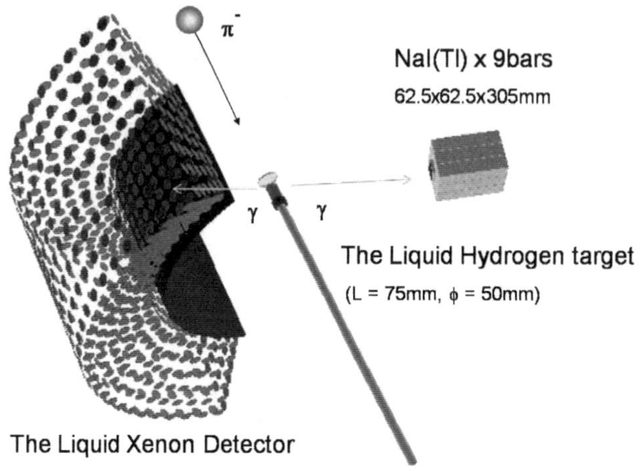

Figure 11. Experimental setup for CEX calibration.

distribution ranging from 54.9 to 82.9 MeV. By requiring 82.9 MeV gamma-ray detection on the NaI counter, semi-monochromatic gamma-rays of 54.9 MeV can be prepared on the liquid xenon detector side.

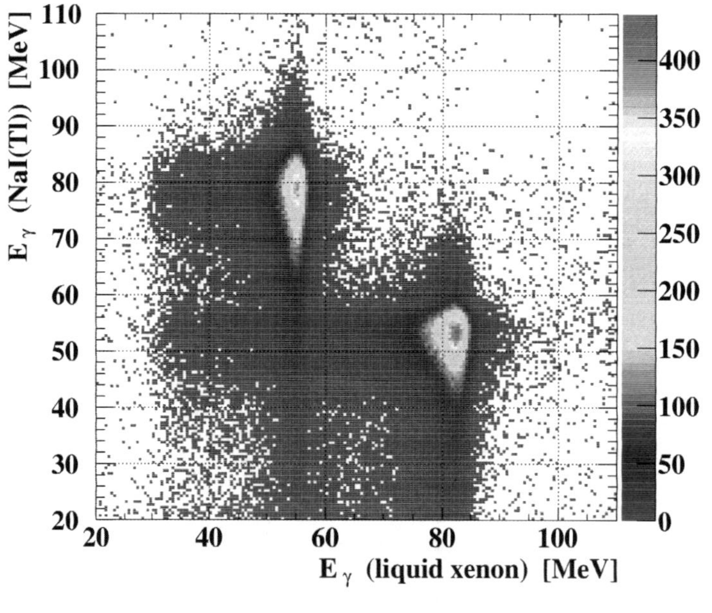

Figure 12. Gamma-ray energy distributions from CEX π^0 decays observed in the liquid xenon photon detector and NaI counter.

MEG data is taken with a trigger requirement both on the liquid xenon detector and positron detector. It is also required that the gamma ray and the positron be emitted back to back in time coincidence. The energy deposition in the liquid xenon detector, measured as the total sum of the observed charge in all photomultipliers, should exceed a certain threshold. The trigger electronics comprises FADCs and FPGAs to realize this. Photomultiplier signals are digitized by the FADCs, and then, arithmetic calculations are performed in the FPGAs to decide whether to record or discard an event. In order to maintain the trigger quality during a long-term data acquisition, the results of detector calibration are reflected to the trigger system as needed.

Energy scale calibration is preformed primarily using CEX data. However, the CEX calibration is time consuming and the setup needs to be modified, which would cause loss of physics data acquisition time. Therefore, another method has been developed to monitor the energy scale around the signal region while simultaneously acquiring physics data. The data acquisition trigger system can generate 32 different types of trigger signals with proper pre-scaling. This enables us to mix the calibration data in the physics data to monitor the detector status at any given time. Among these types, there exists a trigger logic that is generated only using the deposited energy in the liquid xenon detector with a threshold of around 40 MeV. The events recorded using this trigger logic are dominated by muon radiative decays and positron annihilations in flight with a contamination of comic-ray events, as shown in Figure 13. The green and blue lines indicate contributions from radiative decay events and positron annihilations in flight, and cosmic-ray background events, respectively. The effect of detector resolution is taken into account in this description. It can be seen that the event distribution is well reproduced by the combined function shown in red. The end point of the green line can be precisely determined along with the acquisition of physics data, and this can be used to monitor the stability of the energy scale of the liquid xenon detector.

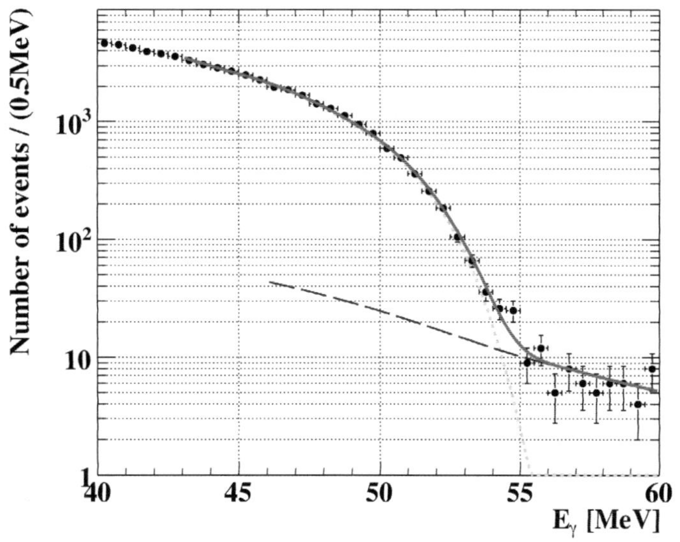

Figure 13. Energy distribution of observed photons around the end point of the radiative muon decay. The green and blue lines indicate contributions from radiative decay events and positron annihilations in flight, and cosmic-ray background events, respectively. The red line indicates the combined function of these two distributions.

Journal of Physics: Conference Series **308** (2011) 012009 doi:10.1088/1742-6596/308/1/012009

3.4. Detector performance

Here, we summarize the best performance of the detector achieved so far. It should be noted that more precise calibration is in progress along with a better description of the detector by a simulation, which provides us with opportunities to improve the detector performance in the future.

3.4.1. Energy measurement The gamma-ray energy is reconstructed by summing up the number of photoelectrons observed by all photomultipliers. A small but significant correction is applied to achieve better resolution by taking the effect of the solid angle into account. The effect of non-uniform coverage of the photocathode is also corrected. Figure 14 shows a distribution of reconstructed gamma-ray energy in a CEX run for an energy deposition of 82.9 MeV on the NaI counter with an opening angle requirement between two photons larger than 175°. It is also required that the reconstructed position in the liquid xenon detector be located deeper than 2 cm from the inner face, so that events with energy leakage are removed from the active volume. The distribution is fitted with a combined function of a gaussian and an exponential

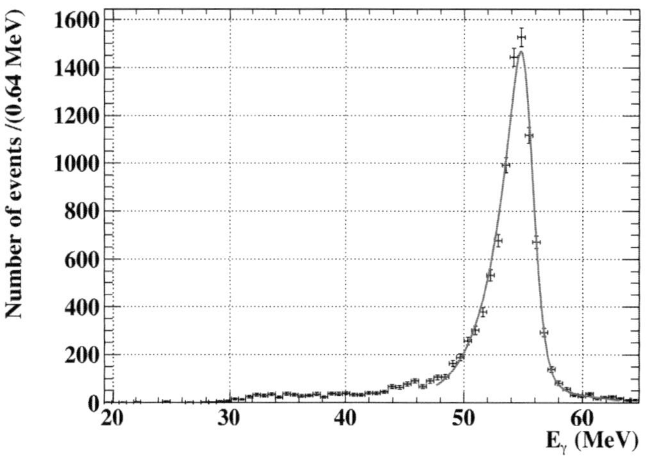

Figure 14. Measured energy spectrum of 54.9-MeV gamma rays.

function taking into account the pedestal distribution measured in the CEX run. An average energy resolution of $(5.8 \pm 0.35)\%$ in FWHM with a right tail of $\sigma_R = (2.0 \pm 0.15)\%$ is obtained.

3.4.2. Position reconstruction The gamma-ray interaction position in the active volume is reconstructed by comparing the light distribution observed by photomultipliers and solid angles covered by them viewed from the reconstructed position. The resolution of the position reconstruction is evaluated using lead collimators having thickness of 1.8 cm and 1-cm slits located in front of the liquid xenon detector. Figure 15 shows a distribution of the reconstructed position along the vertical direction with collimators in the CEX run. Three clear peaks corresponding to the shadow of the three slits are observed. A fitting for the distribution is overlaid in the figure. From these results, the resolution of the position reconstruction is estimated to be 5 mm by taking into account the effects of the slit width and the beam spread at the target.

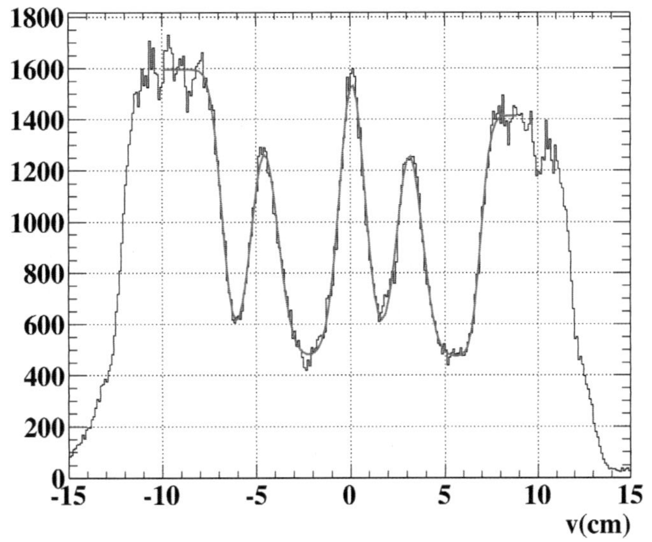

Figure 15. Distribution of reconstructed positions along vertical direction with collimators located in front of photon detector.

3.4.3. Time measurement The gamma-ray arrival time is calculated in the following manner. First, the scintillation photon arrival time at each photomultiplier is calculated by subtracting the propagation time of the scintillation photon in the liquid. Time offsets in signal transfer cables and electronics are subtracted during this calculation. Then, the gamma-ray interaction time is calculated by averaging the measured time at each photomultiplier with a weight of the number of photoelectrons. Figure 16 shows a distribution of the measured time difference between the liquid xenon detector and the plastic scintillation counters located in front of the NaI counter in the CEX run. A gamma-ray energy deposition of around 54.9 MeV is required at the xenon detector during the event selection. The distribution is fitted with a gaussian function, showing a spread of 135 ps in σ. The time resolution of the plastic scintillation counters is estimated to be 93 ps by carrying out an independent analysis. The beam spread of the π^- beam on the hydrogen target is estimated to be 58 ps equivalent in time. By subtracting these two contributions, we obtain a time measurement resolution of 78 ps in σ.

3.4.4. Detection efficiency The detection efficiency of the signal gamma ray is estimated to be 67% for photons with energies larger than 46 MeV by performing the Monte Carlo simulation, in which all the realistic information of materials used in the detector setup is incorporated. The efficiency is also calculated using the CEX run data for the 54.9-MeV gamma rays by requesting the 82.9-MeV gamma-ray observation in the NaI counter; an efficiency of 63% is obtained after subtracting the effect caused by a neutron background event from a radiative capture process ($\pi^- + \mathrm{p} \rightarrow \gamma + \mathrm{n}$) . The discrepancy of 4% between these two estimation is considered to be as an uncertainty. Thus, we conclude that the photon detection efficiency is $(63 \pm 4)\%$ in the range of $E_\gamma > 46$ MeV.

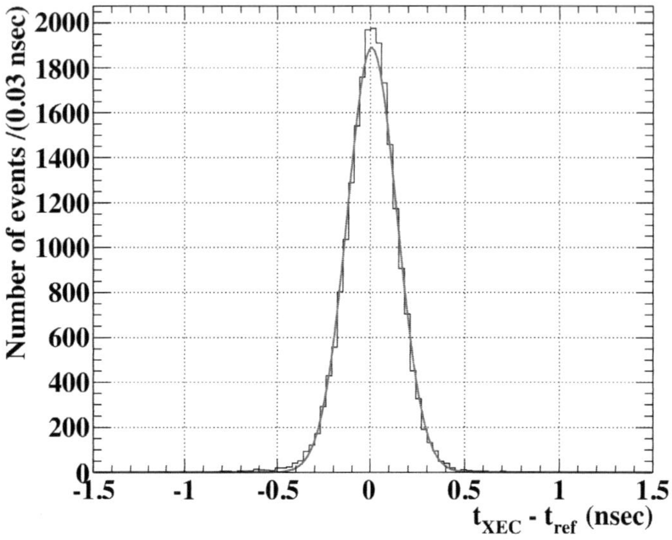

Figure 16. Measured time difference between photon detector and plastic scintillation counters located in front of NaI counter.

4. Summary

The MEG liquid xenon detector has been successfully constructed after extensive R&D activities for more than 10 years and is being operated without any major problems. This detector is the first instance of the application of a large-sized liquid xenon detector to a physics experiment. As expected, the detector shows outstanding features. We intend to further improve the detector performance by carrying out more precise calibration and gaining a better understanding of the detector based on simulation results.

References

[1] Aprile E *et al.* 2006 *Phys. Rev. Lett.* **97** 081302
[2] Yamashita M *et al.* Astro. Phys. **20** 79
[3] MEGA Collaboration Ahmed M *et al.* 2002 *Phys. Rev.* D **65** 112002
[4] Mori T *et al.* 1999 *Research Proposal to PSI* R99-05
[5] Nishiguchi H 2007 *Nucl. Phys.* B **172** 266
[6] Cattaneo P W 2010 *Nucl. Instru. and Meth.* A in press
[7] Ootani W *et al.* 2005 *IEEE Trans. Applied Superconductivity* **14** 568
[8] MEG Collaboration Adam J *et al.* 2010 *Nucl. Phys.* B **834** 1
[9] Ritt S *et al.* 2002 *Nucl. Instr. and Meth.* A **494** 520
[10] Mihara S *et al.* 2002 *IEEE Trans. Nucl. Sci.* **49** 588
[11] Mihara S *et al.* 2004 *Nucl. Instr. and Meth.* A **518** 45
[12] Haruyama T *et al.* 2004 *AIP Conf. Proc.* **710** 1459
[13] Baldini A *et al.* 2005 *Nucl. Instr. and Meth.* A **545** 753
[14] Iwamoto T *et al.* 2008 *Cryogenics* **49** 254
[15] Baldini A *et al.* 2006 *Nucl. Instr. and Meth.* A **565** 589
[16] Mihara S *et al.* 2004 *Cryogenics* **44** 223
[17] Mihara S *et al.* 2006 *Cryogenics* **46** 688
[18] Doke T 1981 *Portugal Phys.* **12** 9

The XENON Dark Matter Search

Elena Aprile

Columbia University, NY, USA, on behalf of the XENON collaboration

E-mail: age@astro.columbia.edu

Abstract. The XENON100 experiment aims at detecting cold dark matter particles via their collisions with xenon nuclei in a two-phase time projection chamber filled with a total of 165 kg of ultra pure liquid xenon. The detector sensitive target mass is about 65 kg, surrounded by about 100 kg of active veto. The detector has been installed underground at the Gran Sasso National Laboratory (LNGS) since 2008 and after a successful callibration, dark matter data taking has started. The current status of the XENON100 as well as future plans for the upgrade are presented.

1. Introduction

There is an increasing number of astrophysical and astronomical observations pointing to the existence of a non-luminous, non-baryonic and cold (i.e. non-relativistic) matter component in the universe, called Cold Dark Matter (CDM) [1, 2, 3]. The most appealing candidates for CDM are Weakly Interactive Massive Particles (WIMPs) predicted by supersymmetric theories (SUSY), models with extra dimensions and little Higgs models [4, 5, 6, 7]. Among the different WIMP detection approaches that are used in order to determine the characteristics of such particles are experiments at LHC looking for their appearance in collisions, high energy astroparticle physics experiments that are looking for WIMP annihilation signatures from the center of the Sun or from the center of the galaxy, and dark matter direct detection experiments that are looking for the scattering of WIMPs off target nuclei.

The XENON project is currently one the most promising experiments for the direct detection of dark matter. After the successful results of the first 10 kg scale detector XENON10 [8, 9], the collaboration has designed and built a second-generation experiment exploiting the two-phase time projection chamber (TPC) technique based on liquid xenon (LXe). The XENON100 detector features an increase in mass by a factor of 10 and a reduction of the radioactive background level by a factor of 100, as shown by the first measured scientific data. XENON100 sensitivity reach is a factor of 50 better than that of XENON10.

2. Detector Concept

Liquid xenon is appealing as the target material for dark matter direct detection. It is a heavy (A = 131) and dense ($\rho \sim 3$ g/cm^3) medium and an efficient scintillator (80% of NaI). The high mass number provides excellent detection capabilities for the spin-independent WIMP-nucleon scattering ($\sigma \propto A^2$). The abundance of odd isotopes (about 50%) allows the detection of spin-dependent interactions ($\sigma \propto$ J(J+1)). The high density and high Z provide self shielding against external gamma radiation. Xenon has no long lived isotopes and purification of krypton in xenon has been shown down to the level of a few ppt, which is an important advantage in the search for rare dark matter induced events.

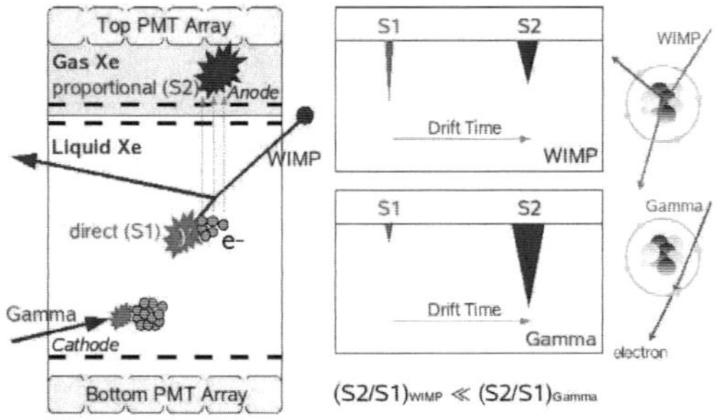

Figure 1. Principle of a two-phase liquid xenon TPC. A particle generates primary scintillation light (S1) and ionization electrons. These are drifted upwards with the field and detected via secondary scintillation light in the gas phase (S2). The S2 hit pattern (xy) and the drift time (z) give complete information for the position of events. Additionally, the ratio S2/S1 allows event discrimination between nuclear recoils (WIMPs, neutrons) and electron recoils (γ, β).

The XENON100 detector is a two-phase (liquid-gas) time projection chamber (TPC). A particle interacting with the target generates electron ion pairs and excited xenon atoms, which produce scintillation light and free electrons in the medium (Figure 1). The primary light (S1) is detected immediately by the two photomultiplier (PMT) arrays above and below the target. An electric field ($\sim$ 0.53 kV/cm) across the TPC drifts the free ionization electrons upwards, where they are extracted into the gas phase by an even stronger extraction field. In the gas phase, the electrons generate very localized proportional scintillation light (S2). The PMT hit pattern of the S2 signal can be used to determine the xy-position of the interaction point and since the z-position is known from the drift time, the event positions can be reconstructed in three dimensions. This allows to select an inner fiducial volume in our target which together with the self shielding capability of liquid xenon drastically reduces the radioactive background from external sources.

The high ionization density of nuclear recoils in liquid xenon leads to larger S1 and smaller S2 signals compared to electron recoils. The simultaneous measurement of charge and light provides a powerful discrimination between signal (nuclear recoils) and background events (electron recoils) via the ratio S2/S1. A discrimination of 99.5-99.9% has been achieved in both XENON10 and XENON100.

3. Experimental Setup

The XENON100 detector consists of 165 kg of liquid xenon divided in two concentric cylinders. The inner sensitive volume contains 65 kg of Xenon and is separated from the outer volume by a PTFE cage on the sides, a diving bell on the top and a PMT array in the bottom. Two electric field regions are created in this volume with one mesh in the bottom and three in the top, near the liquid gas interface. These electric fields allow electrons produced in interactions to drift in the main volume and be extracted to the gas where they produce proportional scintillation light. The PTFE cage acts as a reflector for the UV light from the liquid xenon and also accomodates a set of 40 field shaping rings and their resistive divider chain to improve the homogeneity of the field. The outer volume acts as an active veto reducing the amount of interactions in the inner volume and allowing to identify multiple scatter events. This in fact reduces the DAQ rate and the amount of data collected and improves our ability to reject gamma events as WIMP candidates.

The light readout is based on 1" $\times$ 1" Hamamatsu R8520-06-Al low-radioactive PMTs with quantum efficiencies up to $\sim$ 35%. 98 PMTs in the top array above the anode mesh are arranged in a circular pattern to improve position reconstruction, while 80 PMTs on the bottom are arranged in a compact grid to optimize the light collection. 64 PMTs in the active veto allow to detect energy depositions as low as 100 keVee reducing the overall background by a factor 4.

The detector is surrounded by a passive shield consisting, from the inside to the outside of 5 cm copper, 20 cm polyethylene and 20 cm lead and 20 cm of water. All the systems associated with the cooling and the purification of the xenon are placed outside this shield in order to minimize the background level.

The cryostat and detector production was completed in early 2008 and the detector was installed underground at LNGS. An extensive calibration of the detector systems has been performed in 2009 and blind dark matter data taking started in January 2010. A picture and the schematic of the inner TPC are shown in Figure 2.

Figure 2. The XENON100 TPC. Left: Photo showing the PTFE panels as well as the veto PMTs on the top and bottom. Right: Drawing of the structure to show the location of the inner top and bottom PMT arrays.

3.1. XENON Recirculation and Purification System

An Iwatani PC150 Pulse Tube Refrigerator (PTR) is used to liquefy xenon (170 W cooling power) and keep the detector at cryogenic temperature of -100°C. This provides excellent stability of operation, with temperature variations below 0.1^{o}C and pressure changes of less than 1%.

Successful operation of the xenon TPC requires long VUV light attenuation and effective charge drift (long electron lifetime in liquid xenon). We are using a purification system based on continuous xenon gas circulation with purification through a high temperature metal getter (SAES) to reduce the impurities in commercially available xenon (water, oxygen etc.) to a level below 1 part per billion (ppb) O_2 equivalent.

While xenon has no long lived radio isotope, krypton is present at the ppm to ppb level in commercially available xenon. The isotope ^{85}Kr (E_{max} = 687 keV, $T_{1/2}$ = 10.6 yr) is present in krypton with a ratio ^{85}Kr/Kr=10^{-11} and it represents a significant source of background for any xenon based dark matter search experiment. In order to reduce the ^{85}Kr level to ~100 parts per trillion (ppt) as required by the XENON100 sensitivity goal, we have installed and used underground a cryogenic distillation column made by Taiyo-Nippon Sanso. The xenon has been purified with this system and we have measured with the method of delayed coincidences a ^{85}Kr concentration of ~150 ppt.

3.2. Readout Electronics and Data Acquisition

The XENON100 data acquisition system (DAQ) is used to generate a trigger signal for the TPC, digitize the waveforms of the 242 PMTs, and store this data. The PMT signals are first amplified by a factor of 10 (Phillips 776 amplifiers) and then digitized by CAEN V1724 Flash ADCs with a sampling rate of 100 MHz. A typical XENON100 waveform from a ^{137}Cs calibration is shown in figure 3. The DAQ system is fully operational and used continuously to characterize the detector. Detector calibration is performed with several gamma sources, and the PMTs single photoelectron response is measured regularly with blue LEDs.

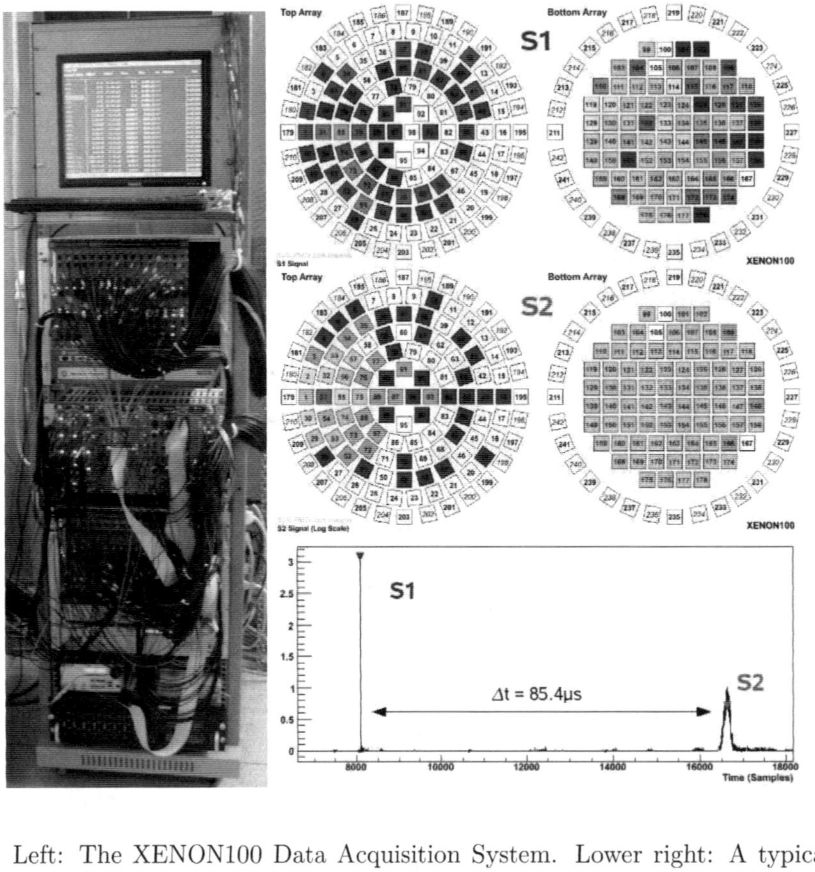

Figure 3. Left: The XENON100 Data Acquisition System. Lower right: A typical electron recoil event in the XENON100 TPC. The S1 and S2 signals are separated by 85μs, yielding the Z position of the interaction point. Upper right: The X and Y positions of the interaction point can be reconstructed using the hit pattern of the S2 signals on the top PMT array.

4. Detector calibration

The XENON100 detector has been installed underground in LNGS since the middle of 2008, and has tested in a series of succesful calibrations during this period. Monitoring of the detector performance has been done on a regular basis using ^{137}Cs and blue LEDs to study the PMT gain, the light collection efficiency and the electron absorption. For dark matter detection one of the most important features of a two-phase liquid xenon TPC is the ability to distinguish between electronic and nuclear recoils. In order to characterize the electron recoil regions several calibrations with a ^{60}Co source have been performed. A high energy veto has been used to acquire only the lower part of the spectrum $\leq$150 keV.

To study the response to nuclear recoils, during December 2009 and for three days a calibration was performed with an AmBe neutron source. This allowed us to collect a large sample of elastic nuclear recoils, but also a rather homogeneous sample of 40 keV and 80 keV gamma-rays from inelastic scatterings which have been used to measure the energy resolution of the detector down to this energy. Figure 4 shows the measured electron and nuclear recoil bands. A discrimination between nuclear and electronic recoils better than 99% is found at all energies for a nuclear recoil acceptance of 50%.

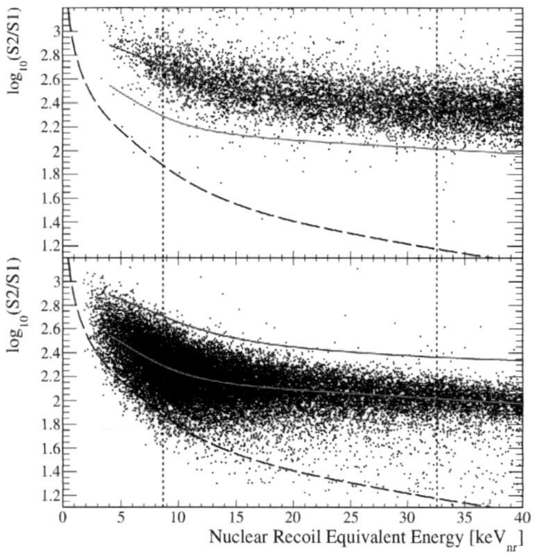

Figure 4. Discrimination between nuclear and electronic recoils using the ratio of charge and light. The red (blue) curve is the median of the nuclear (electronic) recoil band. Top: Electronic recoil band measured with a ^{60}Co gamma source. Bottom: Nuclear recoil band measured with an AmBe neutron source.

5. XENON100 First Results

During the detector callibration period in 2009 some data were taken when no source was present and in ideal detector conditions. While this data were not originally blinded, we decided to perform a blind analysis on them by defining selection cuts only on the callibration data [11]. 11.2 days of data were analyzed with a fiducial volume of 40 kg. Only very basic cuts were defined, aiming to remove noisy events, events interacting in the gas or events with multiple interactions in the detector. After applying these cuts a total of 22 events were observed in the fiducial volume and the energy region preselected for the signal, corresponding to a background level of ~7 mdru, which is the lowest measured in a dark matter search up to date. None of the measured events lies in the nuclear recoil band, showing for the first time the succesful background free operation of a liquefied noble gas TPC. Figure 5 shows the limit for spin independent dark matter parameter space established from this result, which already at this early stage of operation is comparable to the best exclusion limit to date.

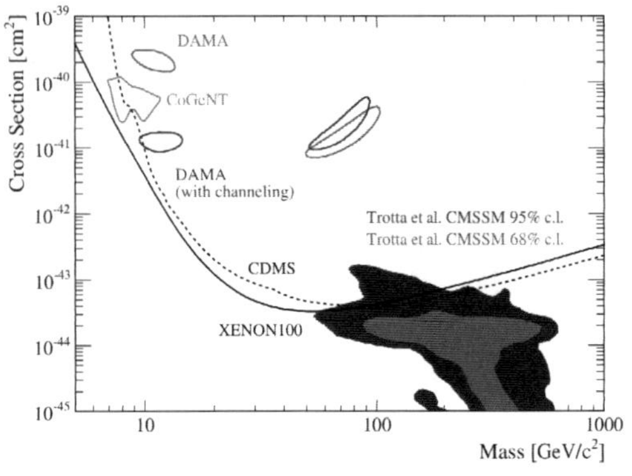

Figure 5. Exclusion plots for spin independent WIMP-nucleon interactions. The current limits are the curves around 10^{-43} cm^2, given by XENON100 (solid, [11]) and CDMS II (dashed, [10]). The projected sensitivity for XENON100 is one order of magnitude lower. The shaded regions are theoretical expectations of CMSSM.

Blind data acquisition started at the beginning of 2010 and to date more than 10 times the exposure used for this analysis has been collected.

6. XENON1T

The XENON100 gamma background is dominated by the PMTs and the PMT bases, followed by the polyethylene of the shield and the stainless steel of the cryostat. The next step in the XENON dark matter project will be XENON1t, with a fiducial mass of $\sim$1.1 ton. In order to achieve reduced background , PMT arrays will be replaced by QUPIDs [12], novel photosensors with an extremely low intrinsic radioactivity, developed by UCLA and Hamamatsu for this experiment. Additionally, a copper cryostat and 10 meter diameter water shield acting also as an active muon veto will be implemented. This detector will bring an improvement in the spin-independent WIMP-nucleon sensitivity of 2 orders of magnitude by 2015.

References

[1] W. Freeman and M. Turner, Rev. Mod. Phys.,**75**, 1433 (2003).
[2] D. Clowe et al., ApH, **648**, L109 (2006).
[3] M. J. Jee et al., http://xxx.lanl.gov/abs/0705.2171
[4] A. Bottino et al., *Phys. Rev.* **D 69**, 037302 (2004).
[5] J. Ellis et al., *Phys. Rev.* **D 71**, 095007 (2005).
[6] H. C. Cheng et al., *Phys. Rev. Lett.* **89** 211301 (2002).
[7] A.Birkedal-Hansen and J. G. Wacker, *Phys. Rev.* **G 69** 065022 (2004).
[8] J. Angle et al., (XENON10 collaboration), *Phys. Rev. Lett.*,**100**, 021303 (2008).
[9] J. Angle et al., (XENON10 collaboration), *Phys. Rev. Lett.*,**101**, 091301 (2008).
[10] Z. Ahmed et al. (CDMS II collaboration), *Science* **327**, 1619 (2010).
[11] E. Aprile et al. (XENON100 collaboration), *Phys. Rev. Lett.* **105**, 131302 (2010)
[12] K. Arisaka et al., arXiv:0808.3968 (2008).

Journal of Physics: Conference Series **308** (2011) 012011 doi:10.1088/1742-6596/308/1/012011

XMASS

Hiroyuki Sekiya, for the XMASS collaboration

Kamioka Observatory, ICRR, University of Tokyo, 456 Higashi-Mozumi, Kamioka, Hida , Gifu, 506-1205, Japan

E-mail: sekiya@icrr.u-tokyo.ac.jp

Abstract. The XMASS detector is a large single phase liquid Xenon scintillator. After its feasibility had been studied using a 100 kg size prototype detector, an 800 kg size detector is being built for dark matter search with the sensitivity of 10^{-45}cm^2 region in spin-independent cross section. The results of R&D study for 800 kg detector, especially ultra low background technologies, and the prospects of the experiment are described.

1. Introduction

Recently, many dark matter search experiments are underway using liquid Xe because of its special features [1]. As sharing many features of liquid Xe with lower cost, liquid Ar is also interesting for dark matter search [2], however Xe still has the advantage of density and mass number A.

Simultaneous detection of both scintillation and ionization in Xe provides the nuclear recoil discrimination power, consequently, most experiments adopted the two-phase time projection detectors. In contrast, XMASS chose a single-phase scintillator, because in principle simple detector setups should easily achieve the high radio-purity that is mandatory for dark matter search. Accordingly, the background suppression technologies should be more critical.

The key idea of the background reduction in XMASS is "self-shielding". The external gamma rays can be absorbed by the liquid xenon itself. Figure 1 shows a simulation of ^{238}U-chain gamma rays injected on the surface of a liquid xenon volume. The outer surface of the liquid xenon volume absorbs the external gamma rays within 10-20 cm of thickness, and a low background environment in the central volume is achieved, while dark matters interact throughout the detector. Therefore dark matters can be observed in a low background environment by extracting the events in the central fiducial volume.

2. The 800kg detector

The 800 kg detector, currently being built, is the first designed XMASS detector dedicated to search for dark matter. Actually, the detector consists of 857 kg liquid Xe in a OFHC vessel, newly developed 642 PMTs, and a 10 m diameter and 10 m height water tank. Figure 2 shows the schematic drawings of the detector and the pictures of the completed "PMT ball" for Xe scintillation light readouts. The PMTs are mounted on a 80 cm diameter pentakis-dodecahedron with 64.2% photocathode coverage of the inner surface. This configuration provide a light yield of 4.4 photoelectrons/keV and the energy threshold below 5 keV (electron equivalent).

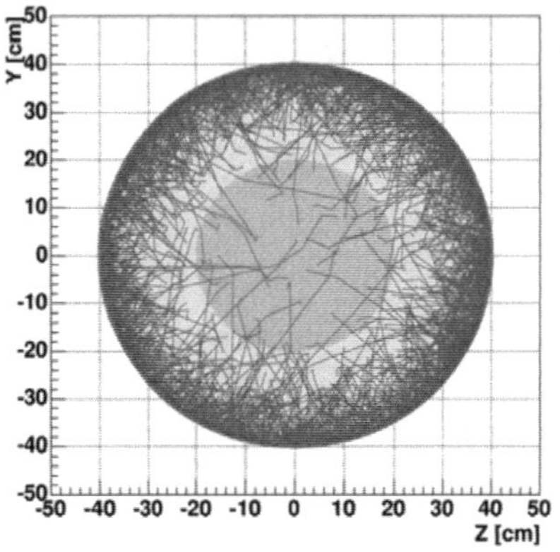

Figure 1. Simulation of ^{238}U-chain gamma rays injected on the surface of the liquid xenon volume. The tracks show the trajectory of gamma rays. The diameter of the liquid xenon volume is 80 cm.

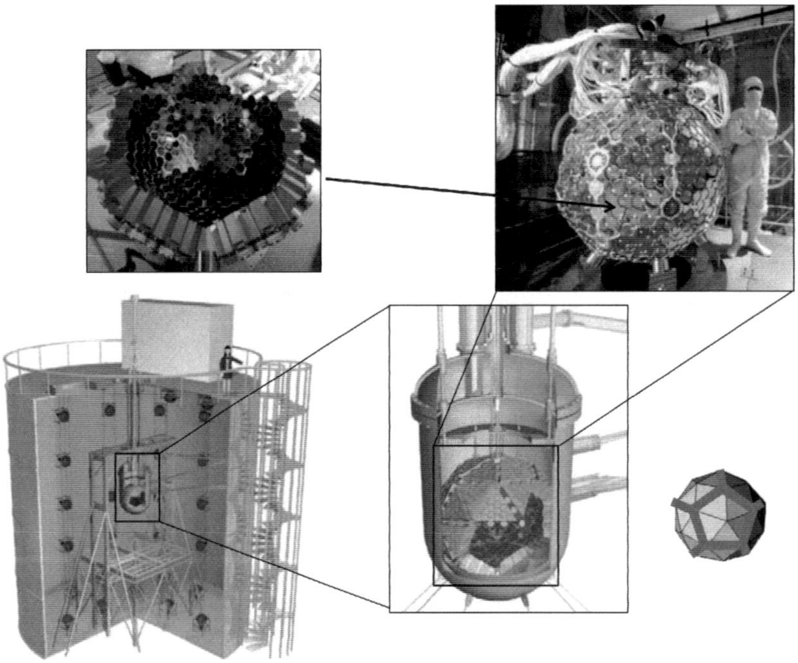

Figure 2. The 800 kg detector.

3. Ultra low background technologies

3.1. Water tank

In order to reduce the environmental radioactive background, the 800 kg detector is put in a water shield. Simulations of gamma rays and fast neutrons show that about a 200 cm thick water layer is enough to reduce the environmental backgrounds to a level below that of the PMT induced backgrounds, however a water tank 10 m diameter and 10 m height is prepared not only for this 800 kg detector but also for future projects. As shown in Figure 2, 72 PMTs (20 inch PMT) are attached to the wall of the tank to detect the Cherenkov light and thus the water tank can also work as the shield for muon induced events.

The water in the tank is circulated and purified with the flow rate of 5 ton/hour by a dedicated water purification system. The Rn concentration in the water is kept below 1 mBq/m^3.

3.2. Low background PMTs

The most serious external backgrounds come from the radioactive contamination in the PMTs. Since 2000, we have developed high radio-pure PMTs with with Hamamatsu Photonics K. K. All the components of PMTs were examined individually and selected by ourselves. In 2002, R8778, a 2 inch diameter PMT for the 100 kg prototype XMASS detector, was developed and it is also used in LUX [1] now. Then, in 2009, we have succeeded in reducing about a factor 10 of radioactivity and developed a hexagonal R10789. The PMTs have rather high quantum efficiency, up to 39% for 178 nm photons at 173 K.

Table 1. shows the contamination of the radioactive impurities of R8778 and R10789 measured by a high purity Ge (HPGe) detector in Kamioka underground laboratory.

Table 1. Radioactive impurities in the developed PMTs measured by HPGe detector in Kamioka. The uranium and thorium chains were assumed to be in radioactive equilibrium.

RI	R8778	R10789
^{238}U(mBq/PMT)	18±2	0.70±0.28
^{232}Th(mBq/PMT)	6.9±1.3	1.51±0.31
^{40}K(mBq/PMT)	140±20	<5.1
^{60}Co(mBq/PMT)	18±2	2.92±1.61

3.3. Materials screening

Not only the components of the PMT, but other components are also screened with HPGe measurement. Although our detector is quite simple, more than 250 materials are used around the liquid Xe region, such as OFHC PMT holder, screws, cables, PTFE gaskets, LEDs for calibration, and so on. Sum of radioactive impurities of the measured materials (upper limits) are listed in Table2.

Table 2. Sum of the measured radioactive impurities in the detector components compared with those in 642 PMTs

RI	compared with 642 PMTs
^{238}U(/PMT)	$< 35\%$
^{232}Th(/PMT)	$< 35\%$
^{40}K(/PMT)	$< 20\%$
^{60}Co(/PMT)	$< 20\%$

3.4. Kr removal

As for the internal background, ^{85}Kr in xenon is one potential source due to the β-decay of ^{85}Kr (τ =10.8y, Q_β = 687 keV). The natural abundance ^{85} Kr/Kr is 1.2×10^{-10} and commercial available "pure Xe" contains about 0.1 ppm level of Kr, as a consequence, to keep the background from ^{85}Kr below 10^{-5} counts/day/keV/kg, Kr concentration should be below 1ppt, i.e 5 order reduction is required. In order to achieve this, we have developed new distillation system based on our previous system [3]. The distillation technique efficiently removes Kr from Xe because the boiling temperature of Xe and Kr are quite different(178 K for Xe and around 145 K for Kr at the operation pressure of 0.2 atmosphere). The schematic diagram and the picture of the system is shown in Figure 3. The system has 99% efficiency for collecting xenon (i.e. only 1% of original xenon is rejected) with the process speed 4.7 kg/hour.

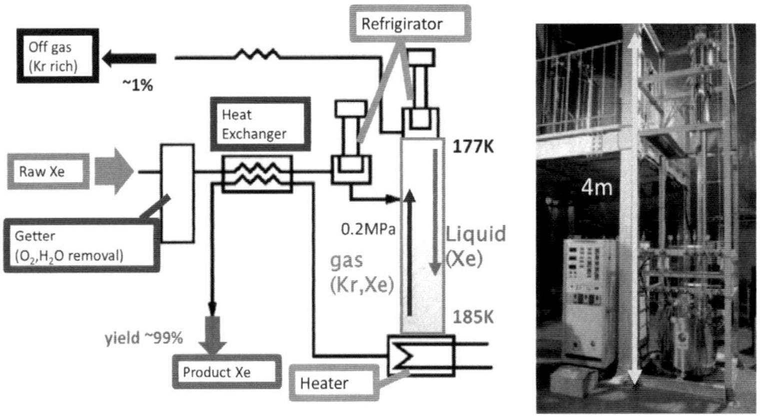

Figure 3. The distillation system

3.5. Rn removal

^{222}Rn is the most difficult background source because β-decay of ^{222}Rn daughters remain in Xe; moreover, ^{222}Rn is emanated from all the materials. During the materials screening, ^{222}Rn emanation rates from the components were also measured using high sensitivity Rn detectors developed for Super-Kamiokande [4]. It is found that up to 15 mBq sources exist in the detector, thus these ^{222}Rn should be continuously removed from the system. Figure 4 shows the purification system.

In order to suppress the background from ^{222}Rn as the level of those from the PMTs, Rn concentration should be kept below 1.2 mBq/ton in liquid phase or 0.8 μBq/m^3 in gas phase. We have succeeded removing ^{222}Rn from gas phase using cooled (−100°C) activated charcoals with more than 90% efficiency, however as the flow rate of gas circulation is limited by the cooling power of the condenser, Rn removal from liquid is preferable. We are testing some setups in liquid phase now.

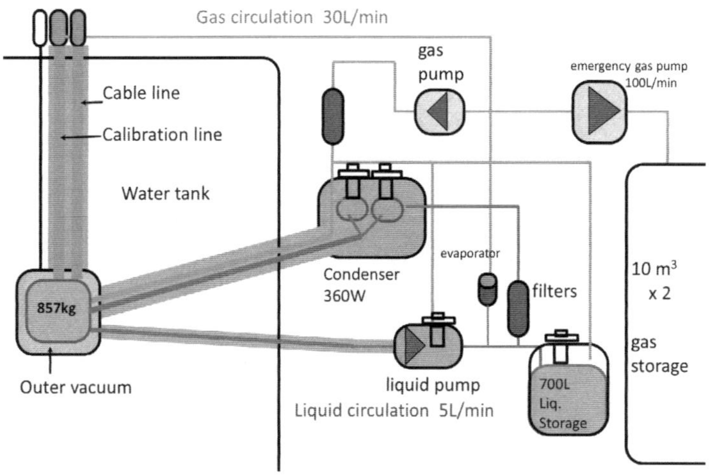

Figure 4. The Xe circulation and purification system

4. Expected sensitivity

By putting above listed background sources into the full detector MC simulation code, the total backgrounds in fiducial volume of 100 kg was evaluated and turned out to be below 10^4 counts/day/kg/keVee[1]. As an example and in order to demonstrate the "self-shielding" effect, the background contribution from ^{238}U-chain in PMTs is shown in Figure 5.

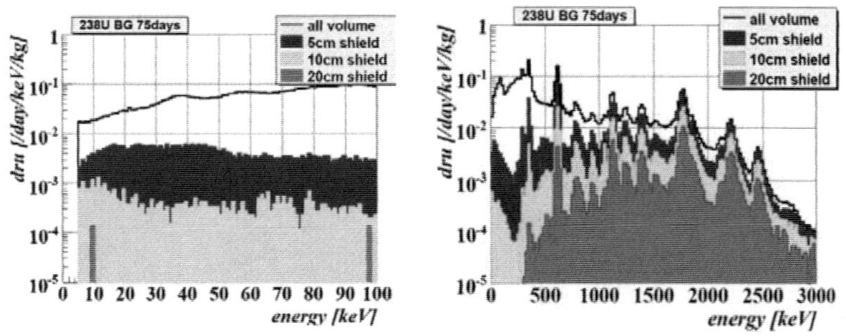

Figure 5. The evaluated background rate from ^{238}U in 642 PMTs. 20 cm of the self-shield layer corresponds to the fiducial volume of 100 kg.

With these background assumptions, the expected detection sensitivity is derived. Figure 6 shows the expected energy spectrum in case of M_χ=50GeV and $\sigma_{\chi-p} = 3 \times 10^{-44}$cm^{-2}, and 90% CL spin-independent sensitivity lines for 10 days operation and 1 year operation are shown in Figure 7. In this plot, we assume that the energy threshold of the detector is 5 keVee and the quenching factor of Xe around this energy region is 0.2.

[1] ee stands for electron equivalent.

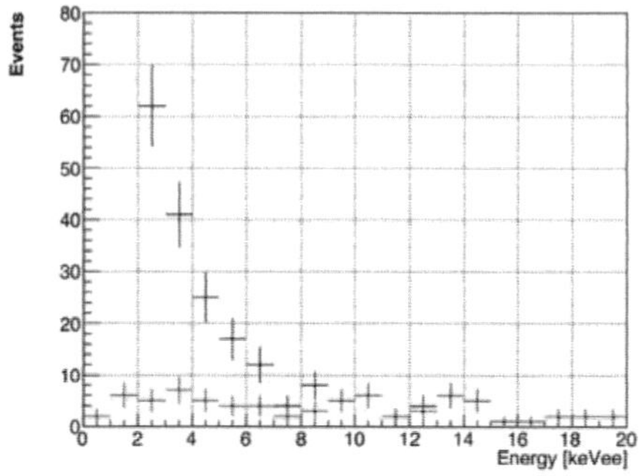

Figure 6. The expected energy spectrum in the 100 kg fiducial volume. Red points show the background events and black points show the background + the expected dark matter signal when $M_\chi = 50$ GeV and $\sigma_{\chi-p} = 3 \times 10^{-44}$ cm^{-2}.

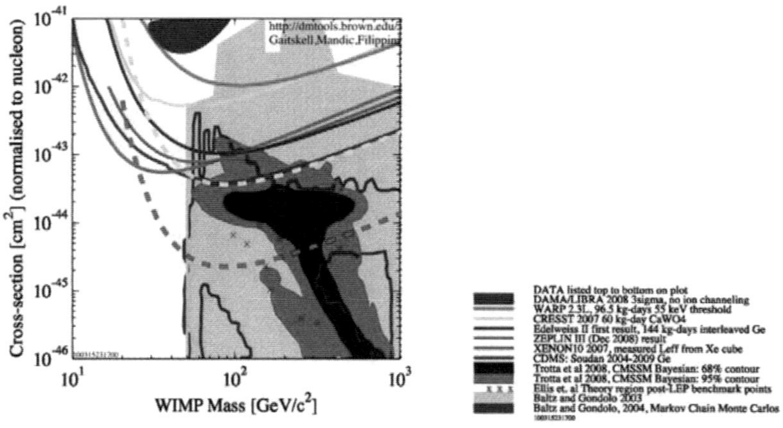

Figure 7. The expected sensitivity of the 800kg detector compared with other experiments. The yellow dotted is the 90% CL spin-independent sensitivity line for 10 days operation and the red dotted is the one for 1 year operation

5. Conclusion

The construction of XMASS 800 kg detector is almost finished in Kamioka underground laboratory. Its sensitivity for spin-independent interaction of $M_\chi = 50$ GeV case is expected to be $\sigma_{\chi-p} = 2 \times 10^{-44}$ cm^{-2} with one year operation. After the commissioning runs for a few months, the dark matter search run will be started in this year, 2010.

References

[1] E. Aprile and T. Doke, arXive:0910.4956.v1
[2] *These proceedings.*
[3] K. Abe et al., *Astopart. Phys.*, **31**, (2009) 290
[4] Y. Takeuchi et al., *Nucl. Instrum. Meth.*, **A 421**, (1999) 113

Journal of Physics: Conference Series **308** (2011) 012012 doi:10.1088/1742-6596/308/1/012012

Supernova Neutrino Detection

Inés Gil-Botella

CIEMAT, Basic Research Department, Avenida Complutense, 22, 28040 Madrid, Spain

E-mail: ines.gil@ciemat.es

Abstract. The neutrino burst from a core collapse supernova can provide information about the explosion mechanism and the mechanisms of proto neutron star cooling but also about the intrinsic properties of the neutrino such as flavor oscillations. One important question is to understand to which extend can the supernova and the neutrino physics be decoupled in the observation of a single supernova. The possibility to probe the neutrino mixing angle θ_{13} and the type of mass hierarchy from the detection of supernova neutrinos with liquid argon detectors is discussed in this paper. Moreover, a quantitatively study about the possibility to constrain the supernova parameters is presented. A very massive liquid argon detector ($\approx$ 100 kton) is needed to perform accurate measurements of these parameters. Finally the possible detection of the diffuse supernova neutrino background in liquid argon detectors is also described.

1. Introduction

Core collapse supernovae are a huge source of all flavor neutrinos. In 1987 astrophysics entered a new era with the detection of the neutrinos from the supernova SN1987A [1], which exploded in the Large Magellanic Cloud at a distance of $\approx$ 50 kpc. The burst of light was visible to the naked eye. Around three hours before the observation of the supernova (SN), an increase of neutrinos was detected by three water Cerenkov detectors (Kamiokande, IMB and Baksan). This observation confirmed important parts of the neutrino supernova theory such as total energy, mean temperature and time duration. However, limited quantitative information on the neutrino spectrum was obtained due to the small statistics recorded.

The flavor composition, energy spectrum and time structure of the neutrino burst from a supernova can give information about the explosion mechanism and the mechanisms of proto neutron star cooling. In addition, the intrinsic properties of the neutrino such as flavor oscillations can also be studied.

Although new data from neutrino detection have contributed to the understanding of the neutrino properties, still the neutrino mixing angle θ_{13}, the CP violation phase and the nature of the mass hierarchy, *i.e.*, normal (n.h.) or inverted (i.h.), remain unknown. On the other hand, detailed supernova simulations (see e.g. [2]) indicate that the average neutrino energy is flavor dependent and hierarchical ($E_{\nu_e} < E_{\bar{\nu}_e} < E_{\nu_{\mu,\tau},\bar{\nu}_{\mu,\tau}}$). However, great uncertainties still exist in the spectral and temporal evolution of the neutrino fluxes [3, 4]. The effects of these uncertainties can change dramatically the characteristics of the events observed in a given experiment. Indeed, this dependence can be used to determine the supernova and mixing parameters.

Several studies in the literature concern water, heavy-water or liquid scintillator targets (see e.g. [5]). This paper is focused on the capabilities of liquid argon Time Projection Chamber (LAr TPC) detectors to investigate supernova neutrinos thanks to the simultaneous observation

Journal of Physics: Conference Series **308** (2011) 012012 doi:10.1088/1742-6596/308/1/012012

of neutrino elastic scattering on electrons and both nuclear charged current and neutral current interactions on argon [7].

Fundamental physics subjects as neutrino mass hierarchy, CP violation, matter stability, astroparticle physics and in particular, supernova neutrino detection, call for a new generation of very large mass detectors able to collect high statistics and perform detailed event analysis. LAr detectors of the order of 100 kton will be ideal devices for these purposes offering the widest physics output (accelerator & non-accelerator) [8].

2. Neutrino oscillation and matter effects on supernova neutrinos

During the gravitational collapse of a supernova, electron neutrinos first and then the rest of flavor neutrinos are produced. It is believed that the 99% of the total binding energy of the star, $E_B \approx 3 \times 10^{53}$ ergs, is emitted in the form of neutrinos. These neutrinos are in equilibrium with their surrounding matter density and their energy spectra can be described by a function close to a Fermi-Dirac distribution. The flux of an emitted neutrino ν_α can then be written as [4]:

$$\phi_\alpha(E_\alpha, L_\alpha, D, T_\alpha, \eta_\alpha) = \frac{L_\alpha}{4\pi D^2 F_3(\eta_\alpha) T_\alpha^4} \frac{E_\alpha^2}{e^{E_\alpha/T_\alpha - \eta_\alpha} + 1} \quad (1)$$

where L_α is the luminosity of the flavor ν_α ($E_B = \sum L_\alpha$), D is the distance to the supernova, E_α is the energy of the ν_α neutrino, T_α is the neutrino temperature inside the neutrinosphere and η_α is the "pinching" factor. We have considered $\eta_\alpha = 0$, $\langle E_{\nu_\alpha} \rangle \approx 3.15\ T_\alpha$ and normalization factor $F_3(0) \approx 5.68$.

The original ν_μ, ν_τ, $\bar{\nu}_\mu$ and $\bar{\nu}_\tau$ fluxes are approximately equal and therefore we treat them as ν_x. An energy hierarchy between the different neutrino flavors is generally believed to hold and imply $\langle E_{\nu_e} \rangle < \langle E_{\bar{\nu}_e} \rangle < \langle E_{\nu_x} \rangle$. However, the specific neutrino spectra remain a matter of detailed calculations. In particular, recent simulations seem to indicate that the energy differences between flavors could be very small and possible collective neutrino flavor conversions could arise for either mass hierarchy depending on the primary fluxes [6].

Neutrino oscillations and matter effects in the supernova will change significantly the neutrino fluxes and therefore, the number of events expected in the detectors.

Before arriving at Earth the neutrino fluxes travel through the supernova mantle undergoing two MSW resonances: one at high density $\rho \approx 10^3$–10^4 g cm^{-3} (H–resonance) and the other one at low density $\rho \approx 10$–30 g cm^{-3} (L–resonance). The propagation through the L resonance is always adiabatic while in the case of the H resonance, the "jump" probability depends on the θ_{13} angle [7].

If the neutrino energy spectra are different, then θ_{13} and the mass hierarchy can be probed. For *small mixing angle (S)* ($\sin^2\theta_{13} < 2 \times 10^{-6}$), there are no effects on θ_{13} and we can not distinguish among mass hierarchies. Only an upper bound on $\sin^2\theta_{13}$ can be set. For *intermediate* θ_{13} ($2 \times 10^{-6} < \sin^2\theta_{13} < 3 \times 10^{-4}$) maximal sensitivity to the angle is achieved and measurements of the angle are possible in this region. For *large mixing angle (L)* ($\sin^2\theta_{13} > 3 \times 10^{-4}$) maximal conversions occur. The mass hierarchy can be probed but only a lower bound on θ_{13} can be put.

Therefore, in order to parametrize the supernova physics, we can consider *five* supernova parameters: the total binding energy $E_B = \sum L_\alpha$, the average energies of the neutrinos emitted from the supernova $\langle E_{\nu_e} \rangle$, $\langle E_{\bar{\nu}_e} \rangle$, $\langle E_{\nu_x} \rangle$ and the relative luminosities of the electron and non-electron neutrinos L_e/L_x. The neutrino oscillations can be studied in terms of two parameters: the sign of Δm^2_{31} and the mixing angle θ_{13}.

3. Supernova neutrino detection in terrestrial detectors

Most of the current and near-future supernova neutrino experiments [9] are water Cerenkov or liquid scintillator detectors and therefore, primarily sensitive to the $\bar{\nu}_e$ component of the signal, via inverse beta decay $\bar{\nu}_e + p \rightarrow n + e^+$. For supernova burst detection, not only statistics but also diversity of flavor sensitivity is needed: neutral current sensitivity, which gives access to the ν_μ and ν_τ components of the flux, and ν_e sensitivity are particularly valuable.

Only two near-future experiments will be mainly sensitive to ν_e's. The HALO detector [10] is under construction at SNOlab and it uses 80 tons of lead blocks instrumented with the unused SNO NCD counters to record neutrons and electromagnetic signals. The ICARUS detector at Gran Sasso [11] is a 600 ton LAr TPC with excellent ν_e sensitivity via ^{40}Ar CC interactions, for which de-excitation gammas will be visible.

All current supernova neutrino experiments participate to the Supernovae Early Warning System (SNEWS) [12], the network of SN neutrino observatories whose main goal is to provide the astronomical community with a prompt alert for the next galactic core collapse supernova explosion.

Very promising for the future are a number of planned mega-detectors exploring essentially three technologies: megaton scale water Cerenkov detectors, like LBNE in DUSEL [13], Hyper-K in Japan [14] and Memphys in Europe [15], 100 kton scale LAr TPC detectors like GLACIER in Europe [16] or LAr LBNE in DUSEL [17] and 50 kton scale liquid scintillator detectors like LENA in Europe [18] or Hanohano in Hawai [19]. Some such detectors can hope to collect individual neutrino events every few years from beyond the Local Group of galaxies (few Mpc), assuming that background can be reduced sufficiently.

4. Supernova neutrino detection in LAr TPCs

In a liquid argon TPC supernova neutrinos (E_ν < 100 MeV) can be detected through four different channels [7]:

(i) Charged current (CC) interactions on argon

$$\nu_e \, ^{40}Ar \rightarrow e^- + ^{40}K^* \tag{2}$$

$$\bar{\nu}_e \, ^{40}Ar \rightarrow e^+ + ^{40}Cl^* \tag{3}$$

The neutrino energy thresholds of these reactions are 1.5 and 7.48 MeV, respectively.

(ii) Neutral current (NC) interactions on argon

$$\overset{(-)}{\nu} \, ^{40}Ar \rightarrow \overset{(-)}{\nu} + ^{40}Ar^* \tag{4}$$

The energy threshold of this reaction is 1.46 MeV.

(iii) Elastic scattering (ELAS) on electrons

$$\overset{(-)}{\nu} \, e^- \rightarrow \overset{(-)}{\nu} \, e^- \tag{5}$$

It is possible to separate the different channels by measuring the associated photons from the K, Cl or Ar de-excitation, or by the absence of them in the case of elastic scattering.

Table 1 shows the expected number of neutrino events from a supernova at 10 kpc in a 100 kton detector. The four detection channels are considered independently and oscillation and non-oscillation cases are computed for normal and inverted hierarchies.

A total of 38550 neutrino events are expected in case of no oscillation assuming $\langle E_{\nu_e} \rangle = 11$ MeV, $\langle E_{\bar{\nu}_e} \rangle = 16$ MeV, $\langle E_{\nu_x} \rangle = \langle E_{\bar{\nu}_x} \rangle = 25$ MeV and luminosity equipartition among flavors (scenario I). CC events are very sensitivie to the change in the energy spectra due to oscillations.

Table 1. Expected neutrino events in a 100 kton detector from a supernova at a distance of 10 kpc.

$\langle E_{\nu_e}\rangle = 11$ MeV, $\langle E_{\bar{\nu}_e}\rangle = 16$ MeV, $\langle E_{\nu_x}\rangle = \langle E_{\bar{\nu}_x}\rangle = 25$ MeV and luminosity equipartition					
	No	Oscillation (n.h.)		Oscillation (i.h.)	
Reaction	oscillation	Large θ_{13}	Small θ_{13}	Large θ_{13}	Small θ_{13}
ELAS	1330	1330	1330	1330	1330
ν_e CC	6240	31320	23820	23820	23820
$\bar{\nu}_e$ CC	540	1110	1110	2420	1110
NC	30440	30440	30440	30440	30440
Total	38550	64200	56700	58010	56700

A clear increase in the number of events is expected, being a factor 4-5 for ν_eCC events and a factor 2-4 for $\bar{\nu}_e$CC interactions. The main variations with the θ_{13} angle are expected for the ν_eCC channel in the case of normal hierarchy and for the $\bar{\nu}_e$CC channel for inverted hierarchy. NC events are not affected by oscillations and they constitute an excellent probe for the supernova properties.

Figure 1 shows the oscillation effects in the neutrino energy spectra of the expected ν_eCC and $\bar{\nu}_e$CC events in a 3 kton detector. Due to the total conversion $\nu_{\mu,\tau} \to \nu_e$ for the n.h.-L case, the ν_e energy spectra is harder and this leads a huge increase of the expected events due to the quadratic dependence of the CC cross section with energy. The same effect can be seen for $\bar{\nu}_e$ events and i.h.-L case.

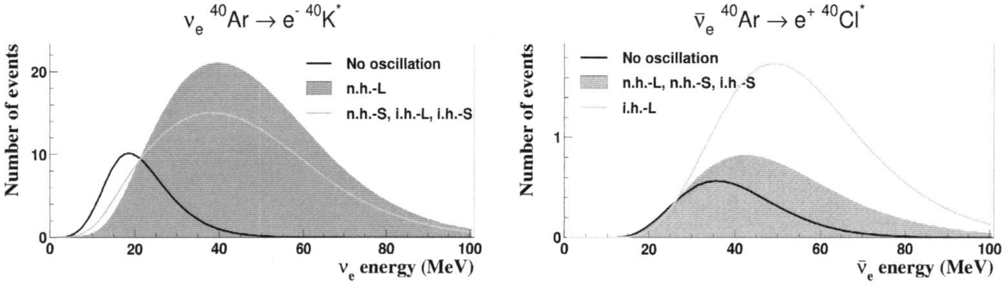

Figure 1. Expected number of ν_eCC (left) and $\bar{\nu}_e$CC (right) events in a 3 kton LAr detector from a supernova at 10 kpc.

5. Decoupling supernova and neutrino oscillation physics with LAr TPC detectors

The understanding of the supernova explosion is still plagued by uncertainties which have an impact on the precision with which one can predict time, energy and flavor dependent neutrino fluxes. On the other hand, the neutrino mixing properties are not fully known and in fact large uncertainty exists on the prediction of the actual effect of neutrino oscillations in the event of a supernova explosion. The main question is to understand to what extent the supernova and the neutrino physics can be decoupled in the observation of a single supernova.

In the case of LAr TPCs, the combination of the four independent detection channels can over-constrain the features of the supernova fluxes and the mixing parameters. The capability of

LAr TPC detectors to simultaneously measure the supernova and the oscillation parameters is quantitatively analyzed in next sections. A global fit to the event energy distribution for elastic and charged current events and to the rate for the neutral current events is performed.

5.1. Study of the oscillation parameters

First we investigate the sensitivity to the mixing angle θ_{13} and the type of mass hierarchy without any assumption on the supernova parameters. We exploit the information from the four neutrino detection channels on argon and we minimize a χ^2 function letting free the five supernova parameters.

In order to understand the uncertainties on the knowledge of the supernova physics, we consider two different sets of reference values for the supernova parameters, corresponding to one hierarchical energy scenario (I) and a non-hierarchical scenario (II). They differ in the assumptions on the average energies of the neutrino flavors. The luminosities of the neutrino flavors are assumed to be equal.

We first analyze the true small and large mixing angle scenarios and study the possibility of distinguishing the mass hierarchy and putting a bound on the θ_{13} value.

Figure 2 shows the variation of the minimized χ^2 value with the $\sin^2\theta_{13}$ parameter and the mass hierarchy for a 100 kton detector. In the top right corner is indicated in bold the "true" scenario considered. The left part of a given plot corresponds to the results of the fit under the assumption of normal hierarchy and the right part for an inverted hierarchy. Solid lines are computed considering the four supernova neutrino detection channels. Dashed lines do not take into account the NC processes. The shaded region illustrates the excluded region by the results of reactor experiments. The one- and two-sigma levels are shown by horizontal lines.

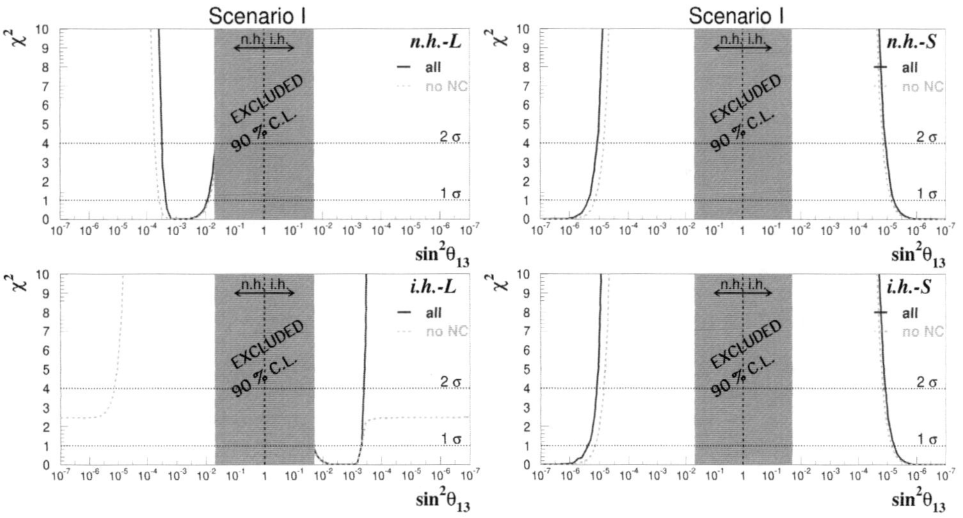

Figure 2. χ^2 value of the fit as a function of $\sin^2\theta_{13}$ for a 100 kton detector. Curves on the left (right) part of every plot correspond to the results of the fit under the assumption of a normal (inverted) hierarchy. Four limiting oscillation scenarios (n.h.-L, n.h.-S, i.h.-L, i.h.-S) are considered as "true" and the reference values of the supernova parameters correspond to scenario I. Solid curves are computed using the information of the four supernova neutrino detection processes and the dashed line does not take into account NC events.

We see that normal and inverted hierarchies are indistinguishable for small "true" θ_{13} mixing angle. The results of the fit are similar and only an upper bound on $\sin^2\theta_{13}$ can be set. If the "true" mixing angle is large, we are able to distinguish among hierarchies. Assuming the "true" hierarchy is normal and the value of θ_{13} is large (n.h.-L), we could put a lower limit on the θ_{13} angle. The same is possible for the case of inverted hierarchy. The wrong hierarchies are excluded. We illustrate the importance of the NC events. If we do not include them, we loose sensitivity in the large mixing angle cases and i.h.-L could be misidentified as n.h.-S.

The sensitivity to the oscillation parameters is almost similar for hierarchical (scenario I) and non-hierarchical scenarios (scenario II: $\langle E_{\nu_e}\rangle = 13$ MeV, $\langle E_{\bar{\nu}_e}\rangle = 16$ MeV, $\langle E_{\nu_x}\rangle = \langle E_{\bar{\nu}_x}\rangle = 17.6$ MeV) considering a 100 kton detector. Table 2 shows the results on the determination of the θ_{13} mixing angle for the different oscillation cases.

Table 2. Estimated limit on the θ_{13} mixing angle at 1σ for different "true" oscillation cases with a 100 kton detector. We compare the results considering as reference values for the supernova neutrino parameters scenarios I (hierarchical) and II (non-hierarchical).

Determination of $\sin^2\theta_{13}$ and mass hierarchy (SN parameters free)		
True osc. case	SN scen. I	SN scen. II
n.h.-L	$> 4.5\times10^{-4}$ n.h.	$> 1.8\times10^{-4}$ n.h.
	excluded i.h.	excluded i.h.
n.h.-S	$< 4.3\times10^{-6}$ n.h.	$< 1.2\times10^{-5}$ n.h.
	$< 5.9\times10^{-6}$ i.h.	$< 4.3\times10^{-6}$ i.h.
i.h.-L	excluded n.h.	excluded n.h.
	$> 5.0\times10^{-4}$ i.h.	$> 2.3\times10^{-4}$ i.h.
i.h.-S	$< 5.9\times10^{-6}$ n.h.	$< 1.2\times10^{-5}$ n.h.
	$< 4.0\times10^{-6}$ i.h.	$< 4.1\times10^{-5}$ i.h.

If the θ_{13} mixing angle is in the intermediate range ($2\times10^{-6} < \sin^2\theta_{13} < 3\times10^{-4}$), maximal sensitivity to the angle is achieved and measurements of the value are possible. Considering a 100 kton detector, statistically accurate measurements can be performed at 1σ and 2σ levels for scenario I (assuming true $\sin^2\theta_{13} = 10^{-4}$):

$$9.1(8.1)\times10^{-5} < \sin^2\theta_{13} < 1.1(1.3)\times10^{-4} \quad 1\sigma(2\sigma) \quad (\textit{assumed n.h.}) \tag{6}$$
$$8.9(7.4)\times10^{-5} < \sin^2\theta_{13} < 1.1(1.3)\times10^{-4} \quad 1\sigma(2\sigma) \quad (\textit{assumed i.h.}) \tag{7}$$

The wrong hierarchies are excluded in both cases. For scenario II, a measurement of the angle is also possible at the 1σ level and the mass hierarchy is well determined.

5.2. Study of the supernova parameters

In this section we study the information about the astrophysical parameters that can be extracted from the detection of supernova neutrinos considering that the oscillation parameters $\sin^2\theta_{13}$ and the mass hierarchy have been determined from reactor and long-baseline experiments.

Assuming that the angle has been measured with a precision of 10% ($\sin^2\theta_{13} = 10^{-3}\pm10^{-4}$) and the mass hierarchy is also known, we perform a χ^2 minimization letting the supernova parameters free in order to see how well we can constrain them. The expected accuracies at 90% C.L. are summarized in table 3. With the statistics given by a 100 kton detector, all the variables can be measured in a certain range, independently on the supernova scenario. Only the L_e/L_x parameter in the case of i.h. is less well determined for scenario II ($\sim 32\%$) than for scenario I ($\sim 9\%$).

Table 3. Expected accuracies at 90% C.L. in the determination of the supernova parameters using the neutrinos measured with a 100 kton detector. We have assumed that the mass hierarchy is known and the θ_{13} mixing angle has been measured by terrestrial experiments, being equal to $\sin^2\theta_{13} = 10^{-3} \pm 10^{-4}$. Supernova scenarios I and II are tested.

With constraint $\sin^2\theta_{13} = 10^{-3} \pm 10^{-4}$ from terrestrial experiment						
True hierarchy	SN scen.	$\frac{\Delta\langle E_{\nu_e}\rangle}{\langle E_{\nu e}\rangle}(\%)$	$\frac{\Delta\langle E_{\bar\nu_e}\rangle}{\langle E_{\bar\nu e}\rangle}(\%)$	$\frac{\Delta\langle E_{\nu_x}\rangle}{\langle E_{\nu x}\rangle}(\%)$	$\frac{\Delta E_B}{E_B}(\%)$	$\frac{\Delta(L_e/L_x)}{(L_e/L_x)}(\%)$
n.h.	I	~ 14	~ 4	< 1	~ 2	~ 11
i.h.	I	~ 5	~ 9	~ 1	~ 2	~ 9
n.h.	II	~ 9	~ 3	< 1	~ 4	~ 12
i.h.	II	~ 6	~ 6	~ 1	~ 2	~ 32

Another possibility is that future neutrino experiments will not be sensitive enough to measure $\sin^2\theta_{13}$ and will place an upper limit on its value. Assuming the limit $\sin^2\theta_{13} < 10^{-4}$, figure 3 shows the allowed regions at 68%, 90% and 99% C.L. on different supernova parameters planes for a 100 kton detector in the case of i.h. Thanks to the high statistics available with a very massive detector, we can fit very precisely the data and the results are essentially equal for both mass hierarchies. For non-hierarchical scenarios, the uncertainties are bigger, specially for the ratio between the electron and non-electron neutrino luminosities.

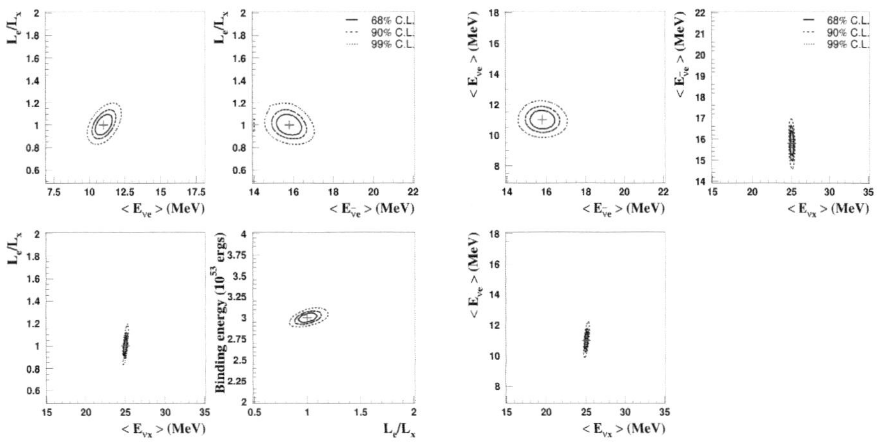

Figure 3. 68%, 90% and 99% C.L. allowed regions for the supernova parameters with a 100 kton detector, assuming that an upper limit on the value of the θ_{13} mixing angle has been set ($\sin^2\theta_{13} < 10^{-4}$) and the mass hierarchy is inverted ($\Delta m^2_{31} < 0$). Crosses indicate the value of the parameters for the best fits.

Finally we investigate the information that can be extracted from supernova neutrinos if a supernova explosion occurred nowadays when the value of the oscillation parameters is unknown. Letting all parameters $\{E_B, \langle E_{\nu_e}\rangle, \langle E_{\bar{\nu}_e}\rangle, \langle E_{\nu_x}\rangle, L_e/L_x, \sin^2\theta_{13}, \text{sign}[\Delta m^2_{32}]\}$ freely vary, we study how well we can determine them in the case of a 100 kton detector. The results show that with large statistics it is possible to decouple the supernova and oscillations physics and determine the value of the parameters with high precision (table 4).

Table 4. Expected accuracies at 90% C.L. in the determination of the supernova parameters using the neutrinos measured with a 100 kton detector. No conditions on the θ_{13} angle and mass hierarchy have been considered.

True osc. case	SN scen.	$\frac{\Delta\langle E_{\nu_e}\rangle}{\langle E_{\nu_e}\rangle}(\%)$	$\frac{\Delta\langle E_{\bar{\nu}_e}\rangle}{\langle E_{\bar{\nu}_e}\rangle}(\%)$	$\frac{\Delta\langle E_{\nu_x}\rangle}{\langle E_{\nu_x}\rangle}(\%)$	$\frac{\Delta E_B}{E_B}(\%)$	$\frac{\Delta(L_e/L_x)}{(L_e/L_x)}(\%)$
n.h.-L	I	~ 17	~ 4	< 1	~ 2	~ 11
i.h.-L	I	~ 5	~ 9	< 1	~ 2	~ 9
n.h.-S	I	~ 6	~ 4	< 1	~ 1	~ 11
i.h.-S	I	~ 5	~ 4	< 1	~ 2	~ 9
n.h.-L	II	~ 21	~ 3	< 1	~ 4	~ 14
i.h.-L	II	~ 6	~ 9	~ 1	~ 2	~ 34
n.h.-S	II	~ 11	~ 5	< 1	~ 2	~ 35
i.h.-S	II	~ 8	~ 4	~ 1	~ 2	~ 37

The expected accuracies are similar to the ones obtained constraining the mixing angle and the mass hierarchy, except for the $\langle E_{\nu_e}\rangle$ parameter and n.h.-L. In this scenario, the knowledge of the θ_{13} value improves the determination of the $\langle E_{\nu_e}\rangle$ energy from 17% to 14% for scenario I and from 21% to 9% for scenario II.

6. Diffuse supernova neutrino background in LAr TPCs

The diffuse supernova neutrino background (DSNB) is the flux of neutrinos and antineutrinos emitted by all core-collapse supernovae occurred so far in the Universe. It will appear isotropic and time independent in feasible observations.

The DSNB has not been detected yet. The Super-Kamiokande experiment established an upper limit on the $\bar{\nu}_e$ flux of $\Phi(\bar{\nu}_e) < 1.2$ cm^{-2}s^{-1} for neutrino energies higher than 19.3 MeV [20], close to the predictions. If Super-Kamiokande is modified with dissolved gadolinium to reduce detector backgrounds and increase the energy range for analysis, then the DSNB could be detected at a rate of a few events per year.

LAr TPCs would be able to mainly detect the ν_e component of the DSNB signal providing complementary information with respect to Super-Kamiokande. The main background sources for these events in the relevant neutrino energy range of 10-50 MeV are solar and low energy atmospheric neutrinos. Depending on the theoretical predictions for the DSNB flux, a 100 kton LAr detector running for five years would get more than 4σ measurement of the DSNB flux [21].

7. Conclusions

Liquid argon TPC detectors have excellent capabilities to study and decouple supernova and neutrino oscillation physics by the observation of four independent channels: elastic scattering on atomic electrons, charged current interactions on argon (independently sensitive to ν_e and $\bar{\nu}_e$ neutrinos) and neutral current interactions on argon.

We have studied the sensitivity to the θ_{13} and mass hierarchy parameters and found that for large mixing angle, the mass hierarchy can be identified and lower limits on the θ_{13} value are set. For small mixing angle, it is not possible to distinguish the mass hierarchy but upper limits on θ_{13} can be obtained. If the mixing angle is in an intermediate range, measurements of its value are possible. The statistics provided by the detection of a supernova at 10 kpc with a 100 kton detector allow to quantitatively determine the neutrino oscillation parameters.

Finally we have studied the ability to measure the supernova parameters with good precision using large LAr detectors. With a 100 kton detector, a single supernova explosion would allow to determine the supernova cooling phase quite precisely. In addition, the diffuse supernova neutrino background signal could be observed in five years at the 4σ level.

Acknowledgments

I would like to thank Prof. André Rubbia and Prof. Takuya Hasegawa for their invitation to this very interesting and productive workshop.

References

[1] Hirata K *et al.* 1987 *Phys. Rev. Lett.* **58** 1490; Bionta R *et al.* 1987 *Phys. Rev. Lett.* **58** 1494; Alekseev E N *et al.* 1987 *JETP Lett.* **45** 589
[2] Takahashi K *et al.* 2003 *Astrop. Phys.* **20** 189-193
[3] Raffelt G G *et al.* 2003 *Preprint* astro-ph/0303226; Takahashi K *et al.* 2003 *Phys. Rev.* D **68** 113009
[4] Lunardini C and Smirnov A Y 2003 *J. Cosmol. Astropart. Phys.* JCAP06(2003)009
[5] Totani T *et al.* 1998 *Astrophys. J.* **496** 216; Minakata H *et al.* 2002 *Phys. Lett.* B **542** 239; Dutta G *et al.* 2001 *Phys. Rev.* D **64** 073011; Bandyopadhyay A *et al.* 2003 *Preprint* hep-ph/0312315
[6] Choubey S *et al.* 2010 *Preprint* arXiv:1008.0308 [hep-ph]
[7] Gil-Botella I and Rubbia A 2004 *J. Cosmol. Astropart. Phys.* JCAP08(2004)001; Gil-Botella I and Rubbia A 2003 *J. Cosmol. Astropart. Phys.* JCAP10(2003)009; Bueno A , Gil-Botella I and Rubbia A 2003 *Preprint* hep-ph/0307222
[8] Rubbia A 2003 *Preprint* hep-ph/0402110
[9] Scholberg K 2010 *J. Phys.: Conf. Series* **203** 012079
[10] Duba C A *et al.* 2008 *J. Phys.: Conf. Series* **136** 042077
[11] Aprili P *et al.* 2001 LNGS-EXP 13/89 add. 2/01, CERN-SPSC-2002-027 (SPSC-P-323)
[12] Antonioli P *et al.* 2004 *New. J. Phys.* **6** 114
[13] Maricic J 2010 *J. Phys.: Conf. Series* **203** 012109
[14] Nakamura K 2003 *Int. J. Mod. Phys.*A **18** 4053
[15] Bellefon A *et al.* 2006 *Preprint* hep-ex/0607026
[16] Rubbia A 2009 *J. Phys.: Conf. Series* **171** 012020
[17] Baller B 2010 *These proceedings*
[18] Marrodan Undagoitia T *et al.* 2006 *J. Phys.: Conf. Series* **39** 278-280
[19] Maricic J 2010 *J. Phys.: Conf. Series* **203** 012137
[20] Malek M *et al.* 2003 *Phys. Rev. Lett.* **90** 061101
[21] Cocco A G *et al.* 2004 *J. Cosmol. Astropart. Phys.* JCAP12(2004)002

1st International Workshop towards the Giant Liquid Argon Charge Imaging Experiment IOP Publishing
Journal of Physics: Conference Series **308** (2011) 012013 doi:10.1088/1742-6596/308/1/012013

Neutrino Interactions on Ar target

Flavio Cavanna

Department of Physics and INFN, University of L'Aquila
Via Vetoio, I-67010 L'Aquila, Italy

E-mail: `flavio.cavanna@aquila.infn.it`

Abstract. The forthcoming experimental effort aiming at precision measurements of the MNSP matrix elements requires a robust control of the systematic uncertainties. A major contribution is presently represented by the large (still persisting) uncertainty associated to the neutrino cross sections. Therefore, precise measurements and modeling of the ν-Nucleus Cross-Section and related Nuclear Effects in the *intermediate* energy range ($\sim$ *0.5-5 GeV*) are today considered as relevant issues.
The LAr-TPC technology is ideal to perform a wide variety of ν-physics studies, thanks to the capability of single particle identification and detailed reconstruction of exclusive topologies, as demonstrated within the ICARUS R&D program.
The ArgoNeuT detector, a 175 lt active LAr-TPC put into operation on the LE NuMI ν-beam line at FNAL in 2009-10, is providing with first data for precision measurements of (quasi-elastic) ν-Ar Cross-Section in the *intermediate* energy range.

1. The Intermediate ν-Energy range

The field of neutrino oscillations, being these conclusively observed by solar, reactor, atmospheric, and accelerator-based neutrino oscillation experiments of first generation, has moved to an era where experiments will seek to precisely measure the oscillation parameters and search for the existence of a subdominant mode of oscillation. In the standard parameterization, within the three neutrinos scheme, neutrino oscillations are described by the MNSP mixing matrix through the squared differences of the neutrino mass eigenvalues, three mixing angles and one CP-violating phase.

The observed oscillation of electron neutrinos and anti-neutrinos from the sun and nuclear reactors, respectively, are primarily driven by the "$i, j = 1, 2$" sector while the "2,3" sector describes the oscillation of ν_μ to other neutrino flavor observed in atmospheric and accelerator experiments. This leads to subdominant oscillations that may manifest itself experimentally as a tiny appearance of ν_e in beam of ν_μ detectable by accelerator experiments if $\theta_{13} > 0$. If the irreducible phase δ is also non-zero, neutrinos will also exhibit CP violation in which neutrino oscillations and the corresponding antineutrino oscillations will have different oscillation probabilities.

In present and future stages, accelerator-based experiments will perform high-sensitivity search for the $\nu_\mu \rightarrow \nu_e$ oscillation associated with non-zero θ_{13} using long-baseline neutrino beams in the "few-GeV" energy range (this lies "intermediate" between the "low" energy domain of solar and reactor neutrinos and the "high" energy range of accelerator neutrinos, e.g. appropriate for ν_τ appearance).

Journal of Physics: Conference Series **308** (2011) 012013 doi:10.1088/1742-6596/308/1/012013

The ability to carry out this ambitious program depends heavily on the full understanding of the neutrino ("standard") interaction mechanisms with the nuclear matter (A nucleus chosen as target in the detector).

An accurate determination of the experimental sensitivity is in fact a key element for a correct design of the experiment and this in turn should be based on the accurate, a priori knowledge of the beam flux and ν-A cross-section for all neutrino types (rather than on effective models, usually - so far - "tuned" up to match data to take care of the differences in fluxes (& detectors) at Near/Far locations).

2. ν Cross Section at the intermediate energies

At the *"intermediate"* energies the charged-current (CC) Cross-Section for ν-interaction on $p, n/q, \bar{q}$ in nuclei/nucleons is usually referred to according to a natural decompositions:

$$\begin{aligned} \sigma_{tot} &= \sigma_{QE} \oplus \sigma_{RES} \oplus \sigma_{DIS} \\ &= \sigma_{0\pi} \quad \oplus \sigma_{1\pi} \quad \oplus \sigma_{n\pi} \end{aligned} \tag{1}$$

where:

- the first term in Eq.(1) refers to the Quasi-Elastic Scattering (QE):

$$\nu_l + n \rightarrow l^- + p \tag{2}$$

  characterized by low Q^2, $x_{Bj} = 1$, $W = M_p$.
  The dynamics can be described by a V-A Current-Current Lagrangian. The hadronic Current is usually defined through the Nucleon Weak Form Factors (FF): the Vector FF's [$F_V^1(Q^2)$ and $F_V^2(Q^2)$], related to the El.M. FF under CVC-hypothesis, the Axial [$F_A(Q^2)$] and PseudoScalar [$F_P(Q^2)$] FF's.
  In particular, the Vector FF's can be (are) determined from e-scattering experiments, while the Axial FF is usually assumed to be in dipolar form and depending on a free parameter M_A, the Axial Mass to be determined from ν-data fits. The Axial Mass describes the nucleon structure and the Axial FF determines a large fraction of the total QE Cross-Section.
  When the n-target nucleon [p for $\bar{\nu}$ interaction in (2)] is bound in the parent nucleus A, the Relativistic Fermi Gas Model (RFG) is usually adopted to describe the nuclear initial state. Final state particles (hadrons) produced at the primary neutrino collision undergo non-perturbative effects (FSI) of strong interactions inside the target nucleus. In this case the absence of well defined models makes the treatment of these *"Nuclear Effects"* the main potential source of systematic uncertainty.
- the second term in Eq.(1) refers to the Resonance Excitation channel (RES):

$$\nu_l + N \rightarrow l + \Delta/N^* \rightarrow l + \pi + N' \tag{3}$$

  characterized by low Q^2, large x_{Bj}, and W.
  From the theoretical point of view this is the most complicated channel. According to the standard FKR model the nucleon N is represented by a 3-quarks system bound by a harmonic potential in ground state. Δ and N^* correspond to excited states, decaying with π production. Each decay channel results from superposition and interference between allowed resonance amplitudes.
  When N is bound in A, the treatment of the Nuclear Effects and of the final state (re)interactions (FSI) are even more crucial for a satisfactory Cross-Section determination. From the experimental point of view, this is the least precisely measured channel.

- the third term in Eq.(1) refers to the (Deep) Inelastic Interaction modes (DIS):

$$\nu_l + N \rightarrow l + X \tag{4}$$

with $x_{Bj} \in (0, 1)$ and large Q^2.
The interaction dynamics is well described by Standard Model propagator (massive $W^{\pm}$). The hadronic current is defined through the Nucleon Structure Functions embedding the standard Parton Distribution Functions (PDF). Precise high-Q^2 DIS data are available from (e, e') experiments for F_1 and F_2 determination, and from ν-N experiments for F_3 fitting. F_4 and F_5 Structure Functions in the ν Cross-Section are proportional to (m_l^2/M_N), i.e. relevant only for $\nu_l = \nu_\tau$.
Nuclear Effects have a limited impact at the DIS regimes.

The available Cross-Section measurements (in particular for the QE and RES channels) from *"1st generation"* experiments ('64-'80) at ANL, BNL, FNAL, CERN and IHEP were affected by large errors, mainly due to low statistics, background contamination and to a limited control of the incoming neutrino flux amplitude and profile (with $\sim$ 40% spread across experiments).

This triggered a renewed interest (*NuInt Workshop* series) [1] and intense activity [2],[3],[4],[5],[6],[7] on precise measurements and modeling of the ν-Nucleus Cross-Section in this *intermediate* energy range ($\sim$ 0.5-5 GeV), and related Nuclear Effects.

Two neutrino beams were/are active in the *intermediate* ν-energy range: at KEK (Japan) and at FNAL-Booster (US) with mean energy of 1.3 GeV and 0.7 GeV respectively and low ν_e contamination. In the framework of the search for ν-oscillation signals, cross-section measurements of *"2nd generation"* ('90-'00) have been performed with the FNAL-Booster beam and at KEK on the LBL beam pointing to the SK detector.
The MiniBooNE experiment at FNAL [8] collected large statistics of ν_μ-CC QE events with a detector of 800 t of Mineral Oil. Measurements of QE Cross-Section on C target have been reported [6].

At KEK the beam pointing to the SK detector (K2K experiment [9]) was monitored with

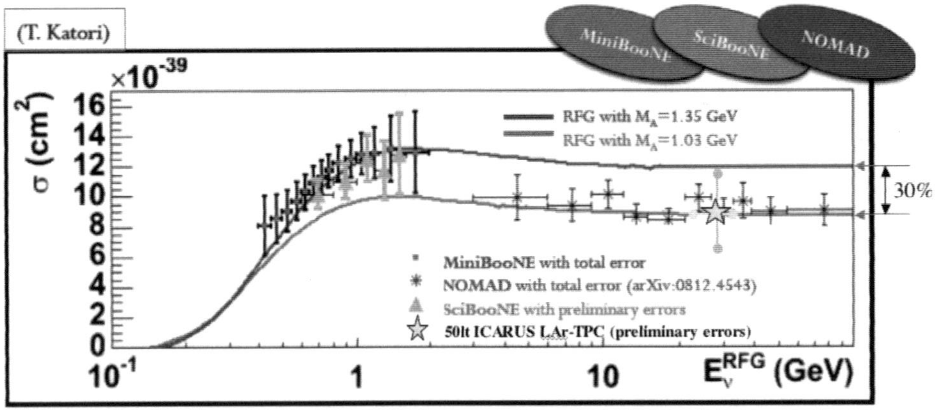

Figure 1. Quasi-Elastic ν_μ CC cross section data from "2nd generation" experiments: MiniBooNE, SciBooNE, NOMAD (T. Katori in [6]). A single point (light blue star) corresponding to the data from the ICARUS 50 lt LAr-TPC prototype has been added as obtained from [14].

a set of three near detectors: the 1 kT-Water Cherenkov detector, the SciFi detector (Water target) and the SciBar fine grained Pb-scintillator calorimeter. From the SciFi detector a new

measurement of the M_A parameter in the Axial Form Factor for Oxygen has been performed. The SciFi detector was successively transported to FNAL and positioned along the FNAL-Booster beam line (SciBooNE experiment [10]). A dedicated run allowed the collection of a new set of data for ν Cross-Section measurements [6].
Finally, a study [11] by the NOMAD Collaboration has recently been published on data collected from the exposure (1995-98) of the active low-Z target of the NOMAD experiment to the CERN wide band beam, though at substantially higher neutrino energies (average 26 GeV). The high statistics and the normalization to the well-known total DIS Cross-Section allowed the first precise QE Cross-Section measurement in the high energy range ($\sim$ from 5 GeV to 80 GeV).

Though the precision of the individual measurements improved substantially compared to the data from the 1st generation, a large ($\sim$ 30%) discrepancy still persists between measurements at low and high energies, as shown in Fig.1 for the quasi-elastic CC cross section.

The need for more accurate measurements thus remains an open issue in view of the next generation ν-oscillation experiments.

3. Perspectives from the *3rd Generation* of ν Cross Section measurements

The NuMI beam-line at FNAL, with its extremely intense ν flux and with the availability of space at the MINOS near detector hall, offers an ideal venue for high-statistics, high-resolution ν and $\bar{\nu}$-nucleon/nucleus scattering experiments.
The MINOS Near Detector itself, operational since 2005, has already recorded many million neutrino interactions on Fe nuclei at neutrino energies around 3 GeV (peak of the LE NuMI flux). Characteristics of the QE CC process have been reported [6].
A fully active and multi-target (A = C, Fe, Pb nuclei) detector, MINERνA (Main Injector Experiment ν-A) [12], is presently under assembly (almost complete). Once operational, the study of ν-A interactions in the *intermediate* energy range would be then performed with unprecedented precision for the most commonly used nuclear targets.

Argon target ($A = 40$) is also of high interest for possible application with next generation ν-oscillation experiments.
The liquid Argon-Time Projection Chamber (LAr-TPC) technology developed within the ICARUS project [13], considered the modern version of the bubble-chamber concept, combines the imaging capability with the additional features of a high resolution calorimetry and of a (virtually) unlimited active mass. Moreover, thanks to the single particle identification and detailed reconstruction of exclusive topologies (e.g. with e-to-π^0 separation, useful for NC vs CC study, and with the detection of the recoiling proton in QE and RES interactions possibly down to the very low threshold of about $T_p \simeq$ 20-30 MeV of kinetic energy), LAr-TPC detector(s) can provide most precise reconstruction of neutrino events.
Direct measurement of ν-Ar interaction cross section at the *intermediate* energies haven't been performed so far. LAr-TPC detectors, even at modest mass scale, definitively represents an ideal tool for the study of ν Cross-Sections, vertex reconstruction and nuclear effects (especially in the "few GeV range"). Final State hadrons from ν-interactions may indeed re-interact inside the (parent) nuclear matter, before propagation in the detector volume, with emission of additional particles (neutrons, protons, π's, γ's, α's,...). These products are usually neglected because not detectable, unless a high quality imaging detector is in use or, if detected, may lead to mis-identification of the final state topology, unless robust PID criteria are at hand.

The capability to identify and reconstruct neutrino interactions in LAr-TPC's was clearly demonstrated by a "50 lt" prototype (ICARUS) with a ν test beam at CERN in the late '90's and also a first indication about the *ν-Ar* QE Cross-Section was extracted from the data analysis.

Possibilities of using LAr-TPC detectors for dedicated Cross-Sections measurements in available ν-beams have been first investigated in 2001 (F.Cavanna and O. Palamara [1]) and further discussed in the following years. In 2007 a small experiment (ArgoNeuT, B. Fleming et al. [15]) to be positioned on the NuMI beam-line at FNAL has been finally approved, as a first step in the vast experimental program of ν-physics with LAr-TPC's in the US. The use for Cross-Sections measurements of a $\sim$ 100 t LAr-TPC detector at the proposed intermediate (2 km) station on the T2K beam line as been also envisaged (A. Ereditato and A. Rubbia in [3] and A. Meregaglia and A. Rubbia in [7]).

In the next sections, a brief overview of the results from the exposure of the ICARUS "50 lt" LAr-TPC prototype to the *high-energy* ν-beam at CERN will be given as well as the perspectives and preliminary results from the exposure of the 175 lt ArgoNeuT detector to the "low-energy" (LE) NuMI beam at FNAL.

3.1. The ICARUS "50 lt" LAr-TPC at CERN

The first exposure of a LAr-TPC to neutrinos (ICARUS-Milano U. R&D program) was performed in 1997-98 at the CERN West Area Neutrino Facility (WANF), where the high-energy wide band beam was available for the CHORUS and NOMAD experiments. The data have been collected with a small LAr chamber located between them. The technical description of the detector is given in [14]. The modest size of the LAr TPC fiducial volume ($32\times32\times47$ cm^3 corresponding to about $\sim$50 lt), coupled with the high energy of the WANF ν beam (mean value 28 GeV), made necessary the use of a muon spectrometer downstream the TPC. To this purpose, a coincidence with the NOMAD DAQ was set up to profit of the detectors located into the NOMAD magnetic dipole as a magnetic spectrometer.
The event reconstruction procedures applied to the collected data allowed to extract the physical information from the TPC wire output signals, i.e. the energy deposited by the different particles and the point where such a deposition has occurred, up to build a complete 3D and calorimetric picture of the event (see Fig. 2 for an example of QE event 3D reconstruction). This was the first time that interactions of (multi-GeV) accelerator neutrinos occurring in a LAr TPC were fully reconstructed.

Over the whole data taking period around 10000 triggers showed a vertex in the LAr-TPC fiducial volume and these were identified as ν_μ CC candidates. A selected QE "$\mathcal{G}olden\ sample$" has been extracted by application of tight cuts. This set consists of events with an identified, fully contained proton and one muon whose direction extrapolated from NOMAD matches the outgoing track in the TPC.
The study of the proton identification and the momentum measurement is particularly important: a negligible $\pi^\pm$/p misidentification probability and a precise interaction vertex reconstruction were achieved with a (conservative) threshold of $T_p \geq 40$ MeV. Indeed, the size of the range of the fully contained protons collected in the QE selected sample offered the opportunity to precisely evaluate the energy loss pattern in the active LAr medium.
The "$\mathcal{G}olden\ sample$" ($N_{\mathcal{G}} = 86$) contains pure QE interactions plus an intrinsic background dominated by RES and DIS production followed by pion absorption in the nucleus (or escaping undetected).
The geometrical detector acceptance, the background subtraction and the QE selection efficiencies have been evaluated using Monte-Carlo simulation and the numerical results rely on the choice of the input MC parameters. The beam intensity and the exposure time were known parameters as well as the mean neutrino energy, estimated to be 28 GeV with an RMS of 18 GeV (all the numerical values are reported in [14]).
The QE ν_μ-Ar Cross-Section at the mean beam energy amounts to $\sigma_{QE} = (0.90 \pm 0.10(stat)) \times 10^{-38}$ cm^2. This value nicely agrees with the result obtained by NOMAD [11] (same ν-beam) as shown in Fig.1.

The estimated total systematic error, evaluated in a separate (yet unpublished) report, is

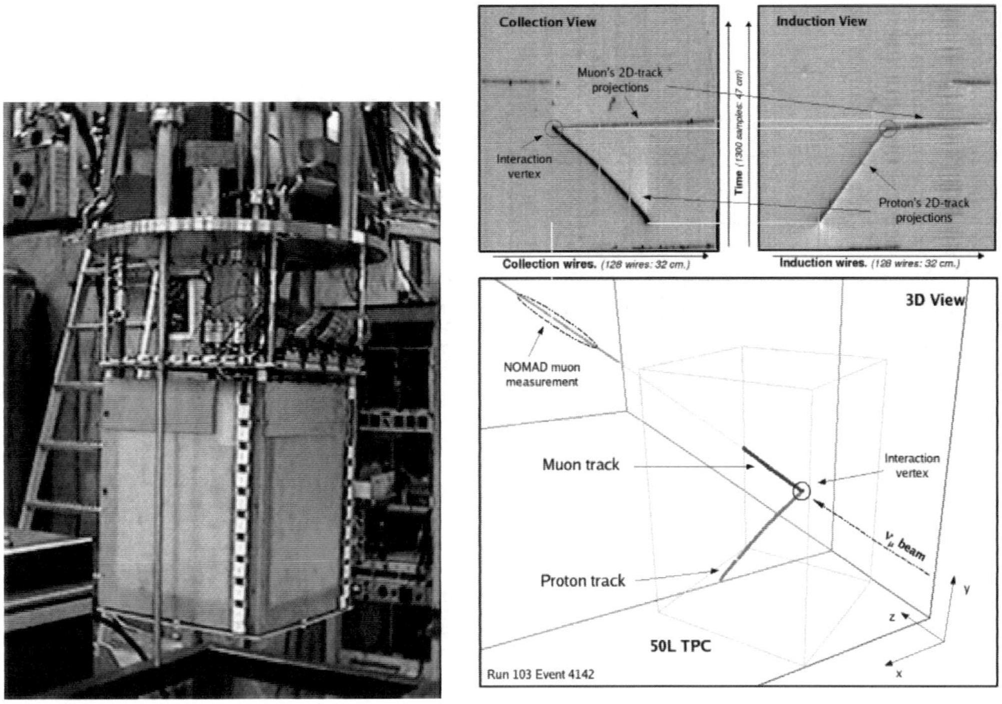

Figure 2. [Left] Picture of the 50 lt prototype. [Right] Example of the 3D reconstruction of a low-multiplicity (QE) ν_μ CC event. The raw image from the two TPC wire planes (on top): Hits and 2D track projections have been identified. Three dimensional view of the reconstructed event (bottom picture).

about 20%. The dominant contribution is from the QE selection efficiency. This is affected by nuclear effects, which modify the topology and kinematics of the final state. Due to the large uncertainties in the modeling of nuclear effects inside the Monte-Carlo, the systematic error on the Cross-Section due to the fraction of $\mathcal{G}olden$ events inside the QE sample amounts to about 16%.

3.2. The "ArgoNeuT" LAr-TPC at FNAL

ArgoNeuT is a joint NSF/DOE R&D project (T962) at Fermilab [15] to expose a small-scale LArTPC to the NuMI neutrino/antineutrino beam. In the LE configuration the NuMI facility produces a nearly pure ν_μ ($\bar{\nu}_\mu$) beam with energies in the 0.5-5 GeV range (peaking at ~3 GeV). ArgoNeuT is a 500 lt (175 lt fiducial) liquid Argon TPC. The TPC dimensions are approximately 0.5 m $\times$ 0.5 m $\times$ 1 m and it is housed in a cylindrical vacuum insulated cryostat. The TPC has 480 active wires in two planes, with individual electronic read-out (analog preamp stage and waveform fast digital conversion). The planes are oriented at $\pm 30^o$ to the vertical. Wire planes separation and wire-to-wire pitch are of 4 mm. Signal feed-throughs and support equipments are mounted on a flange on the top of the vessel. An external "event-flagging" system composed by scintillator pads in front and behind the cryostat along the beam line completes the experimental layout.

Commissioning of the detector and of the cryogenic system in a dedicated experimental facility on surface took place at FNAL in summer 2008 with first cosmic muon events recoding.

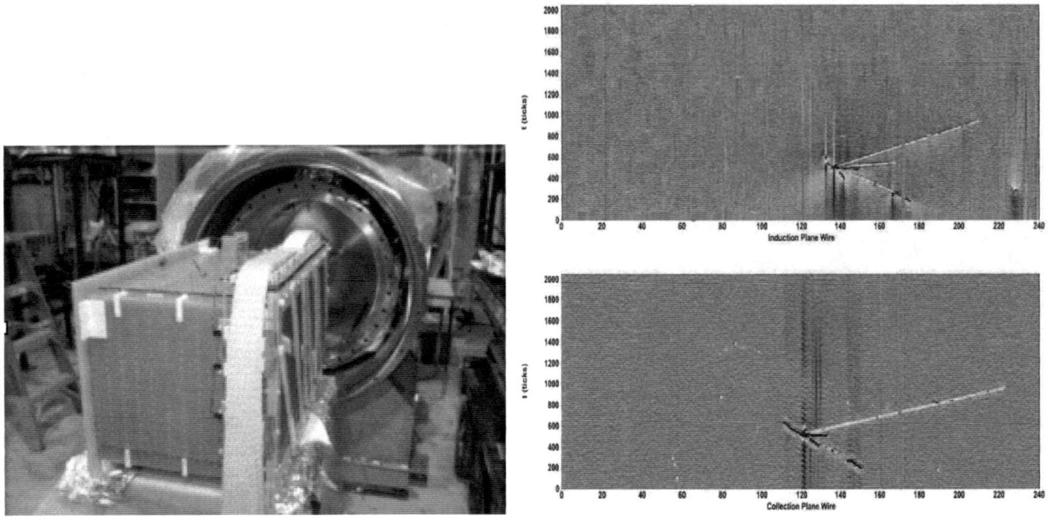

Figure 3. [Left] Picture of the ArgoNeuT detector. [Right] Example of 2D (raw) images of a ν_μ event from the two LAr-TPC wire planes, collected with ArgoNeuT during the run at NuMI-FNAL (Sept. '09-Feb.'10). Two main tracks are visible (one escaping the detector towards MINOS-ND) accompanied by short, densely ionizing tracks at the vertex.

The experimental set-up was then moved to the MINOS Near Detector Hall about 100 m underground, just upstream MINOS-ND, to profit of it as a range stack to measure uncontained long-track muons (and pions) from muon neutrino interactions in the TPC. After remounting and LAr filling, data-taking started in May 2009 with a short technical run. From Sept.'09 up to Feb.'10 a first physics run was successfully completed, most of the time with the beam in antineutrino mode.

Main goal of the ArgoNeuT experiment is to directly measure ν_μ and $\bar{\nu}_\mu$ Cross-Section on Argon in the "few GeV" range, with particular interest on the CC QE contribution and the related nuclear effects.
Considering the amount of $8 \cdot 10^{17} PoT/day$ delivered at NuMI-FNAL, the expected number of ν_μ CC events/day has been evaluated as $N_{tot} = 117$ from the three main channels of Eq.(1) ($N_{QE} = 19$, $N_{RES} = 15$, $N_{DIS} = 83\ evts/day$). These represent the number of events per day where the interaction vertex is found inside the boundaries of the LAr-TPC active volume.
According to MC simulation of QE events, proton tracks are fully contained in the LAr sensitive volume in a fraction of 54% of the total and 86% of muons enter the MINOS NEAR Detector (and the momentum is measured). The incident neutrino energy E_ν can then be rather precisely reconstructed for each QE event from the measured final state muon and proton 4-momentum vectors, assuming the target neutron with off-shell mass and Fermi momentum from RFG model.
During the run period $\sim 1.55 \times 10^{20} PoT$ have been delivered with about 85% live-time of the ArgoNeuT detector, i.e. $\sim 1.35 \times 10^{20} PoT$ acquired, corresponding to about 5500 ν and 3500 $\bar{\nu}$ CC events in the fiducial volume). A picture of the detector and the 2D images of a recorded neutrino event are shown in Fig.3.

Intense activity of data analysis is currently under way (reported in C. Bromberg's talk at this Workshop). In particular, extremely detailed MonteCarlo generators are being developed for simulation and comparison with LArTPC data (FSI in MC codes represent the most difficult present challenge in MC development), as well as accurate off-line reconstruction codes for a full exploitation of the information available (3D imaging, energy/momentum reconstruction and particle identification) from LAr-TPC data.

The high identification capability of ν_μ and $\bar{\nu}_\mu$ QE events [from the collected samples of two-tracks (μ, p) and single track (μ) events possibly associated to vertex activity from de-excitation mechanisms of the intranuclear cascade] may soon shed light on the FSI effects and suppress the systematic errors on the QE CC ν cross section determination on Ar target at the *intermediate* energies.

4. Conclusions

New generation neutrino Cross-Sections measurement is recognized as a well established, necessary step toward the forthcoming second generation of oscillation experiments.

Neutrino beams in the *intermediate* energy range are available in US and Japan, providing an unprecedented richness of experimental opportunities.

The realization of experiments employing state-of-art technologies is considered as necessary for the definitive assessment of the standard neutrino properties.

The LArTPC represents an ideal tool for the study of ν Cross-Sections, vertex reconstruction and nuclear effects in the "few GeV range".
The exposure of the ICARUS "50 lt" LAr-TPC prototype to the *high-energy* ν-beam at CERN in 1997 clearly demonstrated the capability to identify and reconstruct low multiplicity neutrino interactions.

ArgoNeuT, a 175 lt active LArTPC detector exposed to the LowEnergy NuMI beam at FNAL, has recently completed a first physics run (6 months, most in antineutrino mode, with about 10.000 events collected). Intense activity of data analysis and MC simulation is currently under way, aiming at accurate vertex reconstruction, nuclear effects and precise ν-Ar cross section first determination in the *intermediate* energy range.

Acknowledgments

I would like to warmly thank A. Rubbia and T. Hasegawa for the kind invitation to participate to the GLA2010 Workshop and to all the Organizing Committee for the fruitful atmosphere in Tsukuba.

References

[1] NuInt01 Workshop, KEK (Japan), 2001- Nucl. Phys.B - Proc.Suppl. 112, 2002.
[2] NuInt02 Workshop, UCI (US), 2002 - *http://www.ps.uci.edu/~NuInt.*
[3] NuInt04 Workshop, LNGS (Italy), 2004 - Nucl. Phys.B - Proc.Suppl. 139, 2005.
[4] NuInt05 Workshop, Okayama (Japan), 2005 - Nucl.Phys.B-Proc.Suppl. 159, 2006.
[5] NuInt07 Workshop, FERMILAB (US), 2007 - AIP - Conf. Proc. 978, 2007.
[6] NuInt09 Workshop, Sitges (Spain), 2009 - AIP - Conf. Proc. 1189, 2009.
[7] XX M. Born Symposyum, Wroclaw (Poland), Dec. 2005 - Acta Physica Polonica B 37 n.8, 2006.
[8] MiniBooNE: *http://www-boone.fnal.gov*
[9] K2K - KEK: *http://neutrino.kek.jp*
[10] SciBooNe: *http://www-sciboone.fnal.gov*
[11] NOMAD Collaboration, Eur. Phys. J. C63 (2009), 355
[12] MINERνA: *http://www.pas.rochester.edu/minerva*
[13] ICARUS: *http://icarus.lngs.infn.it*
[14] ICARUS-CERN-Milano Univ. Collaboration, Phys. Rev. D74 (2006) 112001
[15] ArgoNeuT, *http://t962.fnal.gov*

Journal of Physics: Conference Series **308** (2011) 012014 doi:10.1088/1742-6596/308/1/012014

Status of ArgoNeuT, a study of ν-interactions using a Liquid Argon TPC in the NuMI beam at Fermilab.

Carl Bromberg
Michigan State University, East Lansing, MI 48824

E-mail: bromberg@pa.msu.edu

Abstract. The Liquid Argon Time Projection Chamber (LArTPC) technology is a leading candidate for a 100-kiloton class detector for electron neutrino appearance in a long-baseline muon-neutrino beam and for nucleon decay within the stable Argon nucleus, both with excellent efficiency and background rejection. The US has taken its first steps toward development of the technology by constructing and operating small LArTPC detectors, the latest being the ¼-ton detector of the ArgoNeuT collaboration. In a recent 6-month run in the low-energy configuration of the NuMI neutrino beam at Fermilab, the collaboration recorded a large sample of neutrino events. This paper reports on the characteristics of that data, the preliminary steps to reduce the data to a sample of neutrino-induced events, and a description of an event consistent with a charged current quasi-elastic electron-neutrino interaction.

1. Introduction

The Liquid Argon Time Projection Chamber (LArTPC) technology developed within the context of the ambitious ICARUS project [1] has not as yet convincingly demonstrated (at least in the US) that the technology could fulfill its promise to provide excellent efficiency and background rejection for neutrino and proton decay physics at a reasonable cost. Over the last 5-6 years has there been a growing community of physicists and engineers who felt that with a staged approach [2] a convincing case could be made for the construction of a large detector [3] in the U.S. After two groups

Figure 1. ArgoNeuT detector in its cryostat with the cathode on the left, wire planes on the right. The signal cables terminate on the bias voltage distribution cards.

Figure 2. ArgoNeuT LAr-containment box, RF-shielding box, and cryogenic purification system, between the MINOS and MINERVA detectors.

successfully tested small LArTPC prototypes [4] the ArgoNeuT collaboration proposed Fermilab T962 [5], a joint NSF and DOE supported R&D project to build and operate a 170 liter (active volume) LArTPC detector, as shown in figure 1, and conduct a program of neutrino cross-section measurements, utilizing Fermilab's 270 kW neutrino beam. Of particular interest are the cross sections for charged current quasi-elastic (CCQE) neutrino scattering.

After testing on the surface the ArgoNeuT detector was installed in the NuMI Near Hall, 100 m below ground, and by the end of the run it had become sandwiched, as shown in figure 2, between the MINERVA and MINOS detectors. The MINOS detector provides momentum and charge sign measurements for energetic forward muons in ArgoNeuT events. Readout was triggered at the beam spill rate (0.5 Hz), thus recording all neutrino interactions in the active region of the detector.

In April 2009 the cryostat was sealed, pumped down and then filled with liquid Argon. A measurement of the statistical noise on the wires yielded an RMS of 1.4 ADC counts with negligible coherent noise. In pure Argon, 4 mm of track length for a minimum ionizing particle should have a mean signal amplitude of ~ 35 ADC counts in the Collection plane and ~ 17 ADC counts in the Induction plane. Thus a signal to noise of at least 10 to 1 was expected. However, during the Spring

Figure 3. Snowflake comprised of tracks from an ArgoNeuT event display, below.

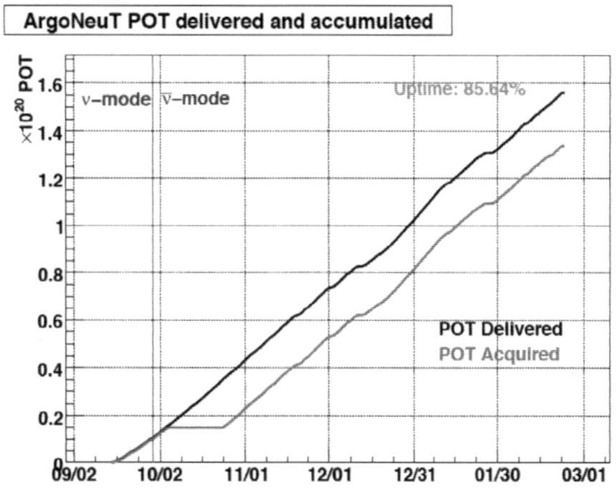

Figure 4. ArgoNeuT beam usage starting in the Fall of 2009. The beam was changed to anti-neutrinos in late Sept, and the majority of the data is in that mode.

2009 running, the electron lifetime was not long enough to approach this level of signal to noise over most of the drift region. Nevertheless, a number of clear neutrino interactions were identified and presented at conferences and meetings [6]. These interaction displays could appear in rather unexpected places, such as on Fermilab's 2009 Season's Greetings card, as shown in figure 3. Though not currently being analyzed, some of this early data may become useful in the future.

During the accelerator shutdown in summer 2009, the electron lifetime was improved, reaching ~ 750 μs as measured in an analysis to be described later but it was not possible to open the cryostat to address the electronics issues to be discussed later. In the subsequent running, except for a 2½-week period in October with a cryogenics problem, the experiment acquired data for 1.5×10^{19} protons on target (POT) in the neutrino mode and 1.2×10^{20} POT in the anti-neutrino mode, as shown in Figure 4. This running should yield in the fiducial volume of the TPC approximately 900 ν Charged Current (CC) interactions in the NuMI neutrino-beam configuration, and 4000 ν and 3500 $\bar{\nu}$ CC interactions in the antineutrino-beam configuration.

2. Electronics and raw data

The analog readout electronics built specifically for the ArgoNeuT detector included a dual jFET front-end integrating preamplifier followed by high- and low-pass filters, implemented on 16-channel

Figure 5. 16-channel preamplifier card with dual jFET front-end, pulse shaping high-pass and low-pass filters, and a cable driver.

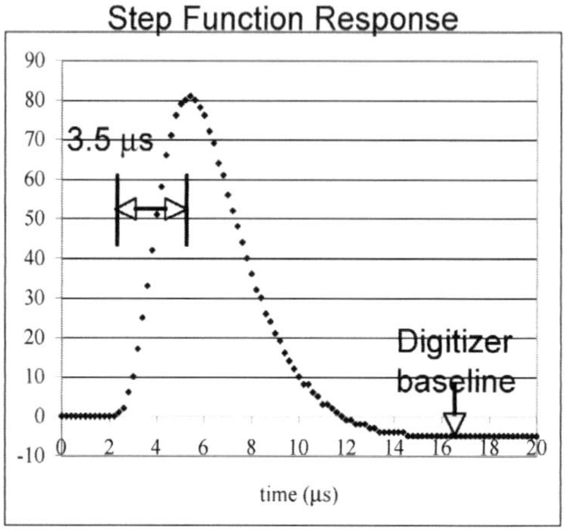

Figure 6. Response of the combined filter and digitizing electronics to a step-function signal at the integrator (a delta function signal on the wire)

cards, as shown in figure 5. These were mounted in a double-walled RF-shield cage enclosing the cards and the signal feed-through. Existing ADF2 32-channel digitizers [7] were borrowed from spares of the D0 experiment. With the limited number of available digitizer channels (480), 4 mm was chosen for both pitch and plane separation of the three TPC wire planes, Shield, Induction, and Collection, with the wire angles, 0°(vertical), and ±30°, respectively. Only the latter two planes were instrumented, each with 240 channels of readout. The signal on each wire was digitized at a 5 MHz rate for a total of 406 µs beginning 10 µs before the start of a 10 µs long NuMI beam spill. It takes ~310 µs to drift the 470 mm distance between the Cathode and Grid wire planes, in the 500 V/cm electric field established between them.

After pulse shaping, the signal for an ionizing track has a rise time of at least a few microseconds, but the capacitive coupling of the digitizer inputs, which could not be modified, were designed for signals ~100 times shorter. In addition, a series resistor compensating for gain differences between ADF2 channels modulated the effects of this capacitive coupling. For the combined shaping circuit and digitizer input, the step-function response, as shown in figure 6, had a peaking time of ~ 3.5 µs with a negative baseline shift of 7-10% of the signal amplitude with a decay constant of 50–100 µs.

There were a number of unique features of the electronics implementation for this LArTPC. Bias voltages were distributed via 1 MΩ resistors to each wire (Shield at -298 V, Induction at -18 V, Collection at +338 V) and decoupled with capacitors feeding the low-voltage signal cables via TPC mounted cards (see figure 1), submerged in the liquid Argon. Through testing in Fermilab's Materials Testing Station [8] we know that these components will not contaminate liquid Argon.

Without the need to carry a bias voltage, the cables and a new feedthrough could use mass-terminated connectors. We were disappointed to later find that the feedthrough had a number of flaws that caused ~5 dead or noisy channels, which could not be repaired. Also, there were no signals from 5 sequential wires in the Collection plane resulting in event displays that exhibit a noticeable gap in tracks crossing this region. The wires neighboring this region collect more charge than expected, indicating that these dead Collection plane wires did not receive bias voltage. No signal shape distortions were seen for the corresponding ionization in the Induction plane. Finally, the bias voltage cards had small un-insulated spots on traces where charge could be collected from outside the active TPC volume. This could generate, primarily on the last 48 wires of the Induction plane (right hand corner in displays), a strange signal with the wrong polarity. Though distracting, these signals are rare in events of interest, and are not present in the other view and thus do not reconstruct in 3D.

After the summer shutdown, coherent noise became more noticeable and reached a peak of about 1.0 ADC counts. As the data to be presented below will clearly indicate, none of the problem areas described here are serious enough to prevent ArgoNeuT from achieving its physics goals

3. Constructing hits

On each wire, track signals are separated from noise by application of a hit threshold. In the raw data, the negative portion of a bipolar signal in the Induction plane or the negative baseline shift of a signal in the Collection plane can cause the positive portion of a subsequent signal to drop below a hit threshold. To insure that a hit threshold is uniformly efficient, a Fast Fourier Transform (FFT) deconvolution was employed to minimize these effects. The deconvolution removes the baseline shift from the Collection plane signals and converts the raw signals in the Induction plane from bipolar to unipolar. Also, some improvement in the signal to noise ratio in the Induction plane can be obtained with this procedure.

This FFT deconvolution technique makes the assumption that every true signal on a particular wire can be resolved into components that to first order have the same base signal shape in time. Ionization electrons follow different paths as they drift through the wire planes, contributing different shapes to the overall signal. The quality of the deconvolution, therefore, depends on the details of the signal formation, which are different in the two readout planes. The majority of the signal on a Collection plane wire is due to ionization from a projected track length equal to the wire pitch, centered on the wire, and generated rather close to the wire. Therefore, the invariance of the component shapes contributing to Collection plane signals is a particularly good approximation.

Signals on Induction plane wires, however, contain significant components from ionization that will be closer to a neighboring wire when passing through the plane. That distant charge will make contributions to the signal with a shape different from that of charge passing closer to the wire. These shape differences (the utility of which is minimized by ionization fluctuations) are washed out by the frequency response of the electronic shaping circuits that favor better signal to noise. The variations in the shape of the digitized Induction plane bipolar signals are not large. For example, the ratio of the positive to negative amplitude of the bipolar signal generated by isolated tracks rarely varies by more than ±15% from the base signal shape. These variations can generate deconvolution artifacts, however, they are small enough that a common signal shape is an adequate first approximation for hit finding. A better approximation to the signal shape is possible, once the angles, ionization, and density of the tracks are known.

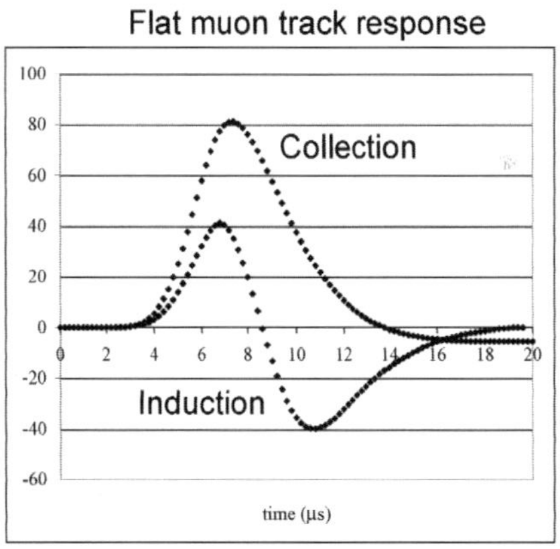

Figure 7. Base signal shapes for the Collection and Induction planes as determined from muon tracks parallel to the wire planes.

The base signal shapes for the Collection plane and Induction plane wires, as shown in figure 7, were derived from thousands of spills containing a long muon track nearly parallel to the wire planes. Track segments with signals with a wider pulse than expected, indicating the presence of a delta ray, were not included in the signal shape determination. As a side benefit of using muon tracks for the signal shape on each wire, the deconvolution can be adjusted to result in signal amplitudes on each

Journal of Physics: Conference Series **308** (2011) 012014 doi:10.1088/1742-6596/308/1/012014

wire normalized to a common ionization scale. These muon track signals should have the shortest rise time of any ionization deposition by a track.

Another aspect of the FFT technique used here is the frequency filter. The filter used in the Collection plane was based on the power spectrum of the base signal. Due to the slowly varying baseline shift, this filter must accept low frequencies. In the Induction plane, baseline variations are quite small so that a filter with a low frequency cut-off could be used.

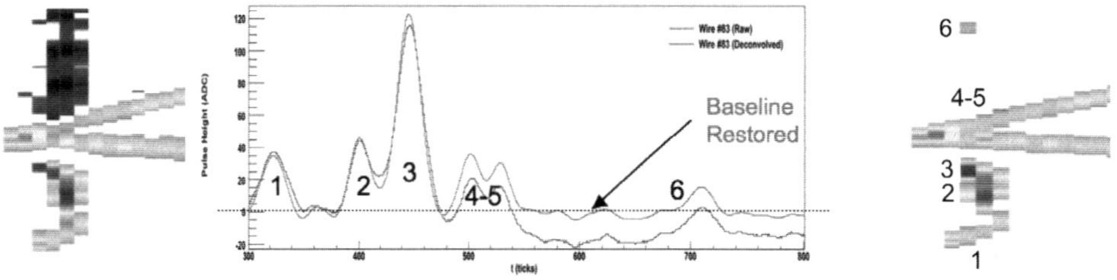

Figure 8. Collection plane signals are shown in the vertex region of a two-prong neutrino interaction. The raw data display (left panel) exhibits pulse heights far below the baseline (dark blue). Signals on wire 83 (center panel) before and after (black/red) the deconvolution, and the display (right panel) following deconvolution, wire 83 indicated by the vertical line.

A few examples will illustrate the capabilities of the signal deconvolution procedure. To demonstrate the effects of the signal deconvolution in the Collection plane, the vertex region of a 2-prong neutrino interaction is analyzed in detail, as shown in figure 8. The signal peak #3, likely due to a photon conversion, has ionization ~ 4 times the minimum and the raw data has depressed the baseline, lowering the pulse heights of hits 4-5 and 6. The deconvolution has restored the baseline, raising the pulse heights of peaks 4-5 to a level consistent with minimum ionizing tracks, and hit 6 is

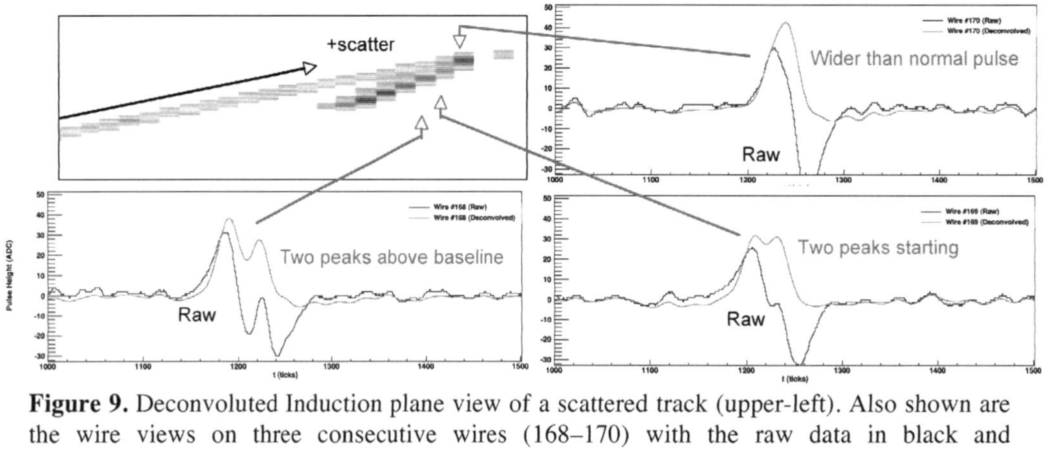

Figure 9. Deconvoluted Induction plane view of a scattered track (upper-left). Also shown are the wire views on three consecutive wires (168–170) with the raw data in black and deconvoluted signals in red. Note: Induction plane pulse heights are typically ½ those of the Collection plane.

now above the hit threshold. In the same event the upper track scatters backward and stops in the active volume, as shown in the upper-left panel of figure 9. Also shown are wire displays for three consecutive Induction plane wires, illustrating that the deconvolution procedure has extracted the signal of the incoming track from the negative portion of the raw bipolar signal of the scattered track.

4. Cluster and track finding

The hit finding described above and the description of the 2D and 3D reconstruction that follows have their software codes packaged in LArSoft [9], a Fermilab supported framework for the LArTPC activities in the US. Once hits are found in the 2D projections, they are grouped in each view into "clusters" using a DBSCAN [10] algorithm. A Hough Transform [11] algorithm applied to the hits within a cluster find those hits that lay along straight-line tracks. The minimum number of missing hits and allowed deviation of a hit from the track are adjustable parameters, initially set at values optimized for efficiency. Both wire planes cover the active volume of the TPC completely, thus a track identified in the one view that enters or leaves the TPC active volume will have a corresponding endpoint with a common drift coordinate in the other view. This provides nearly unambiguous matching of the tracks in the two views. Through-going tracks entering on one boundary and leaving through another have two such corresponding endpoints in 3D.

A neutrino interactions in the material outside the active TPC volume generate all species of particle, however, nearly all of those that enter and then leave the active region of the TPC are muons.

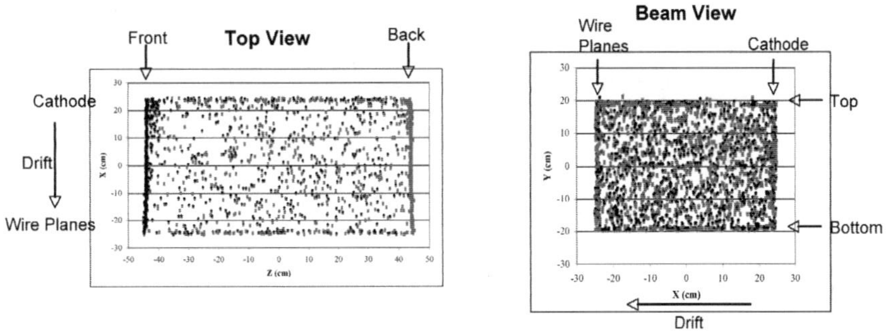

Figure 10. Reconstructed endpoints for muon tracks that enter and leave the TPC active volume, upstream points plotted in black, and downstream points in red. Field distortions (see text) are responsible for the inward tilt for the front and back.

As a first quality check on the reconstruction, about 60,000 beam spills were used to identify a large sample of straight tracks where both endpoints could be reconstructed in 3D. The endpoints should mark the boundaries of the active volume, as shown in figure 10. In the frame labelled Top View, many tracks enter the front (upstream) face or leave the back (downstream) face, at Z = ±45 cm of the TPC. Also, the endpoints show that there are tracks that leave or enter through the Cathode or wire planes at X = ±23.5 cm. Unexpectedly, the boundaries of the front and back surfaces of the TPC look further apart near the wire planes than they do near the Cathode (longest drifts). However, a map of the electric field reveals that some field lines at the edges of the cathode do not end on the wire planes.

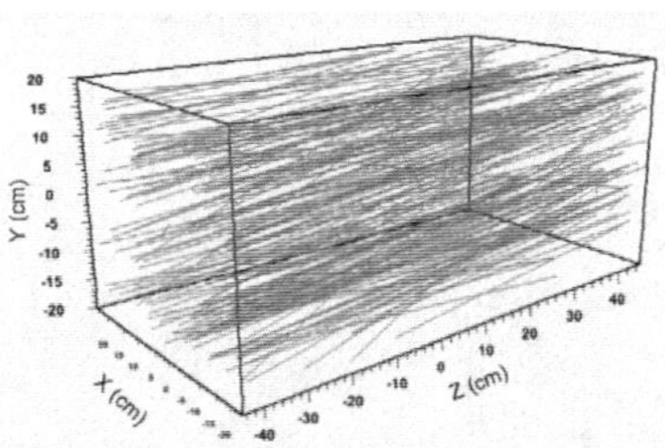

Figure 11. Muon tracks reconstructed in 3D by matching all 2D hits between views, and superimposing tracks from many spills.

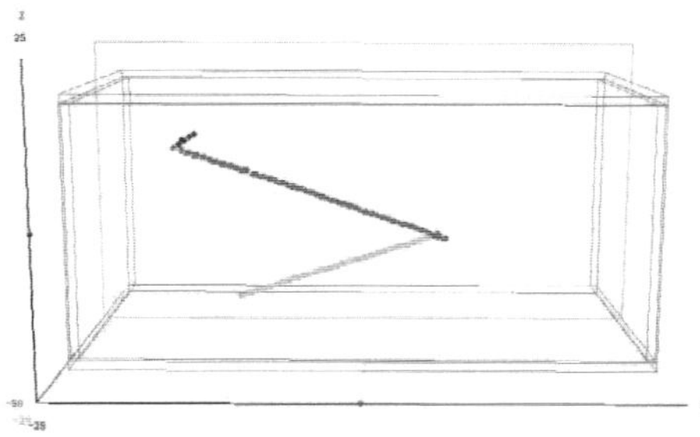

Figure 12. The 2-prong neutrino event shown earlier is displayed with the tracks reconstructed in 3D.

A region about 1 cm inboard from the field-cage is therefore inactive, except close to the wire planes where field lines from the two nearest field-cage electrodes can carry charge to the wires.

The points that lie well within the boundaries of the TPC in the Top View must enter or leave though its top or bottom. This is confirmed in the plot labelled Beam View, where there are concentrations of endpoints on the top and bottom of the TPC. A few reconstructed endpoints appear to lie just outside the top surface of the TPC at Y = + 19.5 cm. The y-coordinate is the least well known of the three coordinates. For endpoints off by 1-2 wires in the two views the y-coordinate can reconstruct up to 16 mm from the true endpoint, consistent with the deviations of those few tracks from the nominal top boundary.

Another approach to 3D reconstruction matches all hits between the two views based on a common drift time. The algorithm works particularly well on isolated tracks such as muons, as shown in Figure 11. Low multiplicity neutrino interactions, such as the two-prong event described earlier in Section 3, are reconstructed with the same algorithm in 3D, as shown in figure 12. Work continues to develop code that will reconstruct tracks in nearly all events in 3D.

5. Electron lifetime from muon tracks.

Using the straight track reconstruction algorithms described above, a selection of minimum ionizing muon tracks at least 40 cm long from ~70,000 beam spills (~ 2 days of running) enabled a measurement of the electron lifetime in the liquid Argon. The deconvolution algorithm yielded an

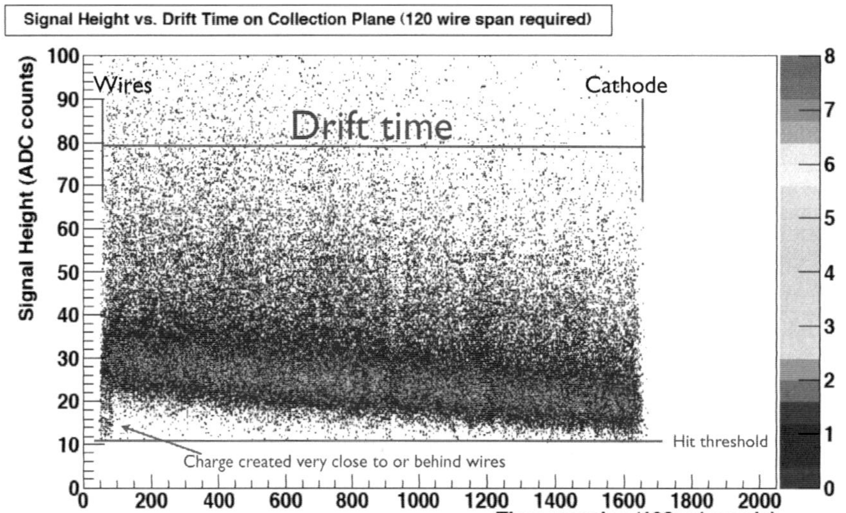

Figure 13. From a single run, the pulse height spectra vs. drift time for tracks reconstructed in the Collection view with at least 120 hits. Electron lifetime effects on the most probable pulse height are clearly demonstrated.

ionization measurement of every hit, normalized in ADC counts to an average of the signal pulse heights in each plane. A plot of the ionization of each hit vs. its drift time, as shown in figure 13, at a fixed drift time exhibits the expected Landau (like) distribution, albeit with smearing induced by instrumental effects and muon track variations in momentum and angle, however, these effects do not vary significantly over the full drift time.

The pulse height spectra in 1.6 μs time slices were selected and excellent fits to a Landau plus a Gaussian-smearing distribution were obtained. From these fits in the Collection plane, excluding drift times near the wire planes or the Cathode, the most probable pulse height plotted vs. drift time, as

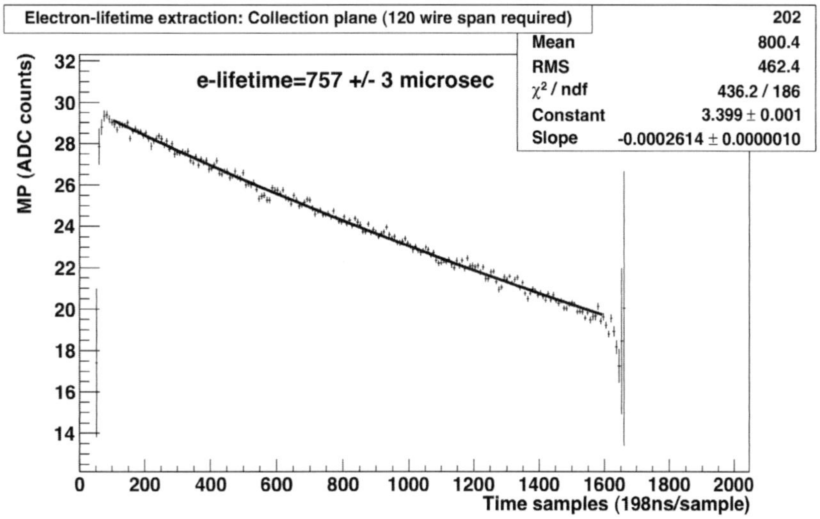

Figure 14. The plotted points are from fits to the most probable pulse height in 1.6 μs slices from the previous figure. The exponential fit for the electron lifetime was restricted to the central time samples where the pulse heights are unaffected by the hit threshold or other instrumental effects. The Slope parameter is the inverse of the lifetime (in time-samples).

shown in Figure 14, exhibits the expected exponential fall with a fitted lifetime of 757 ± 3 μs (statistical error). In the Induction plane a fit yields a similar value when hits affected by a threshold cut are eliminated. The rather poor fit chi-squared, 436 for 186 degrees of freedom, suggests that the fitted mean pulse heights have additional variations not included in the error from the fit in the most probable pulse height from each time slice. One possible source is coherent noise correlated with the common start for the beam spill and the drift time clock.

6. Identification of neutrino interactions

With the muon track 3D reconstruction as described above, a filter removed spills that did not show data consistent with a neutrino interaction. Spills were eliminated if they showed no tracks or just a single straight muon track that enters and leaves the fiducial volume. After removing the ionization of any single muon track, spills were retained for later scanning with more than the ionization equivalent of a 40 MeV energy deposition. During nominal neutrino beam conditions, the filter selects ~3% of the original beam spills. Applying additional cuts in a visual scan, about 25% of these spill records, or less than 0.7% of the original sample, were identified as having a candidate neutrino interaction, and about half of these have unmistakable neutrino interactions with more than 1 charged track at an interaction vertex. The additional cuts applied in the visual scan address cases where there are multiple tracks, most which intersect the front or sidewalls of the TPC, and likely originate outside the TPC. Many of the criteria applied in the visual scan can be coded to reduce the scanning load while others would require more sophisticated reconstruction algorithms to maintain the efficiency in the neutrino event selection. These will be implemented in the near future.

The discovery of CP violation in electron neutrino appearance experiments will require the excellent pattern recognition and particle identification capabilities of the LArTPC technology. With the ArgoNeuT data we hope to make the first measurement of the electron neutrino content of the NuMI beam in the neutrino and anti-neutrino modes. Over the course of the run, random visual scans identified 1 or 2 events that could be electron neutrino interactions, but they were not definitive. In the visual scan of about 450 of spill records selected by the filter from ~15,000 beam spills (about 12 hours of neutrino beam running) an event was found that displayed all the expected characteristics of a charged-current quasi-elastic interaction of an electron neutrino.

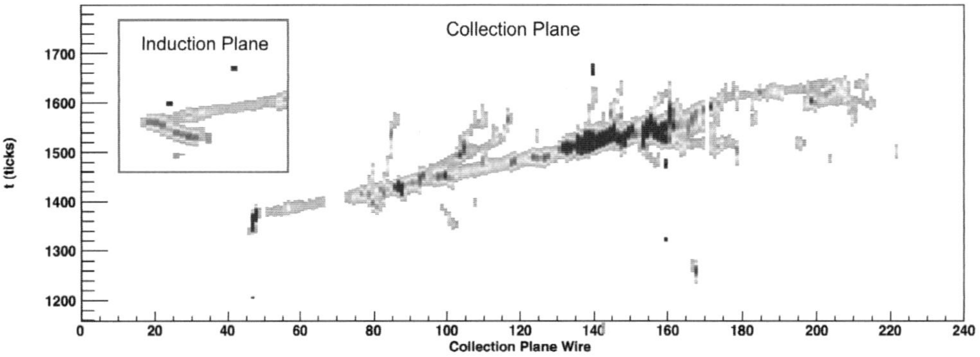

Figure 15. A candidate for an electron neutrino CCQE event shown in the ArgoNeuT Collection plane with one short and heavily ionizing track, and a minimum ionizing track about ~ 11 cm long (big gap is due to 5 dead wires) that develops into an electromagnetic shower. In the Induction plane (inset), the short (~ 5 cm) stopping track and an uninterrupted minimum ionizing track are clear.

The event in the Collection plane, as shown in figure 15, has a two-prong vertex with one minimum ionizing track about 11 cm long (gap is due to 5 consecutive dead wires) that develops into a collinear electromagnetic shower, thus is likely an electron, and a short highly ionizing track that stops in the Argon. The stopping track, more easily seen in the Induction plane (inset), is about 5 cm long and is likely a proton with a kinetic energy of about 85 MeV. Also, the event exhibits no other electromagnetic activity that could be associated with an additional photon of significant energy, e.g., from π^0 or η decays. We continue to scan for additional electron neutrino events. Analysis of the ionization will provide a measurement of the electromagnetic shower energy deposited in the active volume and verification that the stopping track is a proton.

7. Conclusion

The ArgoNeuT collaboration has constructed a 175-liter Liquid Argon TPC, readout by 60°-stereo wire planes, each with 240 wires on a 4-mm pitch. The electronics exhibited an average statistical noise of 1.4 ADC counts with minimum ionizing tracks producing at the midpoint of the drift distance a mean signal amplitude per 4 mm of track length of 31.5 ADC counts in the Collection plane, and 15.5 ADC counts in the Induction plane.

The experiment recently completed a 6-month run at Fermilab in the NuMI beamline in a low-energy configuration providing neutrinos with a mean energy of 3.5 GeV. This run should yield within the fiducial volume ~ 900 ν Charged Current (CC) events in the neutrino beam mode, and ~ 4000 ν and ~ 3500 $\bar{\nu}$ CC events in the antineutrino beam mode. Work is in progress to obtain the momentum and sign of most muons by linking tracks in the LArTPC with the MINOS near detector. The energy determination for electromagnetic showers will be investigated through reconstruction of π^0 decays. Events have been seen with the characteristics of ν_e scattering indicating that measurement of the ν_e fraction in the NuMI beam at the ~ 1% level should be possible.

Acknowledgements
The author thanks the GLA2010 conference organizers for the invitation to talk at the conference, and my ArgoNeuT colleagues for help in preparing the data presented here.

References

[1] Cennini P, et al. 1994 *Nucl. Instr. & Meth.* A **345** 230; Arneodo F, et al. 2003 *Nucl. Instr. & Meth.* A **498** 292-311; Amerio S, et al. 2004 *Nucl. Instr. & Meth.* A **527** 329.
[2] Fleming B T 2009 Liquid Argon TPC for LBNE, DUSEL workshop, Homestake, SD, BFleming-DUSELWorkshopFall2009.pdf
[3] Rameika G 2009 Liquid Argon and the Long Baseline Neutrino Experiment (LBNE), Fermilab Director's Review, Rameika_LBNEoverview_LArDirRevNov2009Final.pdf; Baller B 2010 LBNE Liquid Argon option, *in these Proceedings*.
[4] Curioni A, et al. 2008 The Yale liquid argon time projection chamber, *arXiv*:0804.0415; Bromberg C 2009 Review of LAr Electronics R&D at Fermilab, http://lartpc-docdb.fnal.gov/cgi-bin/ShowDocument?docid=412 .
[5] ArgoNeuT: http://t962.fnal.gov/
[6] See for example, Soderberg M 2009 ArgoNeuT:A Liquid Argon Time Projection Chamber Test in the NuMI Beamline, *Proceedings of the DPF-2009 Conference*, Detroit, MI
[7] Bromberg C and Edmunds D 2006 A Modular Readout System for A Small Liquid Argon TPC http://lartpc-docdb.fnal.gov/cgi-bin/ShowDocument?docid=594.
[8] Finley D, et al. 2006 Work at FNAL to acheive long electron drift time in Argon, *FERMILAB-TM-2385-E*; Rebel B 2010 Results from the materials test stand and status of LAPD, *in these Proceedings*.
[9] Rebel B 2009 Introduction to LArSoft,vmsstreamer1.fnal.gov/VMS_Site_03/Lectures/MicroBoone/presentations/091015Rebel.pdf
[10] Ester M, et al. 1996 A Density-Based Algorithm for Discovering Clusters in Large Spatial Databases with Noise", *The Second International Conference on Knowledge Discovery and Data Mining (KDD-96)*, Portland, Oregon, USA.
[11] Hough P V C 1959 Machine Analysis of Bubble Chamber Pictures, *Proc. Int. Conf. on High Energy Accelerators and Instrumentation*.

Journal of Physics: Conference Series **308** (2011) 012015 doi:10.1088/1742-6596/308/1/012015

Neutrino Event Reconstruction in a Liquid Argon TPC

Gary Barker

Department of Physics, University of Warwick, Gibbet Hill Road, Coventry CV4 7AL

E-mail: G.J.Barker@Warwick.ac.uk

Abstract. We present some preliminary findings and results from activities in Europe and the USA working towards an automated, algorithmic, reconstruction of particle interactions in liquid argon time projection chambers.

1. Introduction

The capability of Liquid argon (LAr) to record particle interactions with a density of detail comparable to the bubble chamber experiments is well recognised. The challenge of reconstructing these interactions to extract physics measurements has been partially met over the years but a fully automated, algorithmic, approach has yet to develop. The reasons for this are partly historical e.g.

- Bubble-chamber event rates/volumes were low by modern standards and hence there was no strong need for an automated solution.
- Available computing power limited what was possible.
- Collider experiments with electronic readout began to dominate the subject.

Tracking algorithms have therefore developed, optimised for reconstructing the sparse, but high precision, space points measured by typical collider experiments (see Fig.1).

Other reasons why automated reconstruction in liquid argon is still developing relate to the fact that it is not such a simple task. Reasons for this include:

- Events are topologically rather complicated (see Fig.2). Tracks and showers develop side-by-side in the same detector volume - compare this to the collider experiment where showers develop outside of the inner tracker volumes (see Fig.1).
- Unlike in the collider format, interactions in LAr can be initiated anywhere throughout the volume making knowledge of the start-point of events uncertain. This is a hurdle for any algorithmic approach.
- The high density of energy deposits (or 'hits') in LAr reveal a tremendous level of detail (vertices, kinks, delta rays etc) presenting challenges for efficient reconstruction.
- Multiple scattering occurs continuously throughout the volume and not just at well defined boundaries between detector elements.

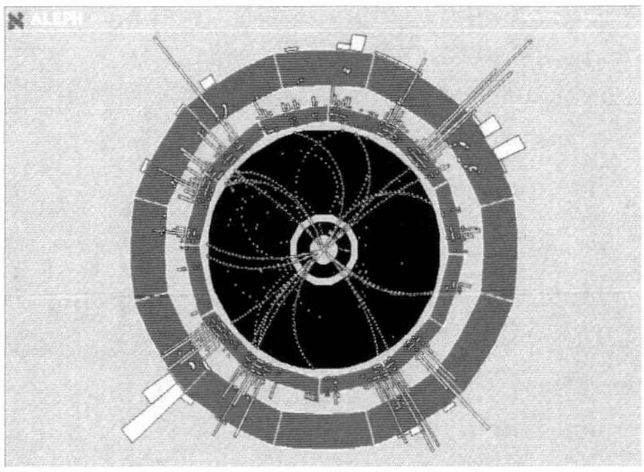

Figure 1. An event display from the ALEPH experiment at the LEP. From[1].

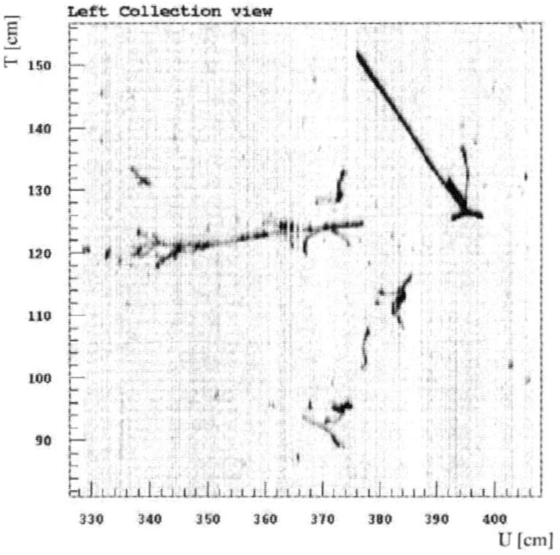

Figure 2. Observation of a π^0 decay, produced in the ICARUS T600 detector. From [2].

2. Previous Work

Early attempts to automate reconstruction in bubble chamber data were led by the development of the Hough Transform technique [3] which later found wide application as a feature extraction method for image analysis. More recently, the full reconstruction of interaction events in a LAr TPC has been demonstrated by the ICARUS programme. The full chain of hit reconstruction, hit-clustering to form track objects and track trajectory fitting in 2D and 3D have been demonstrated (see e.g. [4]). However these results have relied on some degree of visual scanning and selection before applying algorithms. Recently, with the imminent start-up of ICARUS T600 in the CNGS beam and data taking with LAr TPC prototypes in Europe and the USA ramping up (e.g. ArDM and ArgoNeut respectively), interest has again turned to address the issue of a fully automated event reconstruction.

3. Current Status

The physics programme for any future large LAr TPC demands that: (a) hits are classified into 'track-like' and 'shower-like' objects ; (b) particle tracks (e.g. muons, pions, protons) and showers (e.g. electrons and photons) are identified. Reconstructed quantities such as particle trajectories, dE/dx, event topologies, kinematics etc will enable these objectives to be met but their accurate reconstruction depends crucially on the correct assignment of hits to a common event feature i.e. hit-clustering. Recognising this, recent strategies for automatic event reconstruction have centred on the development of clustering algorithms to define groups of hits likely to comprise either a particle track or shower from which a track fitting algorithm or some higher-level shower algorithm (to extract shower energy, shape etc) can be applied as appropriate. As alluded to in the introduction, this task can be significantly simplified if additional information about interaction vertex locations, kinks etc is available. Such 'key-point' detection in event records is an area under active study for use in conjunction with clustering and fitting methods in order to boost performance. In what follows, we briefly summarise some of the studies underway in these areas.

3.1. Density- Based Clustering: DBSCAN

The aim of a hit clustering algorithm is to group together all hits which originate from a common parent while keeping to a minimum, contamination of the group by hits that originate from elsewhere. The DBSCAN algorithm [5] has recently been applied to ArgoNeuT data. This is an example of density-based clustering where clusters are identified as regions of high hit density sitting in an environment of lower density which work by demanding that the region in space around each hit in the cluster (defined by a radius ϵ) contains at least N_{min} other hits. Figure 3 shows two views of a neutrino interaction from the two readout planes of the ArgoNeuT detector. The colours depict the preliminary results [6] of applying DBSCAN – hits of the same colour have been grouped by the algorithm into the same cluster. The performance is promising although close inspection of the event displays shows that sometimes hit distributions that belong to the same cluster can be split into smaller clusters. In order to help optimise the performance studies of the package, Ordering Points To Identify the Clustering Structure (OPTICS) [7], have begun which stores hits in a convenient ordering with which to investigate optimal ϵ-clustering scales [8].

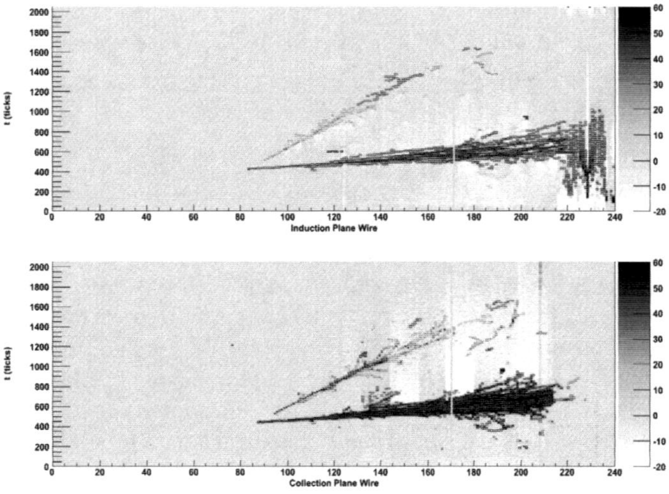

Figure 3. DBSCAN applied to an interaction in the ArgoNeuT detector. Drift time is the y-axis, readout wire number the x-axis and the greyscale represents the size of energy deposit recorded.

3.2. Key-Point Detection

The reconstruction task in LAr should be simplified if the location of certain 'key-points' are known e.g. interaction vertices, start/end points of tracks, delta-electron kinks etc. This is illustrated in Fig. 4 which shows that the DBSCAN algorithm can have difficulty grouping hits into separate track objects around an interaction vertex. If the location of the vertex point was known a-priori, the hits in the immediate location of the vertex could be removed from the clustering step allowing the algorithm to find correctly two, separate, clusters. In addition, Fig. 5 highlights that a straight line Hough-transform track fitting algorithm returns no information on the start and end-points of tracks and so these must be estimated by other means which are subject to errors.

These conclusions have motivated a study based on the Harris-Stephens operator [10] which numerically highlights features by computing the differential of pixel response in the event image and noting that these are large in all directions for something like a vertex, while large mainly in one direction for something like a track end-point. Some preliminary results are shown in Fig 6 and a quantitative analysis of the performance finds that in simulated quasi-elastic

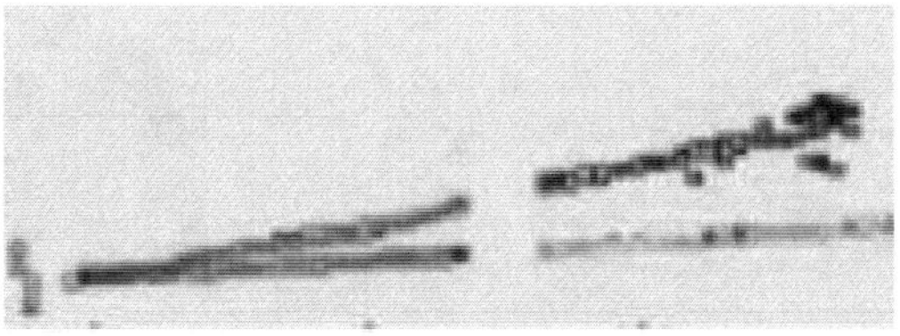

Figure 4. DBSCAN clustering around a vertex point.

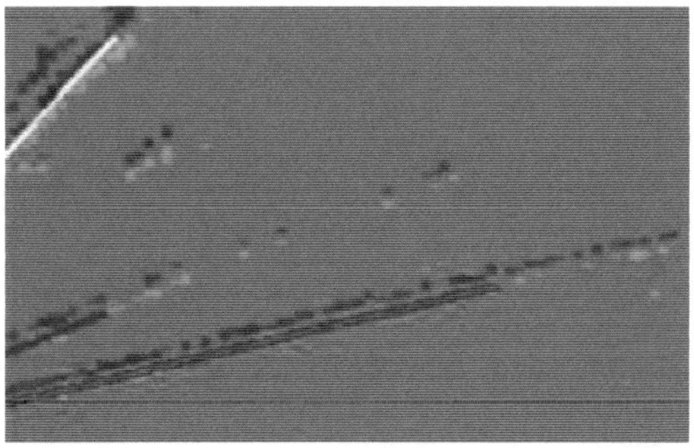

Figure 5. Result of a Hough-transform based track fitting algorithm around the track end-point [9]

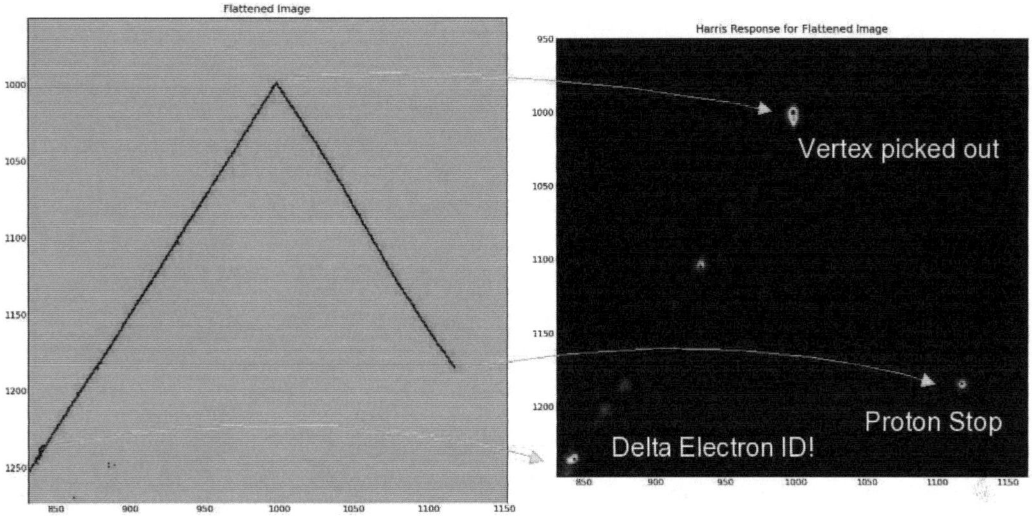

Figure 6. The result of a Harris-Stephens key-point algorithm applied to a simulated charged current neutrino interaction in LAr.

ν_μ interactions in LAr at $E_{\nu_\mu} = 0.77$ GeV, vertex and both the start and end-points of tracks can be identified in 3D with $\sim 85\%$ efficiency [8].

4. Conclusion

Activity towards the goal of automated reconstruction in LAr TPC's has begun in Europe and the USA and preliminary results are beginning to emerge from this effort. A central issue is that of hit clustering, for which the DBSCAN algorithm has returned some promising results and early work has identified the advantages of using this in combination with event key-point reconstruction – for which the Harris-Stephens method has been successfully applied. The work reported on here is still very much in the preliminary stages. Plans for future development include investigating 'trained algorithms' such as neural networks and to continue investigating existing techniques from the disciplines of Machine Vision and Image Processing.

Acknowledgements

I wish to thank the following colleagues for useful discussions and access to results: Andrew Bennieston, Ben Morgan, Yorck Ramachers and Dan Roythorne in Warwick and members of the LArSoft collaboration, especially Brian Rebel at Fermilab.

References

[1] Event display pictures available from: `http://aleph.web.cern.ch/aleph/`
[2] A. Ankowski *et al*, ICARUS Collab. 2008 hep-ex:0812.2373
[3] Hough 1959 *Proc. Conf High Energy Acc. Instr., CERN*
[4] ICARUS Collab. 2006 *Phys. Rev.* D **74** 112001
[5] Sander *et al* 1998 *Data Mining and Knowledge Discovery* **2** 169-194
[6] Results courtesy of Kinga Partyka, Brian Rebel and the LarSoft Collab.
[7] M. Ankerst *et al* 1999 *ACM SIGMOID Int. Conf. on Management of Data* 49-60
[8] Results courtesy of Ben Morgan, Dan Roythorne and the University of Warwick group
[9] Results courtesy of Joshua Spitz, Brian Rebel and the LarSoft Collab.
[10] C. Harris, M. Stephens 1988 *Proc. 4th Alvey Vision Conf.* 147-151

Stable operation with gain of a double phase Liquid Argon LEM-TPC with a 1 mm thick segmented LEM

F. Resnati, A. Badertscher, A. Curioni, S. Horikawa, L. Knecht, D. Lussi, A. Marchionni, G. Natterer, A. Rubbia, T. Viant
ETH Zurich, Institute for Particle Physics, CH-8093 Zurich, Switzerland

E-mail: filippo.resnati@cern.ch

Abstract. In this paper we present results from a test of a small Liquid Argon Large Electron Multiplier Time Projection Chamber (LAr LEM-TPC). This detector concept provides a 3D-tracking and calorimetric device capable of charge amplification, suited for next-generation neutrino detectors and possibly direct Dark Matter searches. During a test of a 3 lt chamber equipped with a 10×10 cm^2 readout, cosmic muon data was recorded during three weeks of data taking. A maximum gain of 6.5 was achieved and the liquid argon was kept pure enough to ensure 20 cm drift (O(ppb) O_2 equivalent).

1. Introduction

The Liquid Argon Large Electron Multiplier Time Projection Chamber (LAr LEM-TPC) is a novel kind of double phase (liquid-vapor) TPC suited for the realization of next generation detectors for neutrino physics and proton decay searches [1], considered as well as for direct Dark Matter detection [2].

The ionization charge is drifted towards the liquid-vapor interface, where it is extracted by means of an appropriate electric field produced by two grids. In the vapor phase, Townsend avalanche takes place in the high electric field regions confined in the LEM holes, similar to the situation of the Gas Electron Multiplier (GEM) [3]. The charge is collected by a set of segmented electrodes, on which the induced signals are recorded.

The resulting charge multiplication yields a significant improvement of the S/N ratio compared to the single phase liquid argon TPC [4]. The very high quality of the images and the dE/dx measurement are key points to address the three-dimensional geometry of the ionizing events and to identify the involved particles. In addition the charge amplification technique is compatible with very long drifts, because it allows to compensate for the diffusion of the electron cloud and the potential charge loss due to electronegative impurities diluted in the liquid argon.

The detector described in this paper was developed during the last three years and different tests of the charge readout devices were performed [4, 5]. Here we present the results obtained with a 1 mm thick Large Electron Multiplier (LEM) with an area of 10×10 cm^2.

2. Description of the detector

The picture of the detector is presented in figure 1. A set of field shapers, spaced every 5 mm, defines the drift volume, 20 cm long and with an area of 10×10 cm^2. An electric field of the order of 0.5-1.0 kV/cm is applied to drift electrons vertically towards the liquid surface. The electrons are extracted from the liquid to the vapor phase by means of a set of two grids parallel to the liquid surface. Each grid is made out of 100 μm diameter stainless steel wires, spaced 5 mm. During operation the level of the liquid argon is precisely set in between the two grids and the voltages are tuned to obtain an electric field of about 3 kV/cm just below the liquid surface (see the schematics in figure 2).

Figure 1. Picture of the detector.

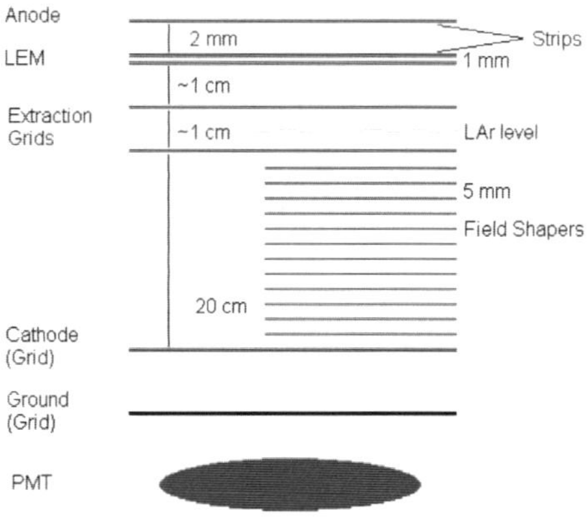

Figure 2. Schematic representation of the detector.

The charge readout apparatus (LEM and anode) is installed in the vapor phase 1 cm above the top grid. The LEM is an electron multiplier with millimeter-size holes and a thickness of about 1 mm. It is a coarser and more robust extrapolation of the Gas Electron Multiplier (GEM) [3]. The LEM is produced with standard Printed Circuit Board (PCB) techniques and the holes are mechanically drilled through the PCB plate. The drifting electrons are driven into the LEM holes where a high electric field (larger than 25 kV/cm) allows to trigger the Townsend avalanche – a detailed description of the working principle can be found in [5]. When the amplified charge leaves the holes, it is collected partially on the anode, installed 2 mm above the LEM, and partially on the top electrode of the LEM. The sharing of the charge between the two electrodes can be adjusted changing the electric field between the anode and the LEM. Each electrode is segmented into 16 strips 6 mm wide and the two sets of strips are orthogonal to allow the reconstruction of the X-Y coordinates of the ionizing event, while the Z coordinate is given by the drift velocity multiplied by the drift time. In table 1 the geometry parameters of the LEM are summarized.

The detector is also equipped with a photomultiplier tube (PMT) that provides the time reference of the events and additional information about the purity of the argon, since the characteristic time of the radiative de-excitation of argon excimer states depends on the concentration (down to the ppm level) of impurities such as H_2O, O_2, N_2 [6, 7]. A more sensitive way to monitor the purity of liquid argon is to measure the charge loss along the drift path (i.e. as a function of the drift time). To ensure a drift as long as 20 cm the liquid argon purity must be of the order of 1 ppb (oxygen equivalent) or less, since the oxygen content ($[O_2]$) is known to be related to the drifting electron lifetime (τ_{e^-}) by $[O_2] \times \tau_{e^-} \approx 300\ \mu$s ppb [8].

Material	copper-cladded FR4
Active area	10x10 cm^2
PCB thickness	1.0 mm
Copper thickness	18 μm
Hole diameter	500 μm
Hole rim	100 μm
Hole pitch	800 μm
Strip width	6 mm
# strips	16

Table 1. Geometry parameter of the described LEM.

Two methods are used to achieve the needed purity: (1) during the filling procedure the liquid argon is filtered with a custom made cartridge loaded with reduced copper-oxyde powder and (2) to maintain the purity during operation, liquid argon is evaporated, pushed by a bellows pump through a commercial SAES getter[1], recondensed through a long serpentine immersed in a liquid argon open bath, and again reintroduced as liquid inside the detector. A proof of the effectiveness of the purification system is given in this paper. For a detailed description of the recirculation circuit see [4].

The detector main vessel was evacuated with a residual pressure smaller than 5×10^{-6} mbar in order to favour the outgassing of the materials and check the vacuum tightness of the system. The filling procedure consists in recirculating pure argon gas through the getter while cooling down the vessel in an open bath of liquid argon. The presence of gas ensures a fast and uniform thermalization of the detector and the recirculation system traps the outgassed molecules into the getter. Finally the detector is filled with liquid argon adjusting the level in the middle of the two extraction grids, that are used also as a capacitive level meter.

3. Cosmic muon reconstruction and argon purity

Cosmic muon events are used to characterize the detector because (1) they release a defined amount of charge per unit path so they can be used as calibration source and (2) they cross the entire volume allowing the evaluation of the liquid argon purity. A typical cosmic muon track with a delta ray emission is presented in figure 3, where the channel number and the drift time are plotted on the X-axis and Y-axis respectively and the gray scale is proportional to the signal amplitude, sampled every 400 ns (2.5 MHz). The two images refer to two orthogonal views of the same ionizing event. We also present as examples two signal waveforms, on the LEM strip number #6 and anode #13.

Cosmic muon tracks are used to determine the effective gain and the drifting electron lifetime. Each event is three-dimensionally reconstructed in order to evaluate the track length (Δl) below each strip and the charge (ΔQ) collected on each channel. The ratio $\Delta Q/\Delta l$, proportional to the energy loss by the muon in liquid argon per unit length, is the relevant quantity used in the analysis.

In order to evaluate the drifting electron lifetime, the drift volume is divided in equal time slices along the drift. A Gauss-convoluted Landau function is fitted to the distribution of $\Delta Q/\Delta l$, as shown in figure 4. The plot of the position of the Gauss-convoluted Landau maximum as a function of the drift time (drift slices) shows the exponential charge attenuation due to electronegative impurities.

The longest drifting electron lifetime (about 500 μs) is achieved at the beginning of the run. The corresponding O_2 equivalent content is about 0.6 ppb. As shown in figure 5 the liquid argon purity degrades with a rate of about 0.6 ppb/day, presumably due to the outgassing of

[1] SAES Pure Gas Inc., MicroTorr MC400.

Journal of Physics: Conference Series **308** (2011) 012016 doi:10.1088/1742-6596/308/1/012016

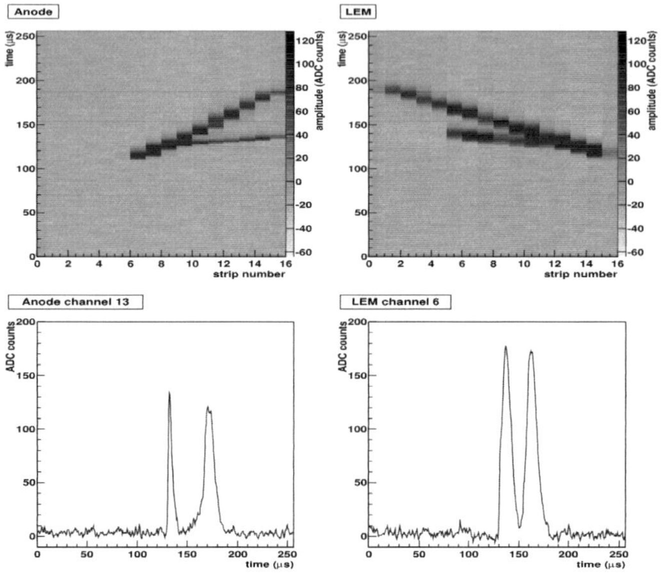

Figure 3. Event display of a cosmic muon producing a delta ray. The two figures on top show the signals from the same event recorded on anode electrodes (left) and on the LEM electrodes (right). Below two examples of signal shapes (on the LEM strip number #6 and anode #13) are shown.

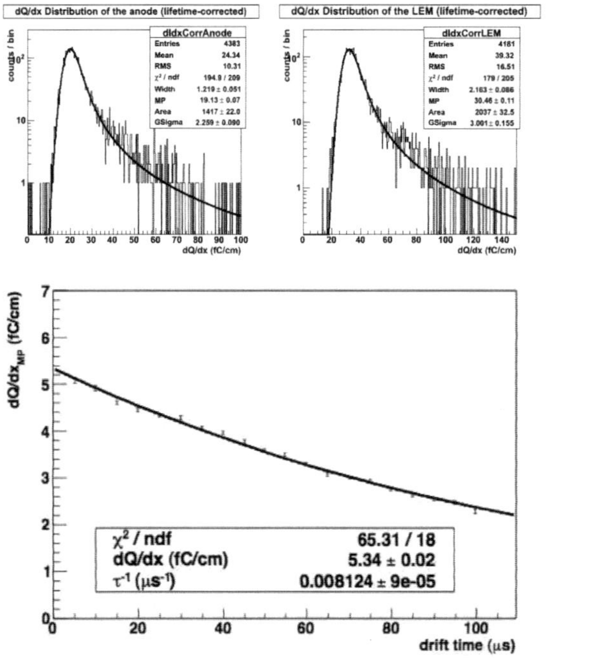

Figure 4. Top: fit of the Gauss-convoluted Landau function to the $\Delta Q/\Delta l$ distribution. Both the anode electrode view and the LEM electrode view are shown. Bottom: most probable value evaluated from the fit of the Gauss-convoluted Landau function versus the drift time.

the materials inside the detector or small leaks to the atmosphere. The purity is re-established by recirculating the argon gas through the purification cartridge and re-condensing the clean argon into the detector volume.

4. Charge amplification and effective gain

The *effective* gain is defined as the sum of the collected charge per unit length on the anode and LEM electrodes (corrected for the drifting electron lifetime) divided by the charge per unit length released in liquid argon by a minimum ionizing muon in an electric field of 500 V/cm.

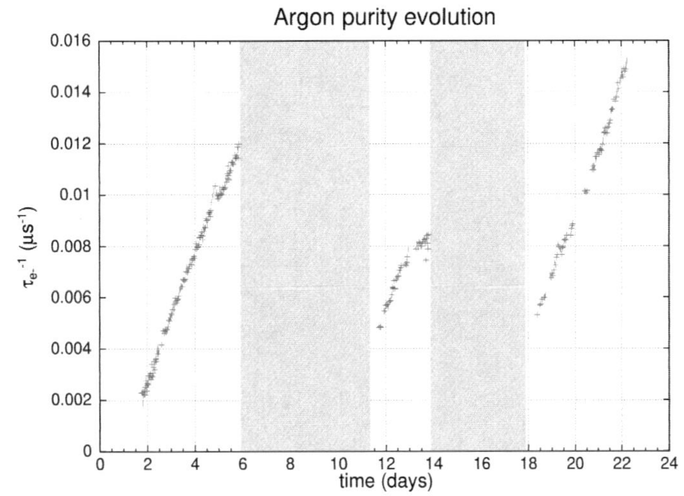

Figure 5. History of the liquid argon purity. The value $1/\tau_{e^-}$ is proportional to the impurities content. While acquiring data the purification system is off and the liquid argon purity degrades. During the purification (highlighted bands) the detector cannot be operated because the liquid argon wets the LEM and no amplification is possible.

The former is estimated from the $\Delta Q/\Delta l$ distribution of the reconstructed muon tracks and the latter is about 10 fC/cm, considering that a minimum ionizing muon in liquid argon releases 2.1 MeV/cm [9]. The effective gain, in addition to the charge multiplication in the LEM holes, takes into account potential charge losses from the efficiency to extract the electrons to the vapor phase and from the transparency of the grids and the LEM (some electrons, in part due to the large diffusion in pure argon, may be collected on the grids or on the bottom electrode of the LEM). For this reason the effective gain is sensitive to the details of the electric field configuration. The working point is found from electrostatic computations of the drift lines and from the optimization of the charge sharing between the anode and the top LEM electrodes. The typical configuration of the electric field is given in table 2.

Anode-LEM	2 kV/cm
LEM holes	36 kV/cm
Grid-LEM	1 V/cm
Extraction	3 kV/cm
Drift	500 V/cm

Table 2. Nominal configuration of electric fields for the run at 6.5 gain.

Figure 6 shows the effective gain as a function of the *nominal* electric field E in the LEM holes, *defined* as the ratio $E = V/d$ of the potential difference V applied to the LEM to the LEM thickness d[2]. The measurements were taken by varying the electric field E, while leaving all the other parameters unchanged, as reported in table 2. Above 30 kV/cm we observe an effective gain larger than 1, but it is eventually limited at 6.5 by the occurrence of discharges, primarily in the LEM holes. The discharge pattern suggests that the voltage break-down are localized around regions of high electric fields, possibly close to the strip segmentation of the

[2] We note that the actual electric field is not constant along the path of electrons inside the holes, and the maximum electric field encountered by drifting electrons (plateau) is slightly less than the nominal electric field.

upper electrode of the LEM. This observation led us to conclude that a promising alternative is to decouple the charge amplification stage from the readout stage with non-segmented LEM electrodes. Indeed, such a configuration was successfully implemented after the completion of this work and improved gain results are published in a separate paper [10]. In the new setup, the X-Y readout feature was retained implementing a 2D projective anode, based on the concept described in [11].

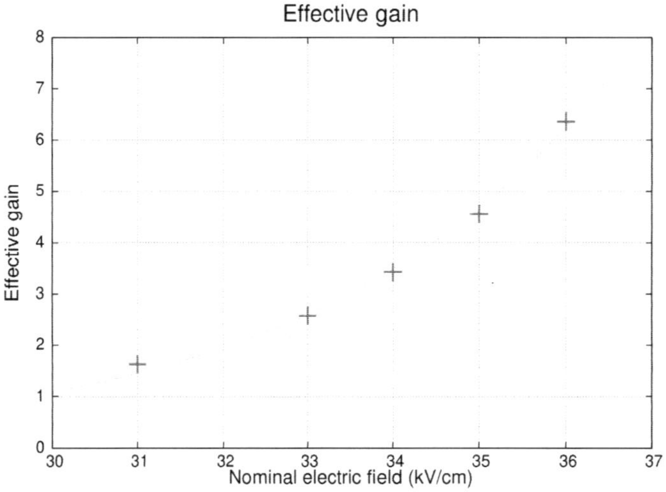

Figure 6. Effective gain as a function of the nominal electric field in the LEM holes. The curve is an exponential fit to the data to guide the reader's eyes.

5. Conclusions

In this paper we presented the successful and stable operation of a double phase LAr LEM Time Projection Chamber over a period of three weeks. The gain was implemented with a 1 mm-thick LEM. Cosmic muons, recorded with very high quality imaging properties, were three-dimensionally reconstructed and used to evaluate the argon purity and the effective charge gain of the detector.

Thanks to a dedicated purification system, pure liquid argon could be obtained, reaching ppb O_2 equivalent level in a few days, for drift paths of 20 cm. The maximum effective gain achieved, measured directly from a fit to the energy deposited by the cosmic tracks, was about 6.5. A possibility to improve the gain performance is to avoid the segmentation of the top electrode of the LEM, since in this configuration, break down discharges are often initiated in the high electric field regions separating the readout strips of the LEM.

The charge amplification greatly improves the imaging capabilities of the TPC making this charge readout technique interesting for next generation proton decay and neutrino experiments and possibly – increasing the maximum gain – for direct Dark Matter search experiments.

References

[1] A. Rubbia, "Experiments for CP-violation: A giant liquid argon scintillation, Cerenkov and charge imaging experiment?," arXiv:hep-ph/0402110.
[2] A. Rubbia, "ArDM: A ton-scale liquid argon experiment for direct detection of dark matter in the universe," J. Phys. Conf. Ser. **39** (2006) 129 [arXiv:hep-ph/0510320].

[3] F. Sauli, "GEM: A new concept for electron amplification in gas detectors," Nucl. Instrum. Meth. A **396** (1997) 531.

[4] A. Badertscher *et al.*, "Operation of a double-phase pure argon Large Electron Multiplier Time Projection Chamber: Comparison of single and double phase operation," Nucl. Instrum. Meth. A **617** (2010) 188-192 [arXiv:0907.2944 [physics.ins-det]].

[5] A. Badertscher *et al.*, "Construction and operation of a double phase LAr Large Electron Multiplier Time Projection Chamber," IEEE Nuclear Science Symposium Conference Record, 2008 (NSS '08), pp.1328-1334, 19-25 Oct. 2008, doi: 10.1109/NSSMIC.2008.4774662 [arXiv:0811.3384 [physics.ins-det]].

[6] R. Acciarri *et al.* [WArP Collaboration], "Oxygen contamination in liquid Argon: combined effects on ionization electron charge and scintillation light," JINST **5** (2010) P05003 [arXiv:0804.1222 [nucl-ex]].

[7] R. Acciarri *et al.* [WArP Collaboration], "Effects of Nitrogen contamination in liquid Argon," Nucl. Phys. Proc. Suppl. **197** (2009) 70 [arXiv:0804.1217 [nucl-ex]].

[8] E. Buckley *et al.*, "A study of ionization electrons drifting over large distances in liquid argon," Nucl. Instrum. Meth. A **275** (1989) 364-372.

[9] S. Amerio *et al.* [ICARUS Collaboration], "Design, construction and tests of the ICARUS T600 detector," Nucl. Instrum. Meth. A **527** (2004) 329.

[10] A. Badertscher *et al.*, "First operation of a double phase LAr Large Electron Multiplier Time Projection Chamber with a two-dimensional projective readout anode," arXiv:1012.0483 [physics.ins-det].

[11] A. Bressan *et al.*, "Two-dimensional readout of GEM detectors," Nucl. Instrum. Meth. A **425** (1999) 254.

Journal of Physics: Conference Series **308** (2011) 012017 doi:10.1088/1742-6596/308/1/012017

Micromegas for charge readout of double phase Liquid Argon TPCs

A Delbart, D Attie, J Beucher, O Besida, E Ferrer-Ribas, F.J Iguaz, A Giganon, A Longhin, E Mazzucato, G Vasseur and M Zito

CEA/CE-Saclay, DSM-IRFU, 91191 Gif sur Yvette cedex, France

E-mail: alain.delbart@cea.fr

Abstract. The 128 μm bulk-micromegas detector is a mature semi-industrial technology well suited for high quality and large scale production of large active area MPGDs as illustrated by its use in the Time Projection Chambers (TPCs) of the ND280 near detector of the T2K experiment. The cooling of a 128 μm bulk-micromegas detector down to 77 K proves its robustness at low temperatures and that is therefore a natural candidate for charge readout in Giant Liquid Argon TPCs.

We propose the use of bulk-micromegas technology as charge readout in the vapour phase of a double phase Liquid Argon TPC. We have made a series of gain measurements with bulk and micro-bulk micromegas at room temperature up to 4.5 bars to predict the behaviour and performances of micromegas in the double-phase Liquid Argon TPC conditions at 87 K. Gas gains of up to 500 were measured in pure Argon at room temperature and 3.5 bars with a 50 μm standard micro-bulk. But the large diffusion of electrons in pure argon demands the operation of the detector with sufficiently low drift over amplification field ratios in order to fully transfer primary electrons in the amplification gap. Gain simulations are in good agreement with these measurements and a gain of 100 in a 128 μm bulk-micromegas is expected in pure Argon at 87 K. Such a large effective gain should be sufficient for charge readout of double-phase liquid Argon TPCs.

1. Introduction

Giant Liquid Argon TPCs are good candidates for next generation rare events experiments, such as neutrino physics or direct search of Dark Matter. A first proof of operation in double phase conditions was reported by A. Badertscher et al.[1] : the ionization electrons released in the liquid phase argon are drifted towards a set of extracting grids to be transferred into the gas phase for amplification in a double stage Large Electron Multiplier (LEM). For such a use, the requirements for the amplification device are :

- sufficient electron multiplication and stable operation in pure argon vapour phase at 87 K: a goal of 100 may suit a large range of physics applications;
- high quality and high yield industrial production for low cost and large scale production;
- ability to cover hundreds of square meters of active area with low dead zones and good uniformity of performances;

We propose the use of the bulk-micromegas technology as charge readout in the vapour phase of a Liquid Argon TPC. After a quick review of the current micromegas technologies, we present a robustness test at 77 K and a series of gain measurements in pure argon and room temperature to prove the capabilities of this detector for such an application. The electron multiplication performances are measured at 3.5 bars since the density of argon at this pressure is roughly equivalent to the expected density of the vapour above liquid argon surface at 87 K and 1 bar.

2. The current Micromegas technologies

Micro-Pattern Gaseous Detectors (MPGDs) are well established and high performance devices widely used in Physics experiments. The international collaboration RD51 gathers 71 institutes on Development of Micro-Pattern Gas Detectors Technologies [2] and a dedicated Working Group is in charge of the development of cost-effective technologies and large scale industrialization through technology transfer.

The MICRO-MEsh GAseous Structure or MICROMEGAS is a MPGD invented in 1996 which amplifies electrons in a typical 100 μm gap defined by a metallic micromesh placed on top of an anode Printed Circuit Board (PCB)[3]. In 2004, a new way to define this amplification gap was introduced and called the bulk-micromegas[4]. A woven micromesh is embedded on top of the segmented anode plane of the micromegas by use of standard photolithographic techniques (fig. 1). This technology was chosen to equip the 3 TPCs of the ND280 near detector of the T2K experiment for its performances in terms of gas gain uniformity, energy resolution and space point resolution, and for its capability to efficiently pave large readout surfaces with minimized dead zones (fig. 2). Eighty six 128 μm gap 34x36 cm^2 bulk-micromegas modules were produced in 18 months. Eighty of them, for an equivalent total surface of 9 m^2 passed the quality and performance tests. The dispersion of gas gain and energy resolution at 5.9 keV within the whole surface of each module is respectively 2.8% and 6% r.m.s. The dispersion of mean gain and mean 5.9 keV energy resolution over the eighty modules is respectively 8% and 3% r.m.s[5]. The cost of the device is about 10 k€/m^2 with 80% of the cost in the 4 layers PCB. These facts illustrate the maturity of the bulk-micromegas technology for a uniform, cheap and high quality mass production.

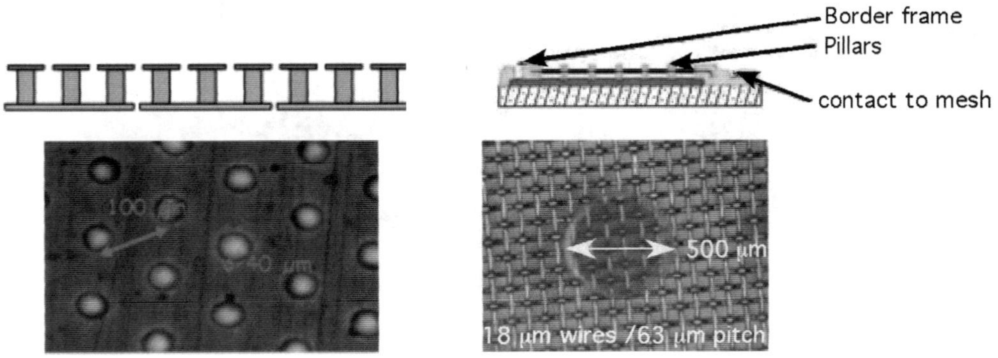

Figure 1. Side view sketch and mesh top view picture of a standard 50 μm micro-bulk (left) and a 128 μm bulk-micromegas (right)

In 2006, a new micromegas structure called micro-bulk was introduced and consists of building the amplification gap within a double side copper-cladded kapton foil [6] (fig. 1) by the same kind of chemical etching techniques as the ones used to make Gas Electron Multipliers

Table 1. Comparison of some bulk and micro-bulk specifications and performances. Gaps of 64 μm for bulk and 12.5 μm for micro-bulk need to be tested.

	bulk	micro-bulk
Standard amplification gap	128 μm	50 μm
Other possible amplification gaps	(64)-100-150-194 μm	(12.5)-25 μm
Standard Mesh pitch	63 μm	100 μm
Standard Mesh openings	45 μm	40 μm
Standard maximum size	50x150 cm^2	10x10 cm^2
R&D maximum size	100x200 cm^2	30x30 cm^2
Best FWHM 5.9 keV resolution	19%	11%
Currently in use in experiments	T2K/TPC	Axion CAST experiment, nTOF
Current R&D programs	ILC/TPC, ILC/DHCAL, SLHC/Muon chambers upgrade, CLAS12 spectrometer, ...	NEXT, MIMAC, ...

(GEM). This more recent technology gives excellent performances especially in terms of energy resolution[6, 7]. A picture of the one used for some measurements done in this paper is shown on fig. 3.

Figure 2. A T2K/TPC readout plane made of twelve 128 μm bulk-micromegas module for a total 1.5 m^2 active surface.

Figure 3. A standard 50 μm microbulk micromegas with a single 35 mm diameter anode.

Table 1 compares these two micromegas technologies for the specifications and performances of interest in double phase Liquid Argon application. The bulk-micromegas is today the more mature micromegas technology for next generation giant liquid Argon TPCs.

3. Cooling of a Bulk-micromegas detector down to 77 Kelvin

The Thermal Expansion Coefficients of the materials used to define the amplification gap of a bulk-micromegas are listed in table 2. They are similar for the Printed Circuit Board (FR4 epoxy and copper anode strips) and the stainless steel woven micromesh. But the pyralux PC1025 which makes the spacers and links the two to define the gap has a 10 times larger coefficient.

We cooled a 10x10 cm^2 128 μm gap bulk-micromegas down to 77 K by gradually plunging it in a Liquid Nitrogen filled dewar (fig. 4). A calibrated Pt100 temperature sensor was stuck on the top side of the detector and the micromesh was supplied with high voltage. At each step of the cooling, high voltage was raised until the first spark occurs to define the maximum amplification electric field. The current drawn through the CAEN 471A high Voltage supply was measured below the 2 nA current sensitivity of the power supply. The results are listed in Table 3.

After these measurements, the detector was dried in an oven at 80 ^oC during one night, and tested in Ar + 5%iC_4H_{10} at few mbars above atmospheric pressure and room temperature with a ^{55}Fe X-ray source (fig. 4). The mesh signal is read out through a 142B ORTEC preamplifier and a 472A ORTEC amplifier. After cooling and plunging into the liquid nitrogen, the detector was not damaged and its gain is comparable to our reference 128 μm gap bulk-micromegas.

These results also show the possible operation of a bulk-micromegas for charge readout in the 87 K Liquid Argon phase.

Table 2. Range of Thermal Expansion Coefficients of the materials used in a bulk-micromegas at 20-200 ^oC .

Material	ppm/^{o}C
FR4	[14 – 16]
Copper	[16 – 17]
Stainless steel	[14 – 17]
pyralux PC1025	[100 – 130]

4. Gain and electron transmission measurements in pure argon at room temperature

4.1. Bulk-micromegas at few mbars above atmospheric pressure

A T2K/TPC bulk-micromegas module was used to measure its gain in pure argon at few mbars above atmospheric pressure with a ^{55}Fe X-ray source. Anode pads were read out with T2K/TPC AFTER based electronics [8]. The 0.2 liter gas chamber is in aluminum with an entrance window in mylar and the drift electrode is a nickel 200 LPI mesh defining a 12 mm drift gap. An argon 6.0 grade bottle fills the chamber with a flow rate of 10 l/h through stainless steel tubes. Gas purity inside the chamber is therefore expected to be high but it was not measured.

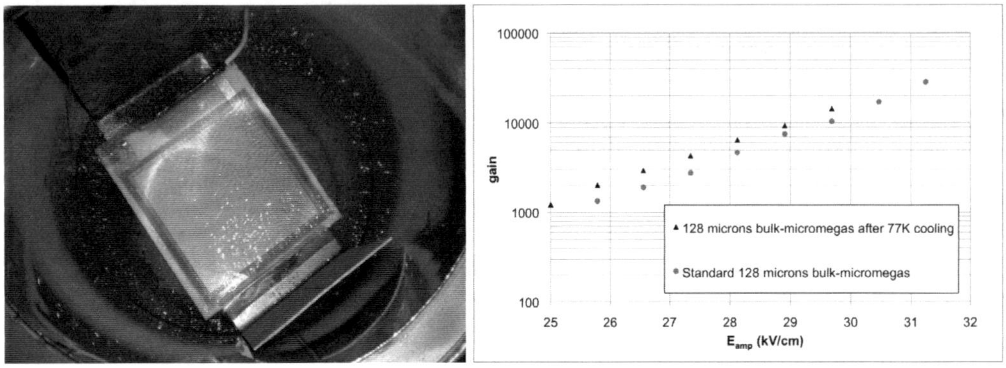

Figure 4. A 10x10 cm^2 128 μm gap bulk-micromegas cooled down to 77 K and plunged into the Liquid nitrogen (left) and its gas gain (right) in Ar + 5%iC_4H_{10} at few mbars above atmospheric pressure and room temperature measured after one night of drying at 80 ^{o}C (black triangles). Gas gain of a standard 128 μm gap bulk-micromegas is also shown for comparison (red circles).

Table 3. Maximum amplification electric field of a 128 μm bulk-micromegas as a function of height above Liquid Nitrogen surface. At -10 mm, the bulk-micromegas is plunged in the liquid.

Height (mm)	Temperature (^{o}C)	Max V_{mesh}(V)	Max E_{amp}(kV/cm)
40	-47	670	52.3
25	-78	750	58.6
8	-119	850	66.4
0	-163	975	76.2
-10	-196	1300	101.6

Electron transmission was measured as a function of the E_{drift}/E_{amp} electric field ratio (fig. 5) and shows a drop for ratios higher than 2.2 10^{-3}. This loss of transmitted electrons from the drift volume into the amplification gap degrades the 5.9 keV r.m.s resolution as shown on fig. 6. A typical ^{55}Fe X-ray source spectrum measured with full electron transmission shows a 10.3% 5.9 keV r.m.s resolution (fig. 6) in contrast with the 9% measured in the T2K/TPC gas[5].

4.2. Micro-bulk micromegas at 3.5 and 4.5 bars

We measured the gas gain in pure argon at 3.5 and 4.5 bars of a 35 mm diameter 50 μm gap standard micro-bulk micromegas (fig. 3) with a $\approx$1 kBq ^{241}Am alpha source in the former HELLAZ setup[9] filled with 6.0 grade pure argon. This setup was recently used at Saclay/IRFU for development of high pressure micromegas detectors within the framework of our collaboration with the Instituto de Fisica Nuclear y Altas Energias at Zaragoza in Spain. Reference [7] describes the setup, reports on measurements performed in argon and xenon mixtures up to 4 bars and shows that the gas purity inside the vessel was high enough for short-term runs in argon mixtures for the 3.1 cm drift distance of alphas. The gas vessel is shown on fig. 7.

The signal was read on the plain anode of the micro-bulk detector through a 142C ORTEC preamplifier fed into an AMPTEK MCA-8000A multichannel analyzer for a counting rate of about 200 Hz.

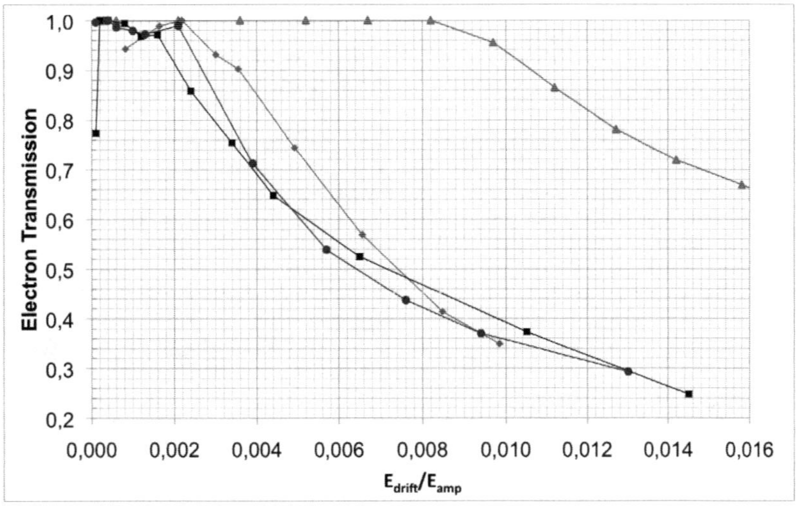

Figure 5. Electron transmission Vs E_{drift}/E_{amp} Electric field ratio for a 50 μm micro-bulk in $Ar + 5\%iC_4H_{10}$ at few mbars above atmospheric pressure (red triangles), in pure argon at 3.5 bars (black squares) and in pure argon at 4.5 bars (blue circles). Electron transmission for a 128 μm T2K/TPC bulk-micromegas module at few mbars above atmospheric pressure is in green rhombs.

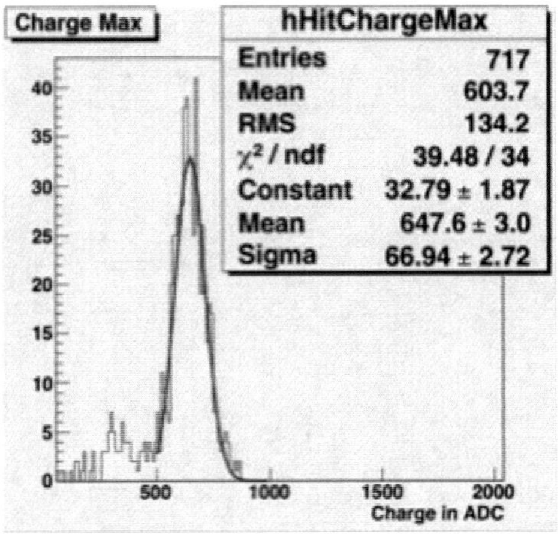

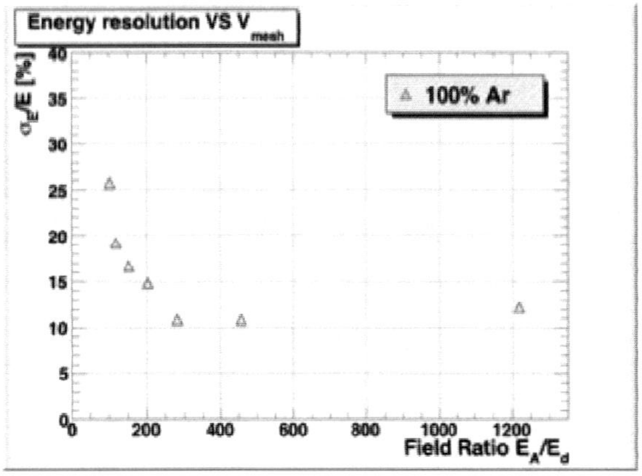

Figure 6. ^{55}Fe X-ray source spectrum operated with $E_a = 32$kV/cm and $E_d = 200$V/cm (left) and FWHM 5.9 keV energy resolution (right) of a 128 μm T2K/TPC bulk-micromegas module at few mbars above atmospheric pressure and room temperature. Drift electric field E_d is varied with a fixed 30.5 kV/cm amplification field E_a ($V_{mesh} = -390$V).

The electron transmission at both pressures are comparable (fig. 5) and show the same drop as the one observed at atmospheric pressure. On this figure, electron transmissions in pure argon are compared to the one measured with the same kind of standard micro-bulk micromegas in $Ar + 5\%iC_4H_{10}$ for which the drop occurs at much higher electric field ratios. The ≈50% lower

Journal of Physics: Conference Series **308** (2011) 012017 doi:10.1088/1742-6596/308/1/012017

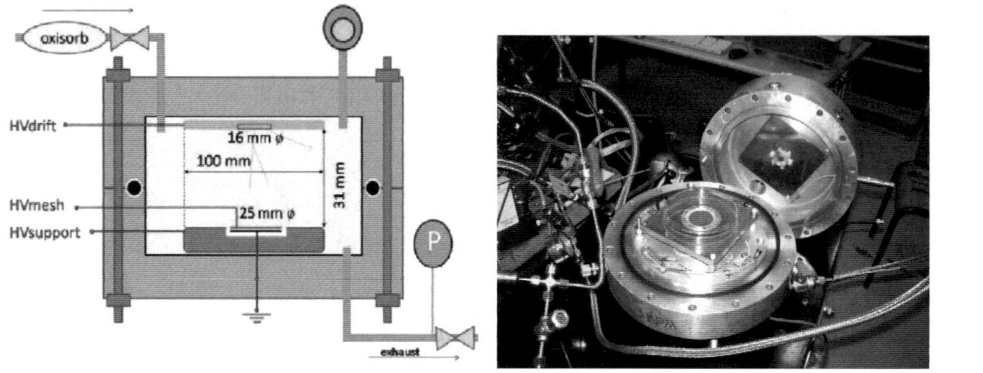

Figure 7. Sketch (left) and picture (right) of the gas vessel used for gain measurements at 3.5 and 4.5 bars.

electron transmission in pure argon compared to Ar + 5%iC_4H_{10} at the same E_{drift} and for the same mesh optical transparency (same micro-bulk micromegas), can be explained by the much larger diffusion of electrons in pure argon (fig. 8)

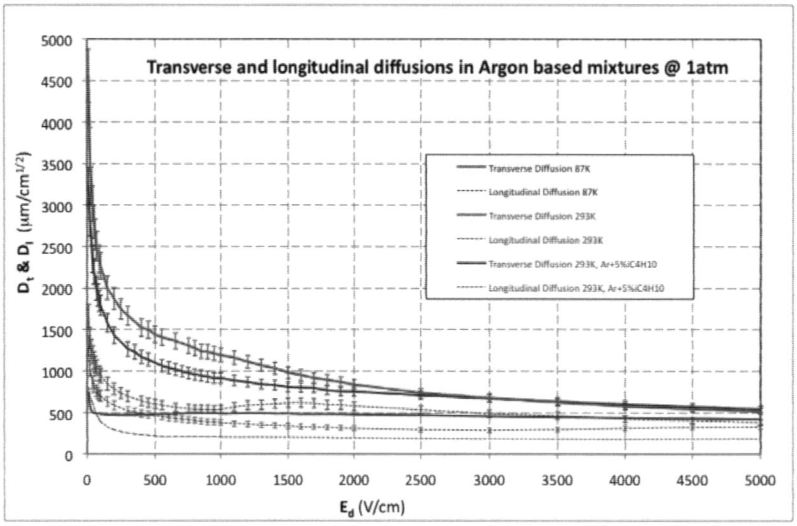

Figure 8. Magboltz simulated longitudinal and transverse diffusions as a function of the drift electric field E_d for different argon based mixtures at 1 atm. Magboltz simulated longitudinal and transverse diffusions of pure argon at 300 K and 3.5 bars are the same as the one for pure argon at 87 K and 1 atm (blue curves).

Gas gains in pure argon at 3.5 bars can reach 500 at 100 kV/cm amplification electric field (fig. 9). For a more stable operation of the detector without sparks, operation at 94 kV/cm (200 gain) should be favoured.

4.3. Gain simulations

A simple modelization of the gas amplification gain G in a micromegas detector of amplification gap x is (at a fixed temperature) : $G = \exp(\alpha \cdot x)$, with $\alpha(E, P) = P \cdot \xi\left(\frac{E}{P}\right)$
with E the amplification electric field and P the gas pressure.
For a noble gas, $\xi\left(\frac{E}{P}\right) = A.\exp\left[-B\left(\frac{P}{E}\right)^{\frac{1}{2}} - C\frac{E}{P}\right]$
Tkachev et al.[10] recently simulated and compared to measurements the Townsend coefficients of pure argon at room temperature : $\mathrm{A} = 43\,[\mathrm{cm.Torr}]^{-1}$, $\mathrm{B} = 27.5\,[\mathrm{V}/(\mathrm{cm.Torr})]^{\frac{1}{2}}$, and $\mathrm{C} = 2.5 \times 10^{-4}\,[\mathrm{cm.Torr/V}]$. Using these coefficients, the calculated gain is in good agreement with the 3 measurements done (solid curves in fig. 9).

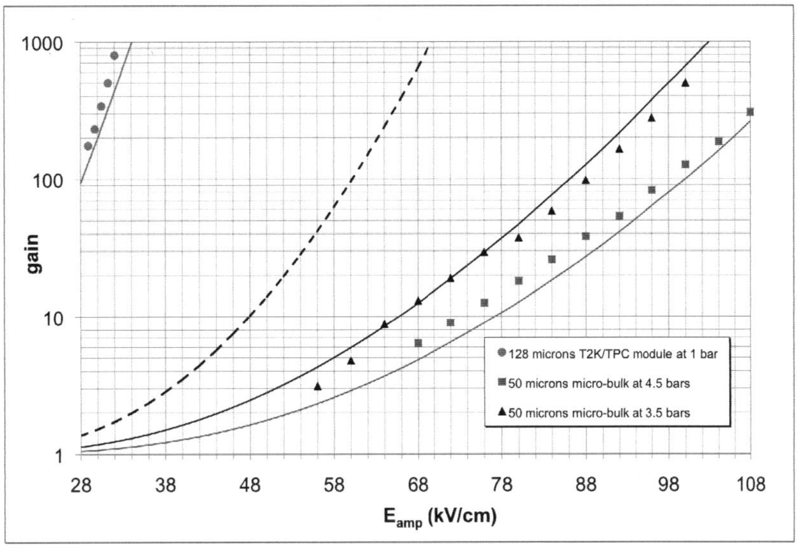

Figure 9. Gas gain in pure argon for a 128 μm T2K/TPC bulk-micromegas module at few mbars above atmospheric pressure (red circles), and for a 50 μm micro-bulk at 3.5 bars (measurements in black triangles, gain simulation in black line) and 4.5 bars (measurements in green squares and gain simulation in green line). Measurements were done with full electron transparency. The black dashed curve is the gain simulation for a 128 μm gap micromegas detector at 3.5 bars.

For a 128 μm gap bulk-micromegas, the black dashed curve of fig. 9 shows an expected 100 gain for an amplification electric field of 58 kV/cm ($V_{\mathrm{mesh}} = -740$V). For full electron transmission, this will require a drift electric field lower than 100 V/cm. For a 500 V/cm drift electric field, only 40% of the primary electrons will be transfered in the amplification gap, and an effective gain of 40 is still reachable and probably large enough for charge readout of double-phase Liquid Argon TPCs.

5. Conclusion and perspectives

Bulk-micromegas was shown to be robust enough to be operated at Liquid Argon temperature. Gas gains of up to 500 were measured in pure argon at room temperature and 3.5 bars with a 50 μm standard micro-bulk. These measurements, those done at 4.5 bars, and those done with a 128 μm bulk-micromegas T2K/TPC detector at atmospheric pressure are in good agreement with gain simulations. A gain of 100 for a 58 kV/cm amplification electric field in a 128 μm bulk-micromegas can thus be expected in pure argon at 87 K, as soon as the drift electric field is lower than 100 V/cm for a full transfer of the primary electrons into the amplification gap. Even if part of the primary electrons are lost in case of the use of a higher drift electric field, such a large effective gain should be sufficient for charge readout of double-phase liquid Argon TPCs.

We are collaborating with the Institute for Particle Physics, ETH Zurich (ETHZ, Switzerland) for the design of a 1D 3mm pitch 64 channels readout bulk-micromegas to be tested in the 3 Liters ETHZ Lar TPC[1] at CERN. 128 μm gap is the baseline design and manufacturing of a 64 μm gap will also be tried. The measurements done in this paper will be the fruitfull guides to tune the operation of the detector in terms of charge collection. The next step will be the design and optimization of a 2D readout bulk-micromegas prototype.

Acknowledgments
We would like to thank our colleagues from the Instituto de Fisica Nuclear y Altas Energias at Zaragoza (Spain) for sharing their experience in the operation of micro-bulk readouts at high pressure. We would like to also thank our colleagues from the Institute for Particle Physics, ETH Zurich (Switzerland) for inviting us to test a Micromegas detector in their 3 liters double phase Liquid Argon prototype at CERN.

References

[1] Badertscher A et al. 2010 Operation of a double-phase pure argon Large Electron Multiplier Time Projection Chamber: Comparison of single and double phase operation *Nucl. Instr. and Meth.* A **617** 259-266.
[2] http://rd51-public.web.cern.ch/RD51-Public/Welcome.html.
[3] Giomataris I, Rebourgeard Ph, Robert J-P and Charpak G 1996 Micromesh Gaseous Structure *Nucl. Instr. and Meth.* A **376** 29.
[4] Giomataris I et al. 2006 Micromegas in a bulk *Nucl. Instr. and Meth.* A **560** 405.
[5] Delbart A et al. 2010 Production and calibration of 9 m^2 of bulk-micromegas detectors for the readout of the ND280/TPCs of the T2K experiment *Nucl. Instr. and Meth.* A (Preprint doi:10.1016/j.nima.2010.02.163).
[6] Andriamonge S et al. 2009 Development and performance of Microbulk Micromegas detectors *JINST* **4** (1st International conference on micro pattern Gaseous Detectors, June 1215, 2009 Kolympari, Crete, Greece, preprint doi:10.1088/1748-0221/4/11/P11016).
[7] Dafni T et al. 2009 Energy resolution of alpha particles in a micrombulk Micromegas detector at high pressure argon and xenon mixtures *Nucl. Instr. and Meth.* A **608** 259-266.
[8] P. Baron et al. 2008 AFTER, an ASIC for the Readout of the Large T2K Time Projection Chambers *IEEE Trans. in Nucl. Sci.* NS-**55**, issue 3, 1744-1752.
[9] Dolbeau P, Giomataris I, Gorodetzky P, Patzak T, Salin P and Sarat A, 2005 Dark matter with HELLAZ *Nucl. Phys. B Proc. Suppl.* **138** 56-58.
[10] Tkachev A N et al. 2007 Townsend Coefficient, Escape Curve, and Efficiency of Runaway-Electron Beam Formation in Argon *Technical Physics* Vol. **52** N^o6 188-192.

Journal of Physics: Conference Series **308** (2011) 012018 doi:10.1088/1742-6596/308/1/012018

Development of Thick-GEMs for a GEM-TPC Tracker

Fuminori Sakuma[1] **and Makoto Tokuda**[2]

[1] RIKEN Nishina Center, RIKEN, Saitama 351 0198, Japan

[2] Department of Physics, Tokyo Institute of Technology, Tokyo 152 8551, Japan

E-mail: `sakuma@ribf.riken.jp`

Abstract. A Time Projection Chamber (TPC) with Thick Gas Electron Multipliers (TGEMs) has been developed for an inner tracker of the J-PARC E15 experiment, a deeply-bound kaonic nuclear states search experiment. The TPC has a cylindrical design with an inner diameter of 170 mm, an outer diameter of 280 mm, and a drift length of 300 mm filled with P10 gas (90:10 argon-methane) at atmospheric pressure. A TGEM is used for the amplification of the TPC in order to realize a self-supporting structure in a limited space. We use a double TGEM structure and the maximal effective gain of more than 10^4 is achieved with 100 × 100 mm TGEM prototypes. Moreover, we have been developing a resistive-electrode TGEM (RETGEM) with graphite electrodes, which has the advantage of being fully spark-protected. The results of our recent studies on the TGEM prototypes with conventional copper electrodes, the RETGEM with graphite electrodes, and the newly developed hybrid-RETGEM with copper and graphite electrodes are discussed.

1. Introduction

The purpose of the J-PARC E15 experiment is to search for the simplest kaonic nuclear bound state, K^-pp cluster, by using an in-flight ^{3}He(K^-,n) reaction [1]. The produced K^-pp cluster will be identified with missing mass spectroscopy using a forward-going neutron with an invariant mass analysis of the expected decay particles from the K^-pp cluster, such as $p\Lambda/p\Sigma^0$ and $p\pi\Sigma$. The baseline detector for the E15 experiment consists of a high-precision beam line spectrometer, a Cylindrical Detector System that surrounds a target to detect the decay particles from the K^-pp cluster and a neutron TOF counter that is located 15 m away from the target position. The detector has been designed to focus on the $K^-pp \to p\Lambda$ channel, which is expected to be the main decay branch of the K^-pp cluster. Therefore, the interaction vertex position resolution is limited to $\sim$ 6 mm along the beam direction (z-direction). However, recent theoretical progresses on the K^-pp cluster indicate the importance of another decay mode, i.e., $p\pi\Sigma$, in reaching an informed conclusion about the K^-pp cluster. In order to realize an efficient measurement of the decay mode $K^-pp \to p\pi\Sigma$, we need to install a new detector in our baseline detector system. The requirements of the new detector system are as follows:

- interaction position resolution in the z-direction must be less than 1mm,
- the new detector system must be compact because of the limited space availability around the target, and

- the material budget of the new detector must be minimized because the detector will be installed as the innermost detector of the tracking system for the experiment.

After a detailed investigation, we chose the Time Projection Chamber (TPC) as the technology for the innermost tracking detector. Moreover, in order to realize the compactness and robustness of the detector, we selected a recently developed technology, a Thick Gas Electron Multiplier (TGEM), for the signal amplification stage of the TPC.

The TGEM is made from a double-sided copper cladding FR4 plate using standard PCB techniques with mechanically drilled holes, and if necessary, with chemically etched rims around the holes [2][3]. The typical parameters of our TGEM prototype are a thickness of 400 μm, hole-diameter of 300 μm, rim of 30 μm, and pitch of 600 μm. Because of the rigidity of the TGEM, it is comparatively easy to construct a self-supporting structure in a limited space. Further, the TGEM has the following advantages: robustness, cost-effective fabrication, high-gain multiplication, and ease of handling. In the recent years, a resistive-electrode TGEM (RETGEM), whose electrodes have been replaced by the resistive electrodes of graphite paint [4] or resistive Kapton [5] has attracted considerable attention as a more robust version of the TGEM. The resistive electrodes protect the detector and the readout electronics from damage by any occasional discharge. We have developed conventional TGEMs with copper electrodes, RETGEMs with graphite-paint electrodes, and newly constructed hybrid-RETGEMs which have copper and graphite electrodes on each side. In this article, we first describe the design of the TGEM-TPC for the J-PARC E15 upgrade and then discuss the results of our recent studies on the TGEM prototypes.

2. Design of the TGEM-TPC

The TPC has a cylindrical design with an inner diameter of 170 mm, an outer diameter of 280 mm, and a drift length of 300 mm filled with P10 gas (90:10 argon-methane) at atmospheric pressure. A photograph of the TPC is shown in Fig. 1. The TPC consists of an end-cap TGEM chamber, an inner field, cage and an outer field cage, which are made from double-sided Flexible Printed Circuits (FPC) with staggered strip electrodes having a width of 8 mm and pitch of 10 mm and connected with 1 MΩ resistors. A drift field of 150 V/cm is applied to the field cages. For amplification, a double-TGEM structure is used as shown in Fig. 2. The TGEM has a nonagonal shape whose one side is subdivided into three sectors and separately connected externally to the voltage supply through high-value resistors in order to reduce both the energy and the propagation probability of the discharge. High voltages are applied to the double-TGEM through connectors penetrating the end cap with a resistor chain. The layout of the readout system is also nonagonal and is divided into 4 $\times$ 4 pads on each side with 4 mm-long and 20 mm-wide pads printed on a standard PCB (the number of total readout channels is 144). For the TPC front-end electronics, preamplifier cards with amplifier-shaper-discriminator ASICs developed for the ATLAS experiment [6] are used, i.e., we obtain only the Z-direction information from the TPC.

3. Results of TGEM prototypes

In order to study the performances of the TGEM, we used 13 types of prototype TGEMs with an active area of 10 $\times$ 10 cm^2; these TGEMs were produced by REPIC Corp., Japan. The TGEM electrodes were made of copper (TGEM#1 - 7), graphite (TGEM#8 - 12), and copper/graphite (TGEM#13). Table 1 summarizes the geometrical parameters of the TGEMs studied in the present work. Measurements were carried out with a double-TGEM configuration using a test bench that consisted of a gas chamber housing, a voltage divider with a resistor chain, and the read out pad with a charge sensitive preamplifier. Figure 3 shows the gas chamber for the TGEM prototypes, and Fig. 4 shows the experimental setup for testing the TGEMs. The effective gain

Journal of Physics: Conference Series **308** (2011) 012018 doi:10.1088/1742-6596/308/1/012018

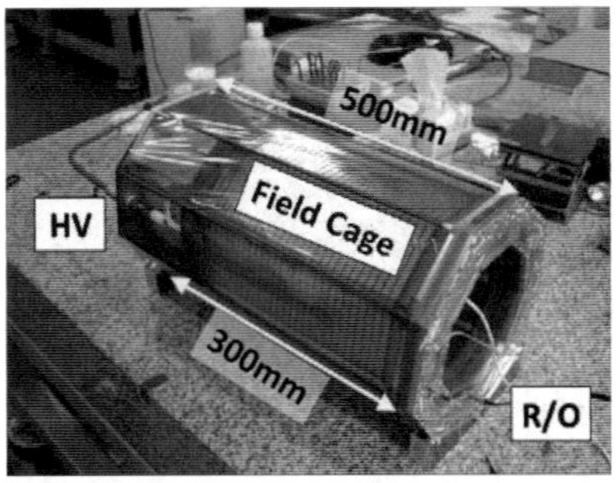

Figure 1. Photograph of the TPC.

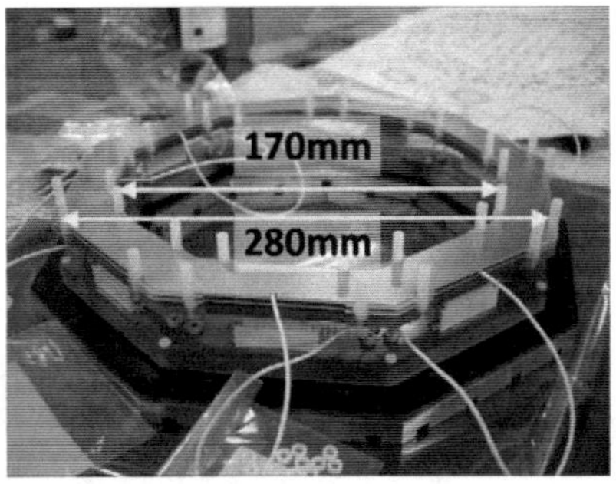

Figure 2. Photograph of the TPC readout with a double-TGEM configuration.

is required to be more than 10^4, and the stability of the gain and the energy resolution should be within 10% per day as per the requirements of the TPC stable operation.

Table 1. A summary of the geometrical parameters of TGEMs studied in this work.

TGEM#	electrode	insulator (mm)	thickness (mm)	hole diameter (mm)	pitch (mm)	rim size (mm)
1	Cu	FR4	0.2	300	0.6	0.05
2	Cu	FR4	0.2	500	0.6	-
3	Cu	FR4	0.4	300	0.6	-
4	Cu	FR4	0.4	300	0.6	0.03
5	Cu	FR4	0.4	300	0.6	0.05
6	Cu	FR4	0.4	300	0.7	0.1
7	Cu	FR4	0.4	500	0.6	-
8	C	FR4	0.4	300	0.7	-
9	C	FR4	0.4	300	0.6	-
10	C	G10	0.4	300	0.6	-
11	C	CEM3	0.4	300	0.6	-
12	C	FR4	0.6	300	0.7	-
13	C/Cu	FR4	0.4	300	0.6	-

3.1. TGEM with copper electrodes

Figure 5 shows the effective gain of the double TGEM measured with ^{55}Fe with four types of the rim configuration (TGEM#3 - 6). Further, the stability of the gain and the energy resolution for a day under the same configurations are shown in Fig. 6. The effective gains shown in Fig. 6 are corrected with the ratio P/T of the absolute temperature and the pressure: a function of the form $f(P/T) = \alpha\exp(\beta P/T)$, where α and β are the fitting parameters. All types of TGEMs achieve the maximal effective gain of more than 10^4, but TGEMs with larger rims require a higher voltage. In the stability plots, the gains of all TGEMs decrease for a few hours after a high voltage is turned on, and the energy resolutions of TGEMs with rims (TGEM#4 - 6) fluctuate during the measurement. Although thus far, such behaviors of the TGEM are not

Figure 3. Photograph of the gas chamber for the TGEM prototypes.

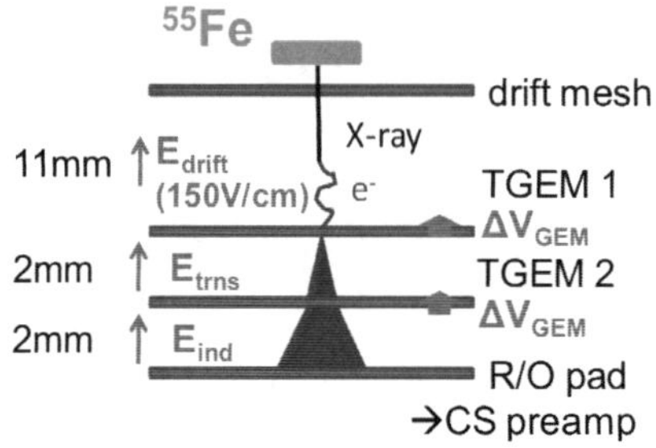

Figure 4. Schematic representation of measurement setup.

understood well, the initial drop effect of the gain is likely to be caused by the charge-up (or polarization) of the overall insulator, and the instability of the TGEM with rims is possibly caused by the charge-up of insulator that is not metalized.

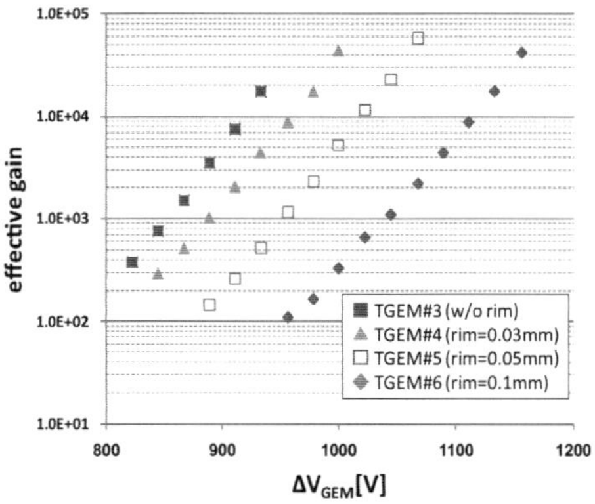

Figure 5. Effective gains measured with double-TGEMs of different geometries.

In order to further investigate the long-term stability of TGEM, we applied a high voltage to TGEM#4 with an ^{55}Fe X-ray irradiation continuously for approximately 10 days. Figure 7 shows the effective gain corrected with the P/T and the energy resolution. It should be noted that the high voltage for the TGEM is turned up manually at interval of around 50 h in order to maintain a high gain of more than 10^4. In the figures, the effective gain is maintained at more than 10^4, but the gain fluctuation is approximately 30% per day. A possible reason for this instability is the charge-up of the insulator that is not metalized.

3.2. TGEM with graphite electrodes

In order to avoid the effects of the rims, we have developed a resistive-electrode TGEM (RETGEM), which has electrodes coated with graphite paint. The RETGEMs have the advantage of being fully spark-protected. We have fabricated and tested several types of TGEMs (TGEM#8 - 12); However, RETGEMs with an FR4 insulator (TGEM#8, 9) were not produced stably because of the discharge from the graphite attachment inside the holes caused by the knot of the glass fiber on the cross section of FR4. Another problem related to the fabrication

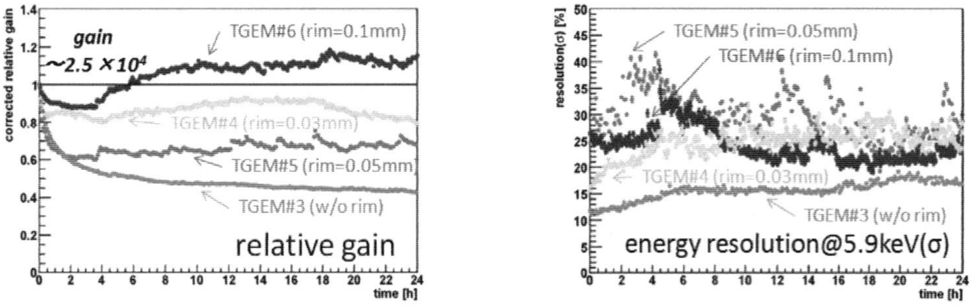

Figure 6. Stability of effective gains (left) and energy resolution (right) measured with double-TGEMs of different geometries for a day. The gains are corrected with the ratio P/T, and unity of the relative gain corresponds to the effective gain of 2.5×10^4.

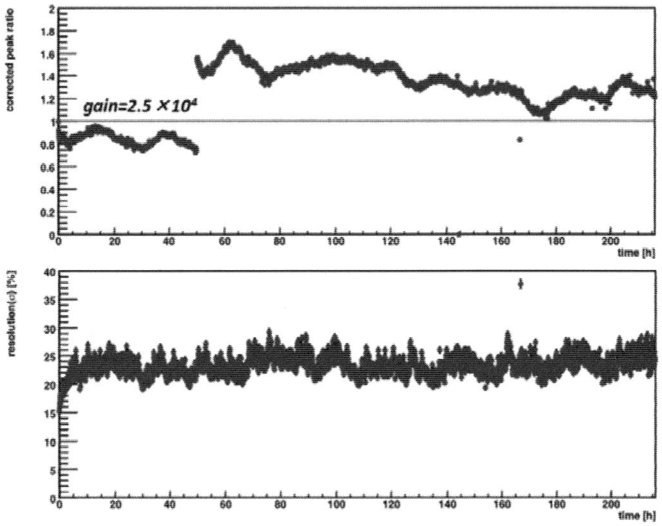

Figure 7. Effective gain corrected with the ratio P/T (up) and energy resolution (down) of double-TGEM#4 for approximately 10 days. Unity of the relative gain corresponds to an effective gain of 2.5×10^4. The high voltage for the TGEM is turned up manually at interval of around 50 h in order to maintain a high gain of more than 10^4.

of the RETGEM with graphite electrodes was the discharge from the burrs arising from the drilling process, but this discharge could be removed by using an antistatic brush. The problem of the graphite attachment could not be solved by improving the drilling procedure. In fact, only the first 2 out of the 11 samples of the RETGEMs (TGEM#8, 9) worked. Now, we have been studying the RETGEM with a CEM3 insulator, which contains less glass fiber than FR4.

3.3. *TGEM with copper and graphite electrodes*

Because of the difficulty in producing graphite RETGEMs, we attempted to fabricate a new type of RETGEM with copper and graphite electrodes on each side (hybrid-RETGEM, TGEM#13). The resistivity of one side of the electrodes would be spark-protected in principle. Further, the

Journal of Physics: Conference Series **308** (2011) 012018 doi:10.1088/1742-6596/308/1/012018

hybrid-RETGEM would have a possibility of a reduction of the graphite attachment inside the holes by an alignment of the drilling process. We fabricated the hybrid-RETGEM with two drilling directions: drilling into the hybrid-RETGEM form the copper to the graphite electrode and vice versa. We found that the two fabrication methods work almost similarly with respect to the gain (and resolution) as shown in Fig. 8, which shows the obtained effective gain of the two double hybrid-RETGEMs. Figure 9 shows the long-term stability of a double hybrid-RETGEM for approximately 25 days with ^{55}Fe X-ray irradiation. The upper figure in Fig. 9 shows the effective gain corrected with the P/T, and the lower figure shows the energy resolution. It should be noted that high voltage for the TGEM is turned up manually at interval of around 40 h in order to maintain a high gain of more than 10^4, and blanks in the plots are absent of the ^{55}Fe source but are kept for turning on the high voltage. The effective gain is maintained more than 10^4, and a gain fluctuation of approximately 5% per day is achieved.

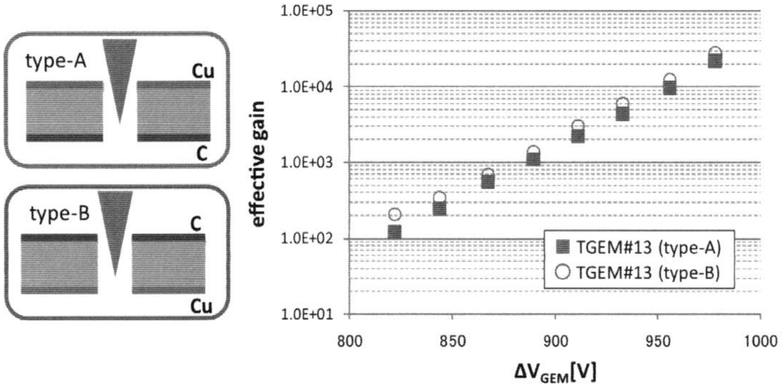

Figure 8. Effective gains of double-TGEM#13s with two different fabrication methods. Type-A indicates drilling into the hybrid-RETGEM form the copper to the graphite electrode, and type-B indicates the opposite.

4. Summary

A TPC with TGEM has been developed for an inner tracker of the J-PARC E15 experiment, a deeply-bound kaonic nuclear states search experiment. The TPC has a cylindrical design with an inner diameter of 170 mm, an outer diameter of 280 mm, and a drift length of 300 mm filled with P10 gas at atmospheric pressure. The TGEM is used for the amplification of the TPC in order to realize a self-supporting structure in a limited space. We have used a double TGEM structure. Further, a maximal effective gain of more than 10^4 is achieved with 100 $\times$100 mm TGEM prototypes with conventional copper electrodes. Moreover, we have developed RETGEMs with graphite electrodes and with copper and graphite electrodes; these RETGEMs have the advantage of being fully spark-protected. Although further investigation is required for a stable production of the RETGEM with graphite electrodes, the newly developed hybrid-RETGEM with copper and graphite electrodes exhibits with good performance: the effective gain is maintained more than 10^4 in the long term, and a gain fluctuation of approximately 5% per day is achieved.

The next challenge will be to establish a stable production method of the RETGEM with graphite electrodes and insulators other than the FR4 insulator.

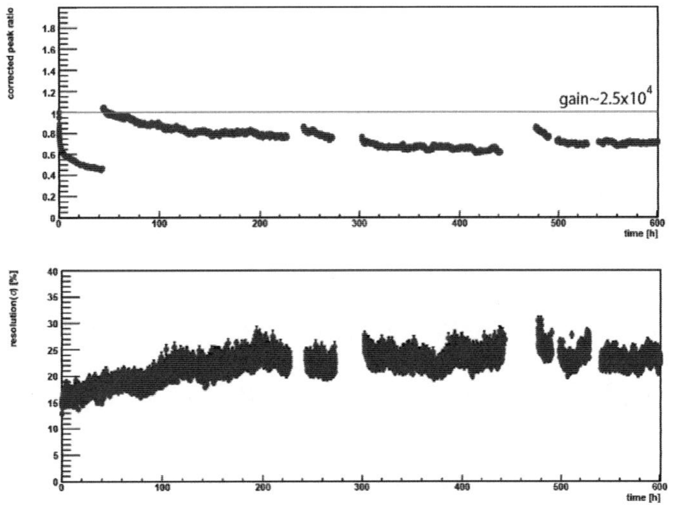

Figure 9. Effective gain corrected with the ratio P/T (up) and energy resolution (down) of double-TGEM#13 for approximately 25 days. Unity of the relative gain corresponds to the effective gain of 2.5×10^4. The high voltage for the TGEM is turned up manually at interval of around 40 h in order to maintain a high gain of more than 10^4, and blanks in the plots are absent the ^{55}Fe source but are kept for turning on the high voltage.

4.1. Acknowledgments
This work was partially supported by Grant-in-Aid for Scientific Research on Priority Areas and RIKEN Special Postdoctoral Researchers Program. Further, we would like to thank Mr. K. Konno at REPIC Corp. and Dr. H. Ohnishi at RIKEN for their assistance.

References
[1] E15 proposal, `http://j-parc.jp/NuclPart/pac_0606/pdf/p15-Iwasaki.pdf`.
[2] L. Periale *et al.*, *Nucl. Instrum. Meth.* **A478** (2002) 377.
[3] R. Chechik *et al.*, *Nucl. Instrum. Meth.* **A535** (2004) 303.
[4] J.M. Bidault *et al.*, *Nuclear Physics B (Proc. Suppl.)* **158** (2006) 199.
[5] R. Oliveira *et al.*, *Nucl. Instrum. Meth.* **A576** (2007) 362.
[6] O. Sasaki, KEK, Japan, ATLAS TGC Collaboration, `http://online.kek.jp/~sosamu/ASD-PRR.pdf`.

Optical readout of liquid argon ionisation

NJC Spooner[a], PK Lightfoot[b], GJ Barker[c], YA Ramachers[c] and K Mavrokoridis[d]
[a]Department of Physics and Astronomy, University of Sheffield, Hicks Building, Hounsfield Road, Sheffield S3 7RH, UK
[b]Department of Physics, University of York, Heslington, York YO10 5DD, UK
[c]Department of Physics, University of Warwick, Gibbet Hill Road, Coventry CV4 7AL, UK
[d]Department of Physics, University of Liverpool, Liverpool L69 7ZE, UK

E-mail: n.spooner@sheffield.ac.uk

Abstract Reading out the charge from a very large liquid argon detector, such as proposed for next generation proton decay and long baseline neutrino detectors, represents a significant challenge. Current proposals suggest using wires in the liquid or a two-phase approach that can provide some gain via amplification in the gas phase. We present here work on an alternative new approach in which the charge is read out by optical means following generation of electroluminescence, such as in a THGEM (Thick Gas Electron Multiplier) mounted within the liquid. This has the potential for significant advantages by providing both simpler readout electronics and significant charge gain, without the need for the complexities of dual phase operation. Tests with a silicon photomultiplier (SiPM) mounted above a THGEM, all submerged in liquid argon, have allowed first demonstration of the technique. Sensitivity to 5.9 keV ^{55}Fe gamma events was observed with an estimated gain of 150 photoelectrons per drifted electron. We review the concepts and results.

1. Introduction

The readout options currently being developed for liquid argon detectors can be divided as follows: (1) Single phase concepts in which readout occurs only in the liquid using either wires, such as in ICARUS [1], or via detection of the primary scintillation using photomultipliers, such as in the dark matter experiments DEAP/CLEAN [2]. The former technique provides essentially no internal gain, while the latter provides only very crude position information and no tracking. (2) Double phase concepts in which readout occurs in the gas phase following extraction of charge out of the liquid. Here, techniques studied include use of amplification of the charge with LEMs, GEMs, THGEMS or Micromegas plus readout with anode plane strips, or conversion to electroluminescence in the gas and readout by photomultipliers. An example of the former is used in the ArDM dark matter experiment, with LEMS [3], and of the latter is in WARP [4] - both dark matter experiments.

None of these concepts look ideal for scale-up to very large mass (>10ktons) for proton decay or long baseline neutrino physics. Here the requirement is for tracking capability with resolution of a few mm and homogeneous calorimetry, with modest gain (say ~x100), effective throughout the large volume, robust enough to stand underground installation over decades at reasonable cost. The development of new optical techniques attempts to address this. To produce a better way to readout charge in the liquid, with gain, noise and stability improved over that obtained with wires, and a better way to control the number of readout channels. The motivation recognises also particular challenges for two phase operation that would be best avoided if possible. For instance, the need for precise, long term, levelling of liquid argon over 100s m^2, the need for extreme cooling stability, for control of the liquid-gas interface, for a very high (~MV) cathode voltage to produce the necessary large drift distance, the need for a design that is not modular, and some complex engineering to allow support of the detector plane above the liquid but within the containment vessel [5].

An optical readout technique in single phase liquid, through making use of electroluminescence generation in the liquid, could solve many of these issues. It would allow modest gain to be produced without the need for the gas phase. This would also allow some degree of internal subdivision of the volume to reduce the requirement for long (>10m) drift distances and hence the complication of very large cathode voltages. This would also reduce the requirements for very high liquid argon purity, already difficult to achieve because the vessel is likely too large to be evacuated.

2. Optical readout concept and SiPDs

The concept under development by our groups to tackle this issue and described here, aims to produce a homogeneous tracking calorimeter based on planar, optical tracking readout with gain in single phase liquid. The particular concept adopted for proof of principle tests is to use THGEMs with a sparse distribution of holes mounted within the liquid argon and observed by cryogenic silicon photodiodes (SiPDs) also mounted in the liquid, positioned above the holes. This concept has been developed by picking the best in current relevant technology. A typical design considered would be to have 36 SiPDs per m^2 of THGEM plane in a 6 x 6 array of row and column readout. Planes of THGEMs could then be built up to reach 10s m^2 readout areas. The array of photosensors behind the THGEMS would be used to image the light signals using the conventional Anger Camera technique to do off-line imaging and track reconstruction, by reconstructing the centroid of the light emission in the XY plane, and using the electron drift velocity in the liquid to give the drift coordinate. Total decoupling of the optical readout from the electronics (that is the drifting charge transfer and readout) promises also superior noise performance. The significant reduction in complexity of this concept allows modular construction of large arrays with readout in any plane and any orientation, not just vertical.

Our concept builds on previous work with various collaborators developing liquid noble gas readout. Notable here is work on readout of charge using GEM structures in the saturated gas phase above liquid argon [6]. Here it was demonstrated with a test device with 1cm thick liquid argon, that charge can be extracted from the liquid phase into the gas and amplified with a triple GEM mounted above. Operation was achieved in single electron counting mode in the gain region from 6,000 to 40,000 with good stability. A parallel experiment using GEMs and Micromegas in liquid xenon was also able to achieve amplification in the gas phase of a two-phase xenon test device [7]. Here a gain of 500 was achieved at 171 K in 1450 Torr. In this case it was found that stability could not be achieved in pure cryogenic saturated xenon gas without the addition of a quench gas. Typically 2% Methane was added to the gas without apparent detrimental effect to charge transport in the liquid. Micromegas has also been successfully operated in various gases, including the negative ion gas CS_2 [8].

Building on these efforts, attempts have been made to obtain charge gain in the liquid using GEM or THGEM structures. However, stability is found to be very hard to obtain, particularly with liquid xenon, given the high electric fields required within the microstructures when using noble liquids. This has led to the alternative concept in which use is made of the potential secondary electroluminescence in the liquid that should be present in the THGEMs, even at lower fields where stability and freedom from breakdown should be more easily obtained. To investigate this requires

Journal of Physics: Conference Series **308** (2011) 012019 doi:10.1088/1742-6596/308/1/012019

precision photon detection in the liquid. In principle, PMTs could be used. However, it is important to understand the behaviour of light production generation in the THGEM holes. For this a new device has emerged that is ideal, the silicon photodiode, for instance as produced by Sensl Ltd. Such a device can be mounted above individual holes and in principle, with a suitable wavelength shifter, be used to study photon production. CCDs where also considered. However, one of the advantages of SiPDs is that the front end electronics and amplification can be positioned away from the sensor and so removed as a potential heat source. This is very difficult for CCDs where the electronics is integral to the chip.

SiPD devices are not specifically designed for cryogenic use, hence extensive studies have been undertaken by us to check operation, stability and gain characteristics at low temperarture [9]. The particular device used had pixel size of 20 μm with 848 cells and geometric efficiency of 43%. Our studies in cold saturated argon gas showed that the device has excellent characteristics at the required temperature of -196 ^{0}C, in fact showing improved gain of up to 3 x 10^6 and reduced dark count of a few 10s Hz, compared to >10^6 at room temperature. The photon detection efficiency was found to be 25% at 460 nm and 11% at 680 nm. Fig. 1 shows typical gain characteristics of the SiPD vs. low temperature. It can be seen from this that for a given bias voltage greater gains are achieved at liquid argon temperatures than at room temperature by a factor 2-5.

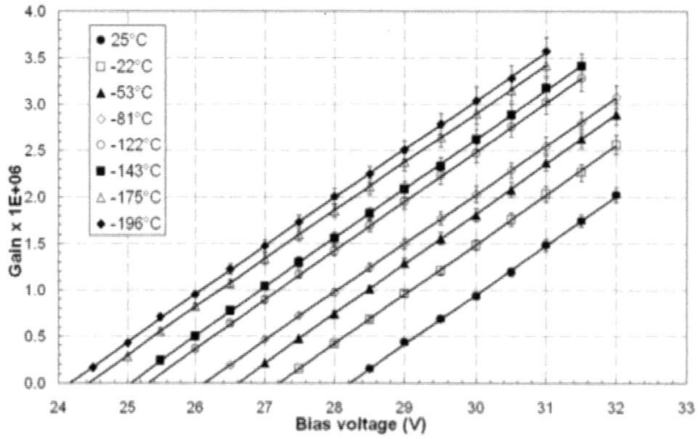

Figure 1 Typical gain characteristics of the Sensl SiPD vs. low temperature.

3. Optical readout test set-up

Based on the Sensl device, a liquid argon stand was constructed that allowed a 1 mm^2 SiPM device to be positioned directly above the centre of a 65 mm diameter THGEM. This was located above a 20 mm drift region defined by a woven steel cathode. Fig. 2 shows a schematic of the test set-up. Further details can be found in ref. [10]. The THGEM was manufactured in-house from a double faced copper clad FR-4 epoxy resin glass reinforced composite plate of thickness 1.5 mm, hole diameter 1 mm, pitch 1.5 mm. Only the central 45 mm diameter region of the THGEM was perforated. No dielectric rims were etched around the THGEM holes. On completion of CNC machining, the THGEM was immersed in 10 M nitric acid to smooth surface irregularities.

Considerable attention was paid to the purification of the liquid argon in the tests to achieve the required level of less than 30 ppb. N6 gaseous argon was first passed from its cylinder through a purification cartridge containing a 1200 g blend of powdered copper and phosphorous pentoxide to remove the bulk of oxygen and water respectively. The argon gas was then passed through a SAES getter at a flow rate of 5 L/min at 1.2 bar to remove oxygen and water to less than 1 ppb. To allow continued removal of impurities such as physisorbed water from within the target during operation, an additional purification cartridge was positioned at the base of the chamber directly below the target assembly. In addition to copper and phosphorus pentoxide, this cartridge contained molecular sieves to remove mineral oil, turbo and rotary pump oil.

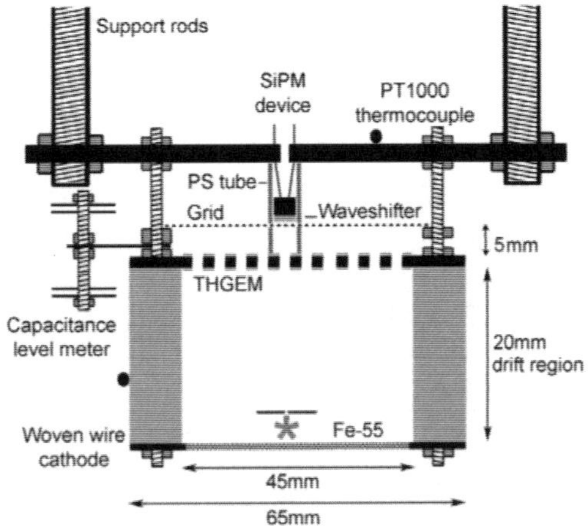

Figure 2 Internal assembly used to detect secondary scintillation in liquid argon.

As seen in Fig. 2 the THGEM was attached to the drift region below, and to the SiPM device above, using long polypropylene bolts. The SiPM device was shielded from the field within the THGEM by a grounded high optical transparency woven steel mesh grid, positioned between the THGEM and the SiPM device in order to deflect electrons passing through the THGEM holes back towards the THGEM top electrode, each hole acting as an independent amplifier. The transparency of the THGEM was 40%. In addition two 5 mm diameter polystyrene cylinders were fitted between the SiPM and the THGEM to isolate the SiPM device from background scintillation light from the argon target. A 50% concentration of the waveshifter tetraphenyl butadiene (TPB) in a mineral oil based diblock copolymer elastomer was applied to the SiPM face.

4. Readout results

The test rig of Fig. 2 was used to investigate the charge and optical response of the THGEM and the SiPD to interactions from ^{55}Fe x-rays in the liquid argon in various configurations. The charge signal, read from the top THGEM electrode, was decoupled from the high voltage line and passed through an Amptek A250 charge sensitive preamplifier and a shaping amplifier to an Acqiris PCI acquisition system, triggered through a discriminator unit. Tests were performed first with two phase argon to establish operation characteristics for the THGEM and SiPD in cold saturated gas above the liquid. Fig. 3 (left) shows an example charge spectrum of ^{55}Fe events taken from cold 1 bar gaseous argon in two phase mode. Here the THGEM was operated at 4.685 kV with a drift field in the liquid of 2.5 kV/cm and in the gas of 4.0 kV/cm. The gain in this test was 300. By adjusting the drift fields it was possible to check that the events were indeed from the ^{55}Fe source.

Turning to the SiPD, we show in Fig. 3 (right) the secondary photon spectrum from electroluminescence generated in the THGEM from the same source as recorded by the SiPD, using the same two phase set-up and the same drift voltages. The SiPM over-voltage was 1 V. Here again a clear ^{55}Fe spectrum is seen with escape peak below 3 keV and evidence for single photoelectron structure at low energy. The main peak was found to correspond to 86 photoelectrons. This demonstrates stable operation of the SiPD at the necessary cryogenic temperatures relevant to liquid argon. Again confirmation of the source detection comes from changing the drift fields and observing changes in the gain and position of the spectral peaks.

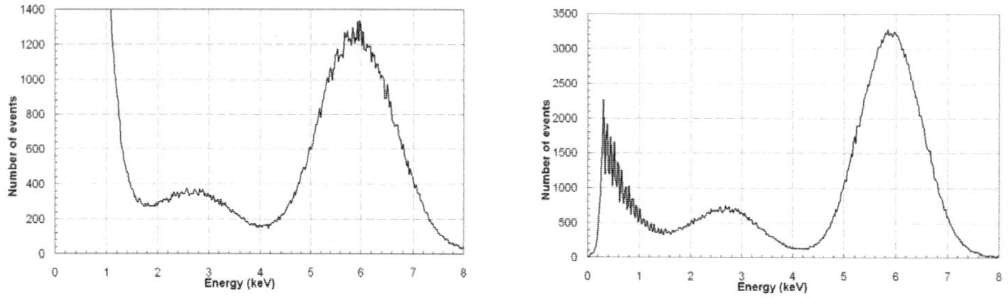

Figure 3 Spectral response from ^{55}Fe x-rays for operation in two phase liquid argon showing (left) charge readout from the THGEM and (right) optical readout of electroluminescence from the THGEM using the Sensl SiPD in the gas phase, all taken using the same drift voltages.

Operation in the two phase mode demonstrated also that the secondary scintillation is initiated in advance of the charge multiplication and consistently produces higher gain at comparable voltages.

The next, and most crucial, test was to increase the level of liquid in the apparatus to cover both the THGEM and SiPD and allow investigation in single phase liquid argon only. Study in this mode first demonstrated that generation of any measureable charge gain from the THGEM submerged in the liquid was not possible. However, as the voltage across the THGEM was raised above 8 kV scintillation light became clearly visible, yielding eventually an observable ^{55}Fe spectral peak. Fig. 4 (left) shows an example of this for a THGEM voltage of 10.15 kV with drift field in the liquid of 2.5 kV/cm. The peak here corresponds to 123 photoelectrons. As the voltage across the THGEM was increased the proportion of events containing discharge sparking also increased and this eventually limited the useful gain. To illustrate confirmation of the detection of the electroluminescence we show in Fig. 4 (right) the result of operation with the same parameters but with zero drift field in the liquid. As expected only noise from the SiPD is now observed, because charge from the signal events is no longer transported to the THGEM.

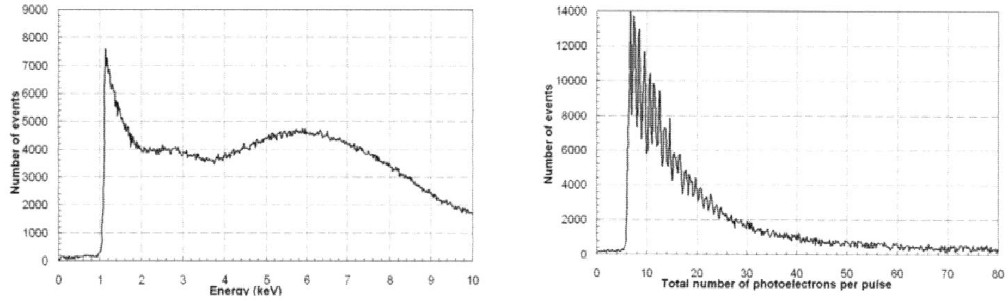

Figure 4 Spectral optical response to ^{55}Fe x-rays for operation in single phase liquid argon where the THGEM, with voltage of 10.15 kV, and SiPD, are submerged in the liquid for: (left) operation with liquid drift field of 2.5 kV and (right) zero drift field.

The results of Fig. 4 demonstrate for the first time the proof of principle that optical readout and gain in single phase liquid argon can be achieved in a stable configuration, at least on a small scale (further details are provided in [10]. The resolution for the optical readout shown at this stage is seen to be degraded somewhat compared to that of the cryogenic gas operation (see Fig. 2). There are several reasons for this, notably the 10μs integration time required to fully acquire the slow component time constant of 1590 ns. Since the software records all photoelectrons generated within this long time window the data will most likely contain in addition to an ^{55}Fe pulse, contributions from primary scintillation due to background interaction with liquid argon contained within the polystyrene tube, effects generated by the SiPM device, and also photons created due to sparking within the THGEM operating close to the breakdown threshold. However, overall, we can state that for a SiPM over-voltage of 1 V, a THGEM voltage of 9.91 kV, and a drift field of 2.5 kV/cm, a total of 62 +/- 20 photoelectrons were produced at the SiPM device per ^{55}Fe event, corresponding to an estimated gain of 150+/-66 photoelectrons per drifted electron.

5. Implications of single phase optical readout and alternative concepts

The new liquid argon detection technique described above opens great potential as a route to cost-effective, large volume, simultaneous tracking and calorimetry targets with excellent performance relevant to neutrino physics. The tests demonstrate that secondary scintillation in the liquid can clearly be observed by the SiPD-THGEM arrangement with gains at least in the region 100-200. At higher gains the resolution degrades in the current set-up. However, there are likely ways to mitigate against this with improved geometry, such as to reduce events from primary scintillation.

Nevertheless, there are issues to address, in particular to improve understanding of the gain vs. light characteristics of the THGEM holes. So far only one hole as been used. There are likely to be effects related to the quality of the manufacturing concerning the dependence on surface treatment in relation to high fields generated on any protrusions. It is also possible that corona discharge from local field instabilities caused by mechanical artifacts in the THGEM hole may be leading to positive photon feedback, resulting in premature breakdown and the narrow operating regime at high voltage. This issue is widely held to be a fundamental reason why significant charge gain has been unattainable using micropattern devices in liquid noble gases. Regarding SiPDs, several improvements are underway by manufacturers. Notably to improve dynamic range and produce larger area devices with more pixels.

The work with SiPDs suggests several routes toward scale-up arrays, for instance using fibre optics and/or the Anger camera technique discussed in Sec. 2. However, alternative optical readout ideas are also being investigated. One possibility is to use new UV sensitive integrated micromegas technology [11]. This detector consists of a CMOS imaging array, a gaseous-detector structure with a Micromegas layout and a UV-photon sensitive CsI reflective photocathode all monolithically integrated using simple post-processing steps. Typical imaging areas of 14 x 14 mm with 256 x 256 pixels have been produced, based on the InGrid technology and Timepix chip variant of the Medipix2 chip. The detector needs to be confirmed for operation in cryogenic liquids and looks costly for large areas.

Another possibility is to use Charge-Coupled Devices (CCDs) [12]. Recent breakthroughs in CCD development toward Charge-Coupled CMOS Pixel devices offer possibility of low noise, high QE and high speed (1-100 MFrames/sec). There is growth to larger area devices and smaller pixels (50 μm) and cryogenic use (e.g. for space). There are several variants, but the key features are: (a) collection of signal charge on a fully-depletable structure (PG or PPD) with large capacitance, (b) shield in-pixel electronics with a deep p-implant, (c) use of sense 'baseline' voltage on the gate of the submicron transistor having minimal capacitance, (d) transfer of the entire signal charge to this gate and sample again, *promptly*. Such devices might offer an effective way of imaging the light from THGEM holes in the gas or liquid as a cost effective way of reading out the light and performing tracking.

The CCD concept itself, essentially of imaging a gaseous TPC (Time Projection Chamber) using CCDs, has been studied for some years for dark matter experiments. For instance, in early work for the DRIFT experiment [13], and more recently by the DM-TPC collaboration [14]. Fig. 5 shows an example of an alpha particle track imaged optically from luminescence generated by a GEM in a TPC with 20 Torr Ar (90%) + CH_4 (10%) + TEA and recorded with a CCD from the DRIFT work [13].

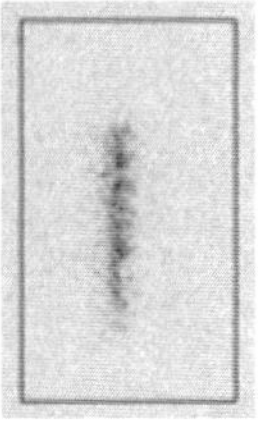

Figure 5 Example of an alpha particle track imaged optically from luminescence generated in a GEM and recorded with a CCD [13].

There are however several potential pitfalls to the CCD optical concept in scale-up operation for liquid argon. One issue is the need for VUV optics required to collect the light, including arrangement for the space required to mount the lenses. This may mean consideration is needed for use of MgF and/or wavelength shifters. Another issue is power consumption, particularly in the case of operation in liquid where it is vital to prevent evaporation. This arises because the CMOS electronics is integral to the CCD. Designs will likely need to minimize pixels numbers and power dissipation. The CCDs will need to be exceptionally fast also, to ensure allow measurement of the drift distance. Finally, recent work by Bondar et al. [15] with Geiger Mode APDs have also now demonstrated readout of secondary electroluminescence light from THGEMs in a two phase device, but achieving this without use of wavelength shifter. This was possible through use of the near infra-red emission, found with 20% photo-detection efficiency. An avalanche gain of 60 was achieved with 1.4 photoelectrons per electron, with ~180 p.e. recorded per 60 keV scintillation event (from an ^{241}Am source) in the liquid.

6. Conclusion

There is evidence now, at proof-of-principle level, that optical readout of secondary electroluminescence in liquid argon could provide an alternative to charge readout. This has the potential to provide the necessary gain required for neutrino physics applications without the need for the complexities of two-phase operation. One solution would be to use an array of SiPDs as the optical sensors in the liquid. Alternative optical readout options may be possible, such as CCDs, and some may be better suited to reading out the light in the gas phase. Although reverting to two phase operation, such concepts would still have the possible advantage of simplification of the electronics.

References

[1] Amerio S et al., 2004 *Nucl. Instrum. Meth.* A **527** 329
[2] Lippincott W H et al., 2008 *Phys. Rev.* C **78** 035801
[3] Boccone V et al., 2009 *JINST* **4** P06001
[4] Mei D M et al., 2009 arXiv:0912.5368v1
[5] Rubbia A et al., 2009 *J. Phys. Conf. Ser.* **171** 012020
[6] Bondar A et al., 2007 *Nucl. Instrum. Meth.* A **574** 493
[7] Lightfoot PK et al., 2005 *Nucl. Instrum. Meth.* A **554** 266
[8] Lightfoot P K et al., 2007 *Astropart. Phys.* **27** 490
[9] Lightfoot P K et al., 2008 *JINST* **3** P10001
[10] Lightfoot P K et al., 2009 *JINST* **4** P04002
[11] Melai J et al., arXiv:1003.2083
[12] Damerell C, 2010 *private communication* RAL, Didcot, UK
[13] Lawson T, PhD Thesis, 2002 University of Sheffield, UK
[14] Roccaro A et al., 2009 *Nucl. Instrum. Meth.* A **608** 305
[15] Bondar A et al., 2010 arXiv:1003.1597

Light Readout Optimisation using Wavelength Shifter - Reflector Combinations

Konstantinos Mavrokoridis

Department of Physics, University of Liverpool, Oliver Lodge Lab, Oxford Street, Liverpool, L69 7ZE, United Kingdom

E-mail: k.mavrokoridis@liverpool.ac.uk

Abstract. The use of reflectors coated with a wavelength shifter (WLS) along with standard bialkali PMTs is an economical method for an efficient readout system for vacuum ultra violet (VUV) light produced in large liquid argon detectors. Various thicknesses of tetraphenyl butadiene (TPB) were deposited by spraying and vacuum evaporation onto both specular $3M^{TM}$-foil and diffuse $Tetratex^{TM}$ (TTX) reflectors. 128 nm VUV light generated in 1 bar argon gas by a 5.4 MeV α source was detected by a 3-inch bialkali borosilicate PMT within a 1 m tube lined internally with a TPB coated reflector. The light collection was recorded as a function of separation between source and PMT for each combination of coating and reflector for distances up to 1 m. Reflection coefficients of TPB coated reflectors were measured using a spectroradiometer. WLS coating on the PMT window was also studied. The optimum coating and reflector combination was TPB evaporated on TTX. Measurements with coating thicknesses of 0.2 mg/cm^2 and 1.0 mg/cm^2 yielded a similar performance. The best PMT window coating is obtained by TPB evaporation of 0.05 mg/cm^2.

1. Introduction

Tonne scale liquid argon (LAr) targets such as that used in the Argon Dark Matter (ArDM) experiment [1, 2, 3] typically require in excess of ten large area PMTs for acceptable light readout. The argon scintillation light due to neutral or charged particle excitation is in the vacuum ultraviolet (VUV) centered at about 128 nm [4, 5, 6]. Currently large MgF_2 windowed PMTs that can detect 128 nm light are commercially not available. An alternative technique is to apply wavelengh shifting chemicals on borosilicate windowed PMTs thus shifting VUV argon scintillation into the visible spectrum, the typical quantum efficiency (QE) of a cryogenic (Pt-underlay) borosilicate windowed PMT being from 10 to 15% at approximately 430 nm. Tetraphenyl butadiene (TPB) powder has been used which has an above average Stokes shift and can absorb 128 nm light, emitting in the required visible PMT region [7, 8, 9, 10]. The argon scintillation light is characterised by two distinct decay times - a slow component, τ_2 (triplet eximer), and a fast component, τ_1 (single eximer)[5, 11]. Determination of the time constants is through a multiple parameters nonlinear least square fit with additional degrees of freedom related to the height, start time and baseline of the pulses. The decay time of the slow component, τ_2, increases with the increase of argon purity and therefore can be used as a measure of the purity of argon. From the literature the purest gas argon has a τ_2 of 3200 ± 300 ns [11]. The analysis of the measurements described in subsections 3.2, 3.3 was based around this property of argon scintillation. Section 2 describes TPB coating methods and sample preparations. Section 3 outlines three experiments aiming for reflector selection, optimisation of TPB coating thickness and deposition technique and presents the results obtained. These results led to the optimisation of the light readout system in the ArDM experiment [12].

2. WLS Coating Techniques and Reflectors

2.1. Reflector type

Two materials, ESR (Vikuiti^{TM} Enhanced Specular Reflector foil) from the company 3M and Tetratex^{TM} (TTX) from the company Donaldson Membranes were investigated. 3M foil is a multilayer specular reflecting polymer film and as such is likely to be of high radio-purity. Its appearance is that of a polished metal although the material is non conducting by its nature. It has a specular reflection coefficient of approximately 100% in a large region of the optical spectrum. TTX is an aligned polytetrafluoroethylene (PTFE) fibrous cloth and is nearly a 100% diffuse lambertian reflector.

Because of the method of manufacture, which relies on extrusion of polymer chains within an oil based emulsion, doubts have been raised concerning both the purity of TTX and its outgassing rate. In order to investigate the radio-purity of TTX, samples were sent to Harwell Scientifics for analysis by inductively coupled plasma mass spectroscopy (ICP-MS). TTX radiopurity was found to be 1.0 ± 0.3, < 0.4 and < 500 ppb for U, Th and K concentrations respectively. The outgassing properties were checked with a dedicated sample of 48 g of 254 μm TTX (equivalent to 1/20 of the total amount required to line the ArDM experiment). The sample was placed within a large chamber which was evacuated. A series of measurements confirmed the good outgassing properties of TTX.

2.2. TPB deposition techniques on substrates and PMT windows

TPB powder can be applied to a reflector or PMT window by vacuum evaporation, spraying, or by dissolving in a polymer matrix [7, 8]. Vacuum evaporation was performed in an Edwards model E308 evaporation chamber. TPB powder, which has a melting point of 207 ^{o}C, was heated electrically in the vacuum chamber by applying 24 A current to a molybdenum sample holder containing up to 3 g of powder. The reflector/PMT window was placed above the TPB powder at a fixed distance and the coating thickness was controlled by varying this distance and the weight of the powder. Sprayed coatings were prepared by dissolving TPB in toluene in a ratio of 1 to 40. This solution was then airbrushed onto the substrate using 1.2 bar argon gas. The polymer matrix coatings were prepared using long chain paraloid or polystyrene plastic fragments dissolved in toluene. An amount of TPB was added and dissolved isotropically. A known amount of liquid was then syringed onto the substrate. The TPB concentration within the solution was varied, as was the amount of liquid applied to the substrate. The solution was left for three hours to allow the toluene to evaporate, forming clear TPB impregnated plastic.

3. Experimental Procedures and Results

3.1. Reflection coefficient measurements of TPB coated reflectors

An Optronics OL750 spectroradiometer with an OL740-70 integrating sphere [13] has been used to measure reflectance and global fluorescence from 200 nm to 700 nm for a range of samples deposited by spray, evaporation, and polymerisation over a variety of thicknesses and concentrations and on a range of substrates. For a full description of the the experimental setup see [12, 14].

The main results of the experiment are shown in Figure 1. The wavelength on the horizontal axis refers to the incident light only, selected by the monochromator. Reflection coefficients greater than one indicate fluorescence. The reflection coefficient for 3M foil decreases approximately by 5% as the TPB coating thickness drops from 4.0 mg/cm^2 to 0.1 mg/cm^2 at 450 nm. The maximum reflectance coefficient for TPB coated 3M foil is 0.95 ± 0.013 at 430 nm

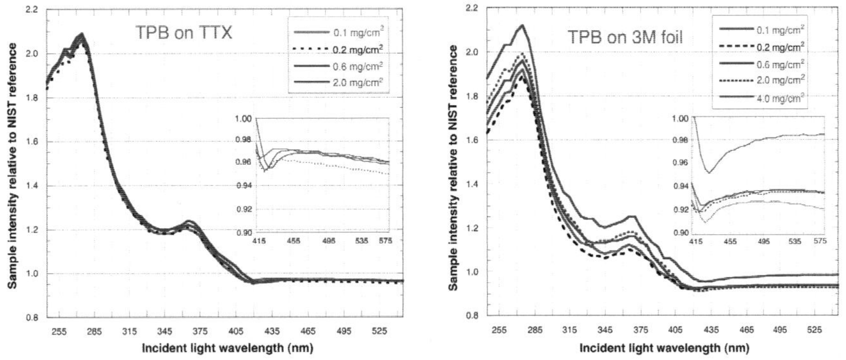

Figure 1. Total reflectance of (left) single thickness TTX cloth; (right) TPB evaporated onto $3M^{TM}$ foil. The insets show an expanded scale in the wavelength range 415 to 575 nm. Taken from [12].

and corresponds to 4.0 mg/cm^2 thickness. Whereas, the reflection coefficient is invariant, within error, with coating thickness for evaporated TTX samples and is 0.97 ± 0.014 at 430 nm.

3.2. Global efficiency of wavelength shifting and reflection with distance

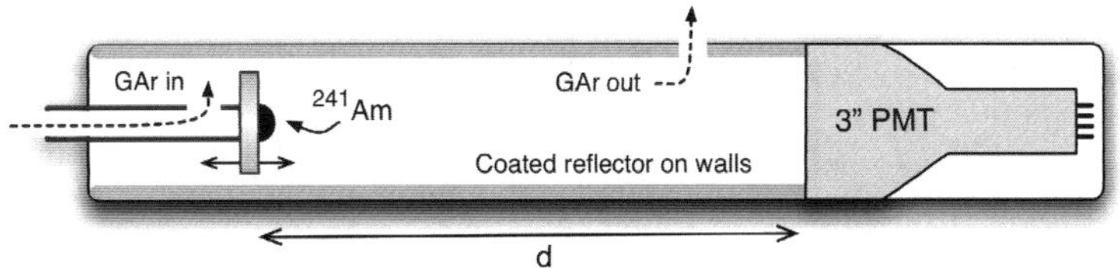

Figure 2. A schematic illustration of the argon gas apparatus used to determine the wavelength shifting and reflection efficiency with distance. Taken from [12].

A schematic illustration of the argon gas apparatus which was used is shown in Figure 2. This apparatus consisted of a sealed polyvinyl chloride (PVC) tube containing a 3 inch uncoated PMT (electron tube type 9302KB). An α source located at the centre of a TPB coated reflector disk was placed within the tube at a fixed distance from the PMT (the total deposition of alpha energy occurs within 4 cm in 1 bar of gaseous argon). Samples of either 3M foil or TTX cloth coated with TPB were placed around the interior walls of the tube. A delivery tube was inserted into the PVC chamber and argon (1 ppm impurity) gas at 1 bar flowed throughout the apparatus. The argon flow rate was used to control the argon purity. Measurements were taken for varied TPB thicknesses between 0.2 mg/cm^2 and 4.0 mg/cm^2, which were deposited both via evaporation and spraying. Additionally, the distance between the α source and the PMT was altered in order to investigate the effect of both the attenuation of light following multiple reflections and the reduction in direct VUV light incident on the PMT. The number of photoelectrons collected at the PMT for each separation (defined as the total area of the light pulse) was then plotted against the slow component decay time (τ_2) for various distances d.

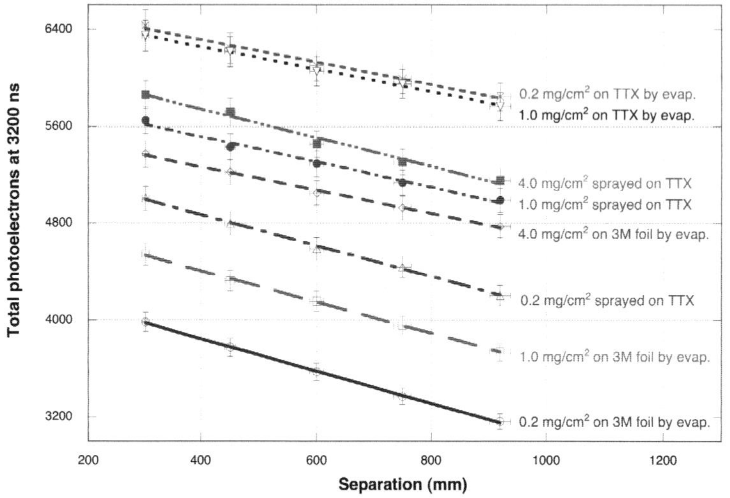

Figure 3. Total photoelectrons for 3200 ns purity against separation from the α source to PMT for TPB coated reflector walled tube. Figure taken from [14].

The results of the analysis are presented in Figure 3. The reduction in the total light collection with increasing distance was found, within errors, to be independent of TPB thickness and substrate. Evaporated coatings on 3M foil consistently underperformed irrespective of coating thickness compared to TTX cloth. Thicker coatings on 3M foil yielded higher light collection whereas light collection from TPB coated on TTX substrates was found to be almost independent of thickness. The $0.2\,\mathrm{mg/cm^2}$ TPB on TTX yielded within errors an identical result to the $1.0\,\mathrm{mg/cm^2}$ coating.

3.3. Wavelength shifting direct light incident on the PMT

Figure 4a shows the argon gas apparatus for direct light measurements. The experiment consisted of a sealed PVC tube containing a 3 inch coated PMT. The PMT window was coated with TPB powder with thicknesses ranging from 0.02 $\mathrm{mg/cm^2}$ to 2 $\mathrm{mg/cm^2}$ via evaporation, spraying and application of a polymer matrix containing TPB. The sides and base of the PVC tube were covered with 3M foil reflector coated with $1\,\mathrm{mg/cm^2}$ TPB powder by evaporation. TPB coated reflector walls were used as the ability of the window coating to shift VUV light is equally important as its ability to allow shifted visible light from the walls to penetrate. An α source was positioned 10 cm away from the PMT window and argon gas was flowed continually. The effect of various PMT window coatings on the total light collection was then recorded by plotting the slow component decay time (τ_2) against the total number of photoelectrons collected at the PMT.

Figure 4b presents the results for the optimum PMT coatings. The optimum thickness was $0.05\,\mathrm{mg/cm^2}$by evaporation, improving the total light collection by $38\,\%\pm3.4\,\%$ at 1000 ns purity compared to that collected with no PMT coating.

4. Conclusions

TPB can efficiently shift the VUV argon scintillation light into the visible blue high quantum efficiency region of low cost borosilicate windowed bialkali photomultiplier tubes. TTX cloth was found to be a preferable reflector when compared with $3\mathrm{M^{TM}}$ foil due to its better light yield and

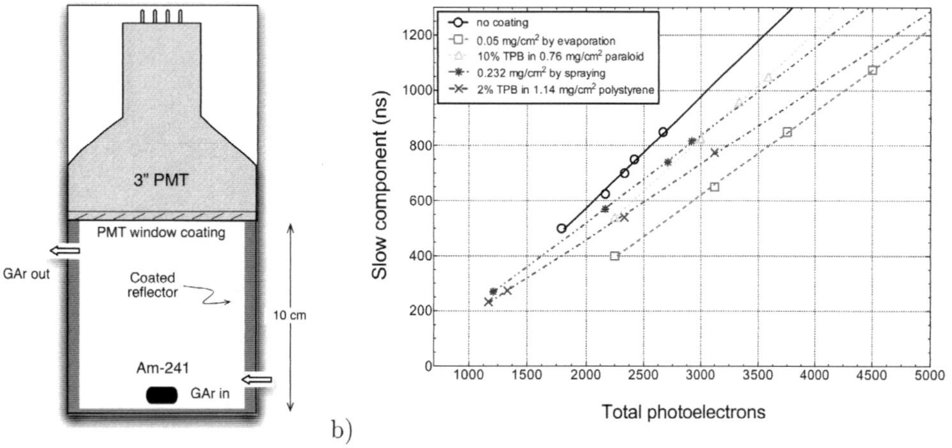

a) b)

Figure 4. a) A schematic illustration of the argon gas apparatus used to determine the wavelength shifting efficiency of direct light incident on a TPB coated PMT. b) Results for the best PMT window coatings (polysterene and paraloid matrices, evaporation and spray). Figure taken from [14].

large tolerance for the TPB layer thickness. Furthermore, the best TPB deposition method was found to be vacuum evaporation which avoids crystallisation and coating inhomogeneities. Based on spectroradiometer measurements, TPB coated TTX was found to have 97% ±1.4% reflectance at 430 nm for all coating thicknesses. The optimum coating and reflector combination was found to be between 0.2 mg/cm^2 and 1.0 mg/cm^2 TPB deposited via evaporation on TTX cloth. The best PMT window coating was found to be 0.05 mg/cm^2 TPB deposited via evaporation.

References

[1] M. Laffranchi and A. Rubbia. The ArDM project: a liquid argon TPC for Dark Matter Detection. *Journal of Physics: Conference Series*, 65:2014, Apr 2007.

[2] A. Rubbia. ArDM: a ton-scale liquid argon experiment for direct detection of Dark Matter in the Universe. *J. Phys. Conf. Ser.*, 39:126, 2006.

[3] A. Marchionni. ArDM. *These Proceedings.*

[4] T. Doke *et al.* Estimation of absolute photon yields in liquid argon and xenon for relativistic (1 MeV) electrons. *Nucl. Instr. and Meth.*, 291:617, Jun 1990.

[5] S. Kubota *et al.* Evidence for a triplet state of the self-trapped exciton states in liquid argon, krypton and xenon. *Journal of Physics C*, 11:2645, Jun 1978.

[6] S. Kubota *et al.* Recombination luminescence in liquid argon and in liquid xenon. *Physical Review B*, 17:2762, Mar 1978.

[7] D. N McKinsey *et al.* Fluorescence efficiencies of thin scintillating films in the extreme ultraviolet spectral region. *Nucl. Instr. and Meth.*, 132:351, Nov 1997.

[8] G. J Davies *et al.* UV quantum efficiencies of organic fluors. *Nucl. Instr. and Meth.*, 117:421, Oct 1996.

[9] I.B Berlman. *Handbook of Fluorescence Spectra of Aromatic Molecules.* Academic Press, New York and London, 1965.

[10] J. M Flournoy *et al.* Substituted tetraphenylbutadienes as fast scintillator solutes. *Nucl. Instr. and Meth.*, 351:349, Dec 1994.

[11] J. W Keto *et al.* Production Mechanisms and Radiative Lifetimes of Argon and Xenon Molecules Emitting in the Ultraviolet. *Phy. Rev. Let.*, 33:1365, Dec 1974.

[12] The ArDM Collaboration. Development of wavelength shifter coated reflectors for the ArDM argon dark matter detector. *JINST*, 4:6001, June 2009.

[13] Optronic laboratories, inc., www.olinet.com.

[14] K. Mavrokoridis. *Characterisation of Liquid Xenon and Argon as Targets for Direct Dark Matter Detection.* PhD thesis, University of Sheffield, 2009.

Journal of Physics: Conference Series **308** (2011) 012021 doi:10.1088/1742-6596/308/1/012021

Cold electronics for "Giant" Liquid Argon Time Projection Chambers

Veljko Radeka[1*], Hucheng Chen[1], Grzegorz Deptuch[2], Gianluigi De Geronimo[1], Francesco Lanni[1], Shaorui Li[1], Neena Nambiar[1], Sergio Rescia[1], Craig Thorn[1], Ray Yarema[2], Bo Yu[1]

[1] Brookhaven National Laboratory, Upton, NY 11973-5000, USA
[2] Fermi National Laboratory, Batavia, IL 60540, USA

[*]Correspondence, e-mail: radeka@bnl.gov

Abstract. The choice between cold and warm electronics (inside or outside the cryostat) in very large LAr TPCs (>5-10 ktons) is not an electronics issue, but it is rather a major cryostat design issue. This is because the location of the signal processing electronics has a direct and far reaching effect on the cryostat design, an indirect effect on the TPC electrode design (sense wire spacing, wire length and drift distance), and a significant effect on the TPC performance. All these factors weigh so overwhelmingly in favor of the cold electronics that it remains an optimal solution for very large TPCs. In this paper signal and noise considerations are summarized, the concept of the readout chain is described, and the guidelines for design of CMOS circuits for operation in liquid argon (at ~89 K) are discussed.

1. Introduction

The Liquid Argon Time Projection Chamber (LAr TPC) is a very promising detector technology for future beam-based neutrino research and nucleon decay research. This technology offers extraordinarily precise event reconstruction and particle identification, as well as scalability to very-large detectors. To go beyond the sensitivities of current experiments for neutrino physics and proton decay, the next generation of liquid argon TPCs are envisioned to be in the range of 20-100 kton, and must be placed deep underground. Detectors of this size pose many engineering challenges. In order of difficulty and cost these are: 1) mechanical and thermal design of the cryostat including the cryogenic system: 2) design, construction and installation of the signal and drift electrode system; 3) electronic readout system.

The number of sense wires, i.e., input signal channels is expected to be in the range of 5-7 hundred thousand for a 20 kton scale TPC (up to several million for 100 kton). To achieve a good signal to noise ratio, while scaling to the large sized detectors, we are developing CMOS ASICs with high degree of multiplexing (128 – 1024) operating inside the cryostat. These front end circuits will be placed close to the sense wires, eliminating the need for long cables which would present major contributors to electronics noise (due to their capacitance and resistance) and to contamination of the high purity liquid. The number of output signal links will be greatly reduced due to the multiplexing, greatly reducing the number of cryostat penetrations (feedthroughs) necessary to read out all the sense wires. The resulting small number of feedthroughs can be placed anywhere on the cryostat (near the end), as there is no need to place them near the wire frames as would be the case with external (warm) electronics). *This gives the cryostat designers complete freedom to choose the optimum size, shape, configuration of the cryostat and the mechanical structure of the top of the cryostat.* It also greatly reduces the risk of leaks in the feedthroughs.

It should be noted that an *a priori* assumption that warm electronics would be the only solution in almost all TPC concepts discussed so far at various workshops and conferences has compelled the designers toward detector concepts with ***very long drift distances (up to 20 meters)*** resulting in very high purity requirements, **very *long sense* wires or *wires from several wire frame modules connected in series*** *and* ***long cables,*** all in an effort to avoid large numbers of signal channels exiting the cryostat. The ***signal-to-noise ratio*** (i.e., the detector sensitivity) suffers twofold, from the ***attenuation of the signal charge*** during a long drift and from the ***increased electronic noise*** due to the capacitance of very long sense wires and long cables.

Cold electronics makes possible an optimum balance among various design and performance requirements.

Pioneering work on LAr TPCs has been done by the ICARUS collaboration [1]. The chamber sizes realized (<1kton of LAr) allowed the use of warm electronics. Truly "giant" detectors require a different approach to the readout, as well as to the overall detector design.

In this paper we summarize the basics of signal formation and electronic noise, present a functional outline of the readout chain and an outline of the front end CMOS ASIC design. We discuss briefly some principal design guidelines to ensure a long life of MOS transistors.

2. TPC electrodes and signal formation

One of the detector concepts under study for the LBNE experiment is based on two or more units, each consisting of a cryostat, with approximate dimensions of $70x15x15\ m^3$, containing ~20,000 tons of liquid argon (LAr) instrumented with time projection chamber (TPC) modules. In each TPC module, ionization electrons created by charged particles produced in neutrino or nucleon decay interactions are drifted up to 2.5 m in a uniform field of 500 V/cm to the readout electrodes.

The simplest electrode configuration which allows track reconstruction with dE/dx measurements consists of three planes of sense wires, arranged one in the vertical direction and two at angles of $\pm 45^{\circ}$ to the vertical, as illustrated in figure 1. These three planes of wires are biased such that electric field lines terminate only on the last plane in the drift direction, so that they collect electrons, while wires in the other two planes (at $\pm 45^{\circ}$) observe induction without collection, allowing two dimensional coordinate determinations in the plane of the wires. This configuration is based on a square cell, where the sense wire pitch is equal to the spacing between the wire planes, figure 2. Under a constraint that each wire is read out, variations in the ratio of these dimensions (i.e., departure from the square cell) offer no overall advantage. More ideal electrode configurations would all require either more wires (to form denser grids), or pad/pixel electrodes in great numbers, where the costs and mechanical construction requirements would become prohibitive.

The total number of wires in each cryostat unit would be on the order of 500,000 to 650,000, for 4 and 3 mm wire spacing respectively. Since each wire spans half the height of the cryostat, each wire approaches ~10 m in length, corresponding to a capacitance of about 200 pF. Each wire is connected directly to one channel of an electronic circuit located at the wire frame in the LAr (at ~89 K).

Journal of Physics: Conference Series **308** (2011) 012021 doi:10.1088/1742-6596/308/1/012021

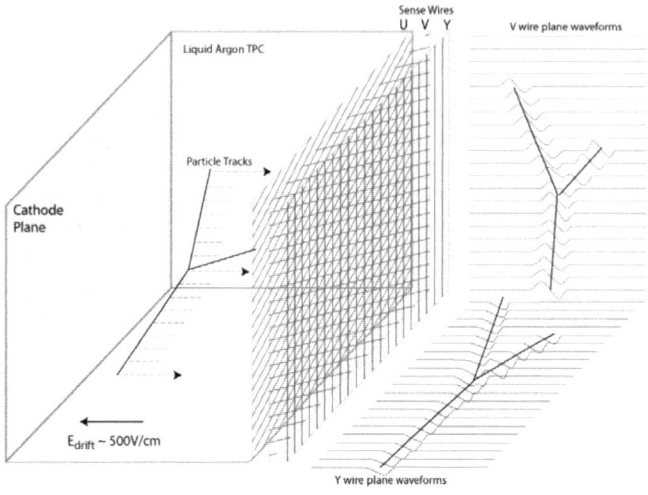

Figure 1. Schematic illustration of a liquid argon TPC: Ionization electrons produced by the passage of charged particles through the liquid argon are transported by a uniform electric field to three wire planes, where they induce electric signals on the sense wires. By recording and analyzing the waveform on each wire, the original track can be precisely reconstructed.

An estimate of the charge signal size on the *collection wire plane* for a track perpendicular to the sense wire and parallel to the wire plane is as follows:

- A 3 mm MIP track should create *210 keV/mm x 3 mm /23.6 eV/e = 4.3 fC.*
- After a 1/3 initial recombination loss (at 500 V/cm): *~2.8 fC* . This would be the maximum signal for tracks close to the wire planes.
- It is expected that the TPC design will maximize the drift path to equal or exceed the charge life time, thereby reducing the signal to 1/e≈0.368.
- The expected signal for *3mm* wire spacing is then *≈1 fC=6250 electrons.*
- The expected signal for 5 mm wire spacing is $\approx 10^4$ *electrons.*
- The ***induction signals*** on the first and second wire plane are smaller as shown in figures 2 and 3.

These signal sizes are for a track perpendicular to the sense wire and parallel to the wire plane. For track inclinations α and β with respect to this orientation, the signal has to be multiplied by $1/(\cos\alpha \cdot \cos\beta)$ and there is a change in the shape of the signals as shown in figure 3. The expected maximum induced charge on a single wire due to a shower would be ~300 fC.

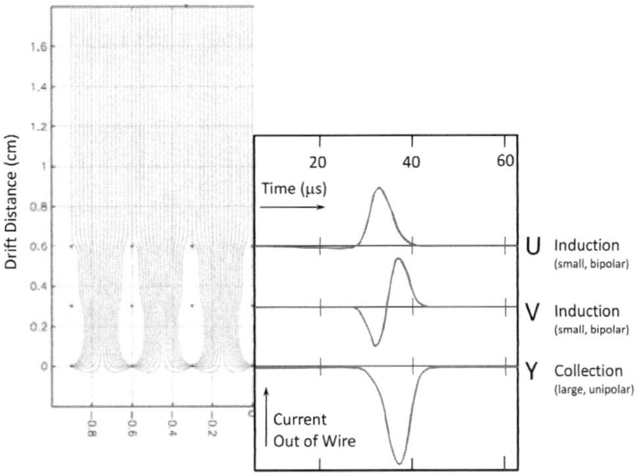

Figure 2. A planar illustration of electric field lines (i.e., electron trajectories) and the signals induced by an ionizing track at 90 degrees to the wire direction and at 0 degrees to the wire planes. The wires in the induction planes *u* and *v* are inclined at +, - 45 degrees with respect to wires in the *y* plane (the wires in this figure are shown all in the same direction for simplicity).

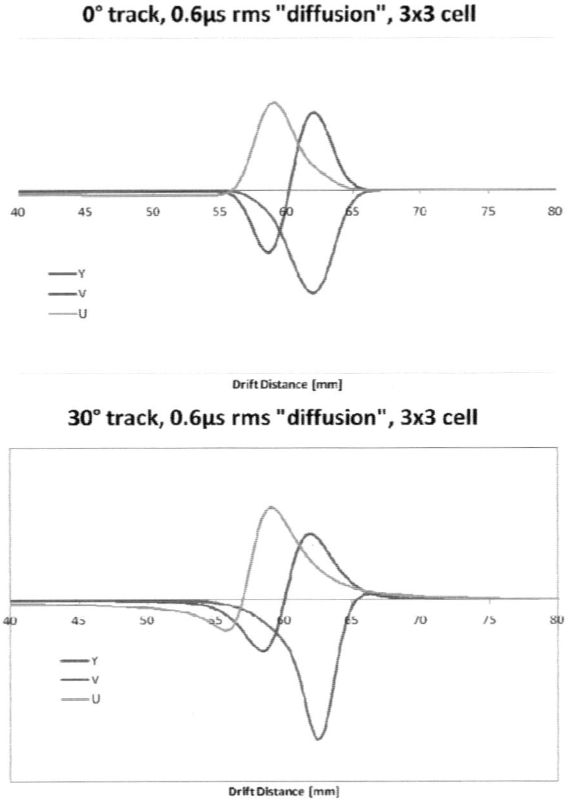

Figure 3. Induced current waveforms for an ionizing track at 0 and 30 degrees to the wire planes. ***Wire plane spacing*** and ***electron drift velocity*** determine the ***time scale*** of the signals and of all the subsequent signal processing. The transit time between the two planes is about 2 and 2.7 microseconds for the 3mm x 3mm and 4 mm x 4 mm cells respectively, for an assumed drift field of 500 V/cm. Both sets of signal waveforms have the same horizontal and vertical scale.

The waveforms in figures 2 and 3 were calculated using Ramo's theorem, as described in [2]. The results of the calculation were then convolved with a gaussian to simulate smoothing and broadening due to charge diffusion. Standard deviation of 0.6 μs was assumed, corresponding to longitudinal diffusion over a drift distance of ~ 0.5m.

3. Electronic Noise

We consider here only the unavoidable sources of noise: 3.1.The ***first transistor noise*** associated with the physics of amplification, and 3.2. The ***thermal noise from the sense wire*** and connection leads (signal cable). We assume that any resistors connected to the amplifier input, such as the feedback and sense wire bias resistors, are chosen so that their thermal noise is negligible.

3.1. The ***first transistor noise*** contribution to the measured signal charge is due to the total capacitance, comprised of the sense wire capacitance, cable capacitance and input transistor capacitance. ***This total capacitance limits the signal-to-noise ratio and it is the one dominant factor on which the feasibility and scalability of a LAr TPC design critically depends.*** This noise contribution is well understood, and the result has been well established analytically and experimentally. The equivalent noise charge (ENC_A) due to the first transistor noise (thermal noise in the case of field effect transistors) can be expressed as,

$$\begin{aligned} &ENC_A^2 = e_n^2 C_{in}^2 / t_p \\ &where: e_n^2 = 4k_B T R_{sn} = transistor\ series\ white\ noise\ spectral\ density\ in\ \left[V/Hz^{1/2}\right] \\ &\quad R_{sn} = transistor\ equivalent\ series\ noise\ resistance \\ &\quad C_{in} = C_{wire} + C_{Cable} + C_{ampl} \\ &\quad t_p = weighting\ function\ peaking\ time \approx 1/bandwidth \\ &\quad k_B = Boltzmann\ constant \end{aligned} \tag{1}$$

Derivation of equation (1) is summarized in figure 4. See also reference [2].

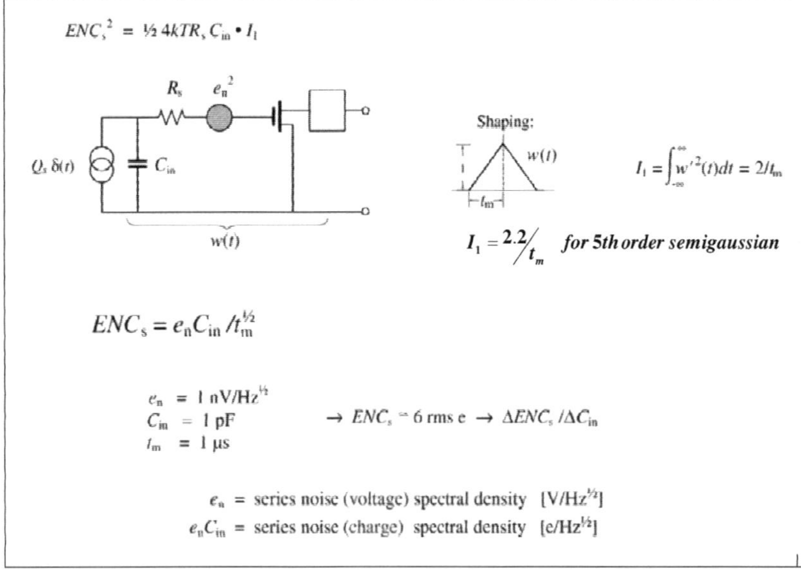

Figure 4. The simple relation for the equivalent noise charge (ENC) due to series white noise requires knowledge of only three parameters: noise spectral density $\boldsymbol{e_n}$, total input capacitance (detector + amplifier) $\boldsymbol{C_{in}}$, and peaking time $\boldsymbol{t_p = t_m}$ of the triangle approximating the weighting function as in figure 5. In a preamplifier design, the expected $\boldsymbol{e_n}$ can be determined from the operating conditions (transconductance) of the first transistor.

3.2 ***Thermal noise from the sense wire*** is calculated based on the transmission line model of the sense wire. The equivalent noise charge can be expressed as,

$$\begin{aligned} &ENC_w^2 \cong 4k_B T C_w^2 \left(R_w/3 \right)\left(1/t_p \right) \\ &where: C_w = sense\ wire\ capacitance \\ &\qquad R_w = sense\ wire\ resistance \end{aligned} \tag{2}$$

The sense wire noise contribution is equivalent to that of a resistor equal to one third of the wire resistance inserted between the receiving end of the wire and the amplifier. This contribution becomes significant for long sense wires (10 to 20 meters) if stainless steel is used. It can be made negligible by using Cu+Au coated stainless steel, or Cu+Be [3].

Both expressions (1) and (2) for ENC contain a weighting function width parameter $t_p = \tau$ (the inverse bandwidth). We assume here the simplest weighting function, illustrated in Fig. 5. This function is optimal for transistor white noise and an impulse signal.

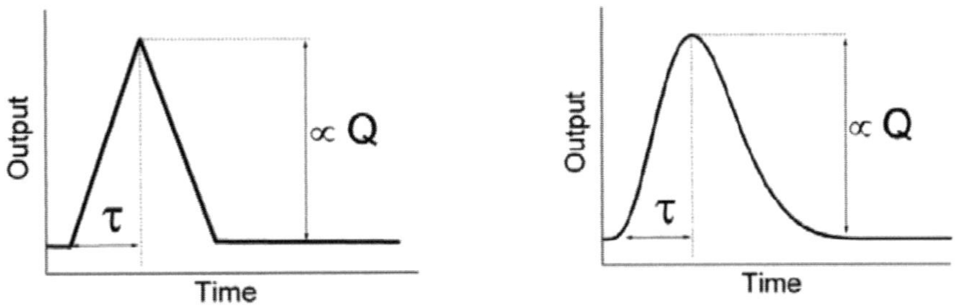

Figure 5. Signal processing weighting functions: ***triangle*** is the theoretical optimum for amplifier white series noise and impulse signal; ***semigaussian*** is easily realized in practical circuits. $t_p = \tau$ in equations (1) and (2), and it results in an ENC within few percent of the triangular one.

In a large TPC we will use different weighting functions to optimize separately the measurements of dE/dx, track segment position coordinates, timing and double track resolution. A function of the type in Fig. 5 will be used for "prefiltering" or "band limiting" prior to sampling. This function will be realized in the front end (preamplifier and pulse shaper) analog circuits. Desired weighting functions for the specific measurements listed above will be realized by weighting multiple samples in subsequent digital processing. For the electron transit times from one wire plane to another of 2-3 microseconds, we choose t_p=1 μs to define prefiltering and the single-sample *ENC*, corresponding to a bandwidth of 1 MHz. This bandwidth requires a sampling frequency of 2 megasamples/second to avoid noise aliasing.

Input transistor: The best one can reasonably achieve with JFETs at 300K for external (warm) readout and with PMOS at 90 K for electronic readout in the cryostat is e_n= 0.5 nV/Hz$^{1/2}$, which corresponds to a resistance R_{sn}=15 ohms. Trying to reduce these values further for warm electronics would bring a diminishing return as the resistance values of the signal cables are approached (~a few ohms). For cold electronics, power dissipation limits any further reduction in the noise.

With these assumptions, two plots of *ENC vs* sense wire length and signal cable length are given in figures 6 and 7, to cover the likely range of parameters to be encountered in a very large TPC, for either external (warm) or internal (cold) electronics.

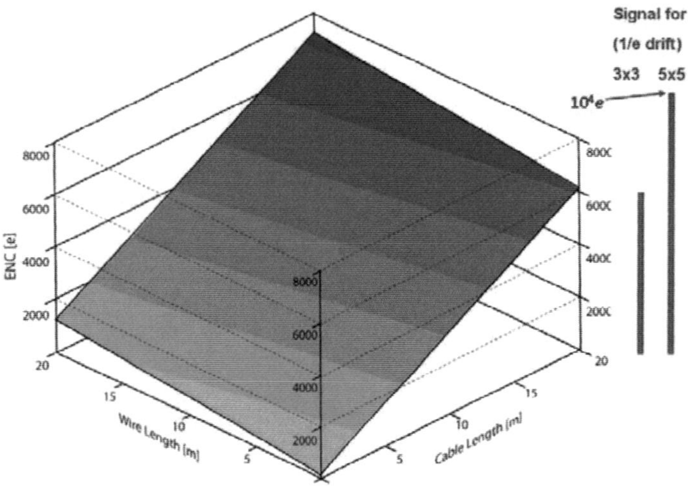

Figure 6. Contour plot of equivalent noise charge (*ENC*) *vs* sense wire length and signal cable length due to amplifier noise with a spectral density of 0.5 nV/Hz$^{1/2}$ and weighting function peaking time t_p=1 μs. Sense wire capacitance is ~20 pF/m and signal cable capacitance ~100 pF/m is assumed (for a different cable capacitance, the cable length scale should be changed). Note that for cold electronics, the cable length is zero. Low resistance sense wire is assumed, figure 7.

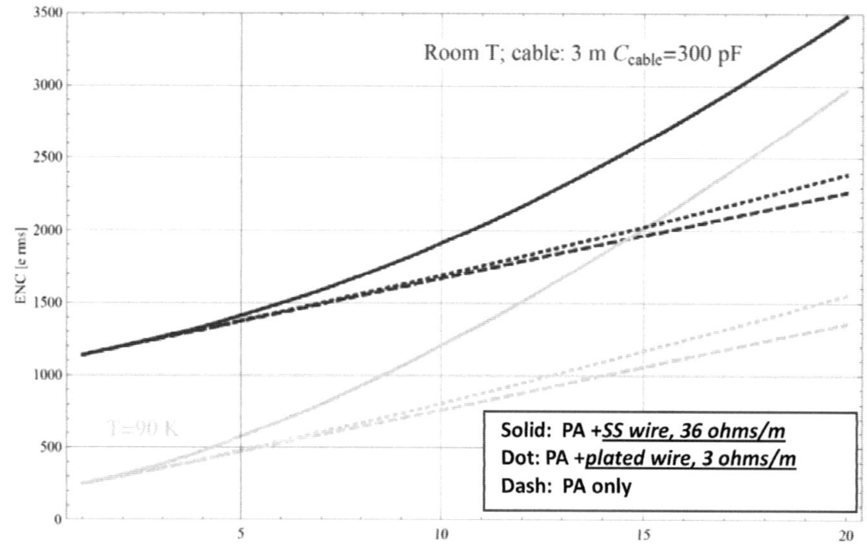

Figure 7. Noise *vs* sense wire length with internal (cold) CMOS electronics in comparison to external (warm) electronics. For calculation purposes, the warm electronics is assumed to have, in the best case, a 3 meter cable. The cold electronics has no cable (only a few cm long connections from the sense wires to the ASIC). Shown is *ENC vs* sense wire length due to amplifier noise with a spectral density of 0.5 nV/Hz$^{1/2}$ for two different sense wire materials, and for weighting function peaking time t_p=1 μs. Sense wire capacitance is ~20 pF/m and signal cable capacitance is assumed ~100 pF/m. For longer wire lengths, low resistance sense wire (such as Cu+Au plated SS wire) offers a significantly lower noise. PA=preamplifier.

4. The readout chain

The readout electronics must digitize signals with 12-bit resolution at a rate of 2 mega-samples per second (MS/s), limited by the longitudinal diffusion of electrons, and compress, buffer and multiplex the data. The power dissipated by each channel must be limited to a few mW to minimize local boiling of the LAr. The expected dynamic range from the signal and noise estimates in sections 2 and 3 would be, ~300 fC/0.1 fC=3000, i.e. ~12 bits. Essentially all the readout electronics should be placed on the sense wire-frame structure as illustrated in figure 12. This would place front-end ASICs next to the sense wire connections and minimize the cabling from the wire frames to the feedthroughs.

A challenging aspect is to bring the data out of the cryostat. For a ~20 kton LAr TPC with half a million sense wires (4 mm wire spacing) the raw data rate would be: *(500 k wires x 2 Ms/s x 12 bits)~* ***10^13 bps.*** Some data compression will be possible without loss of particle track information. The degree of compression and the compression schemes are being studied. Assuming a compression by ~1/8, the data rate would be 1.25x10^12 bps. Both copper links and fiber optic links are being investigated and both are indicated in the figure with their respective data rates.

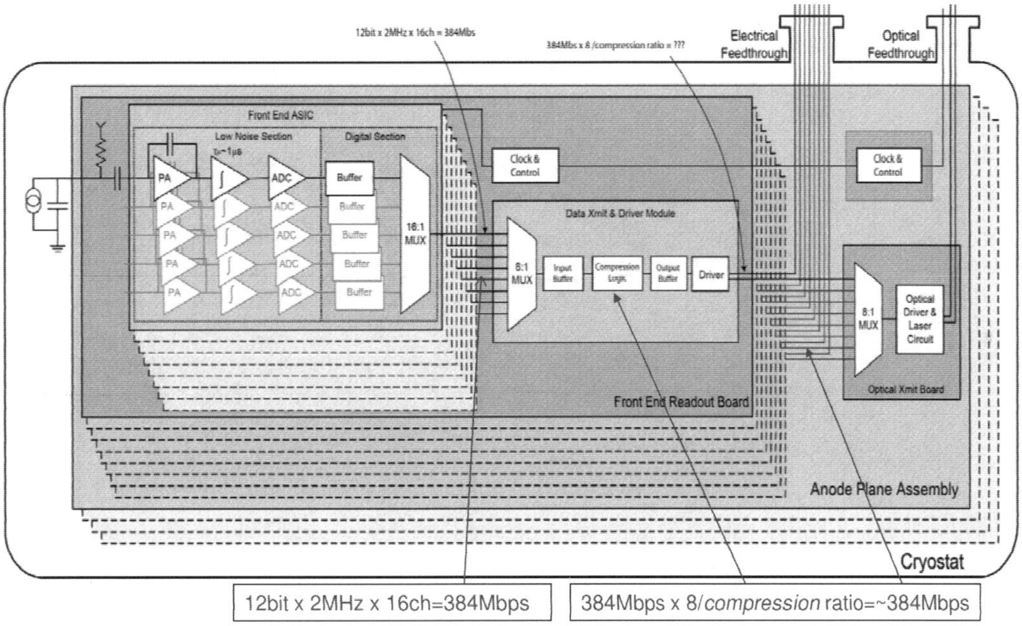

Figure 8. Functional Outline of a Multiplexed Readout Chain for Very Large LAr TPCs. Multiplexing will be performed in two to three steps by 1/128, or by 1/(512-1024). Two options are shown for digital links leading out of the cryostat: copper links at ~400 Mbps, and fiber optical links at 1.6-3.2 Gbps. For each data link (after multiplexing by 1/128, or more), a ***redundant link*** with auxiliary circuits (e.g. driver etc.) will be implemented.

5. MOS transistors at low temperature

For design and operation of CMOS circuits at liquid argon temperature (~ 89 K) two critical aspects must be considered. The first is accuracy of the device models used in the transistor and circuit design in terms of operating point, signal response (transconductance, output resistance and capacitance), noise, interconnects and passive components (resistors and capacitors). The second is the long-term lifetime, since the LAr TPC is required to operate for over 20 years without access to the electronics, short of a very major overhaul.

Concerning the device models, we have characterized a complete mixed-signal CMOS 0.25 μm ASIC, originally developed for room temperature applications [4], at temperatures down to 40 K, observing encouraging results (see for example figure 9). A comparison of the signal response to the simulation suggests that the models are to a first order reliable, and that they may only need some minor adjustments. We are also in the process of fabricating in a CMOS 0.18 μm technology a dedicated analog front-end ASIC (described in the next Section) and a number of test structures.

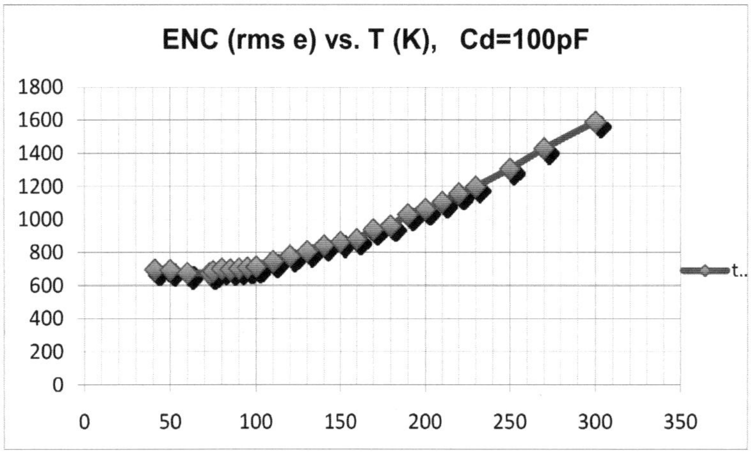

Figure 9. Noise (ENC) vs temperature measured on a mixed signal ASIC designed for room temperature operation. CMOS in LAr has less than half the noise as that at room temperature, higher mobility and higher transconductance/current ratio. Peaking time =0.5 μs

With regard to the ***long-term lifetime***, the major failure mechanisms such as, *electromigration (EM), stress migration (SM), time-dependent dielectric breakdown (TDDB) and thermal cycling (TC) scale with temperature in favor of cryogenic operation* [5]. The only mechanism that affects the lifetime adversely at cryogenic temperature is *degradation due to impact ionization*, which introduces trapped charge in the MOSFET gate oxide [6, 7]. Fortunately *the substrate current is a good quantitative monitor of impact ionization*, which makes possible ***accelerated tests and lifetime extrapolation***.

This is illustrated in figure 10. Impact ionization depends on the electric field near the FET drain. Therefore the lifetime depends strongly on the drain-source voltage. Operation close to liquid nitrogen temperature requires about 10% lower drain-source voltage than at room temperature.

Conversely, the lifetime is decreased or increased by an order of magnitude if the voltage is changed by plus or minus ~10%. This makes accelerated lifetime tests practical.

Some simplified design guidelines for long lifetime at low temperatures are: design both analog and digital circuits with transistors operated at Vds≤2/3Vdd, at low current density and with non-minimum channel length *L*.

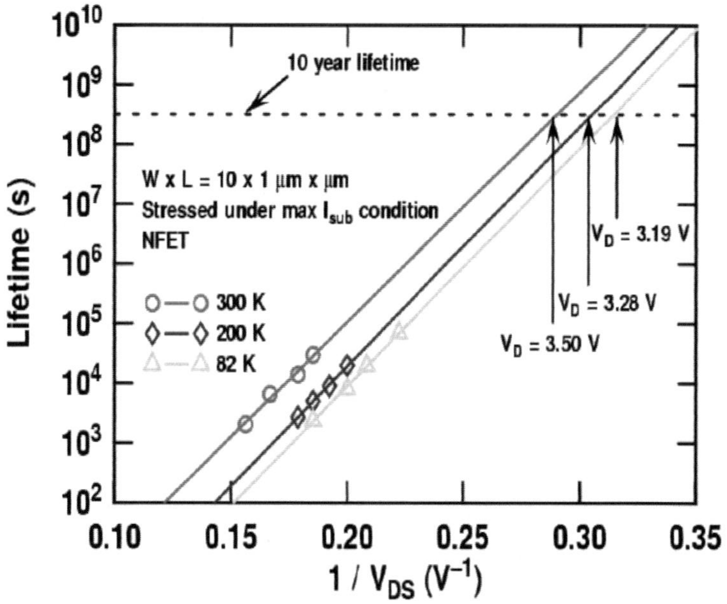

Figure 10. n-channel MOSFET lifetime due to impact ionization as a function of drain-source voltage and temperature.
From: J. Cressler

6. Front end ASIC design concept

In this section an outline is given of some of the circuits leading to the design of the front-end ASIC. In figure 8, a conceptual block diagram of a 16-channel front-end ASIC is shown. Each channel implements a charge amplifier with adjustable gain (4.7, 7.8, 14 and 25 mV/fC), a high-order shaper with adjustable peaking time (0.5, 1, 2, 3 µs), optional AC coupling, baseline adjustment for operation with either the collecting or the non-collecting wires (Y and U, V in figures 1 and 2), a 12-bit 2 MS/s ADC and a data compression stage. Shared among the 16 channels are the bias circuits, registers, a temperature monitor, the digital multiplexer (16:1), the FIFO and the digital interface. The ADC, currently being designed, aims at 12-bit resolution, converting in less than 500 ns, and dissipating about 2 mA (3.6 mW). The ADC design is based on the current mode concept described in [8]. The analog section of the channel (charge amplifier and shaper) has been fabricated and it is being tested and characterized at this writing.

This first prototype of the front-end ASIC is being developed in a commercial CMOS 0.18 µm platform (available through MOSIS). The final choice of the technology will be determined based on the results from the characterization of several test structures from different platforms and sources. For a project of this magnitude, an established technology will be chosen that must be available from more than one source.

A functional diagram of the charge amplifier is shown in figure 11. The input MOSFET is a p-channel biased at 2 mA with an L/W ratio of 0.27 μm/10 mm, followed by a dual cascode stage. The charge amplifier is composed of two stages and it provides a charge gain of up to 20x16 prior to the fifth order shaper. The overall weighting function will be semigaussian as shown in figure 5. The circuits used here are based on well proven concepts developed previously, [4], and [8].

Charge sensitivity calibration.

Charge calibration is accomplished by the charge injection method well established in nuclear and x-ray spectroscopy. A voltage step is generated in every ASIC from an accurately known dc voltage. A simple uniform calibration of numerous channels has been made possible by advances in CMOS technology: the spread in the values of capacitances on a single wafer used for charge injection is remarkably small, $\sigma<0.25$ %, and the lot-to-lot spread is <0.5 %. A check on the capacitance for a single chip could be included as a step in a standard automated test procedure. The capacitances in CMOS are very stable.

The whole TPC with hundreds of thousands wires would thus have well known (~1 %) charge sensitivity, allowing accurate dE/dx measurements calibrated with cosmic muons and determination of the attenuation distance and lifetime.

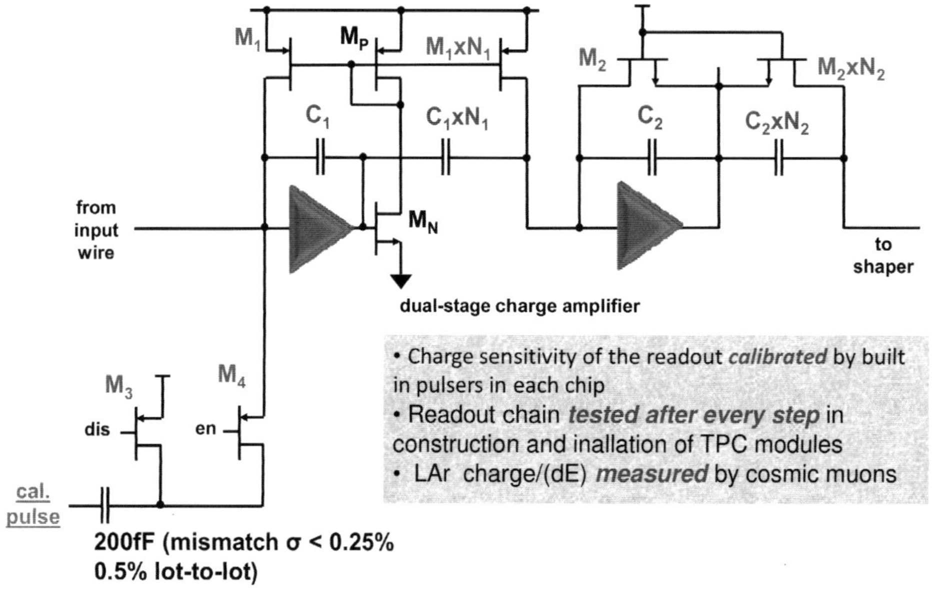

Figure 11. Two-stage charge amplifier with charge calibration. The design concept is described in reference [4]

7. Summary

Cold electronics in very large LAr TPCs allows almost unlimited freedom in the design of the cryostat and anode wire frames. This is an essential condition to enable economical solutions to the cryostat structural and thermal design, an interface between the cryostat walls and the cavern walls not hampered by numerous signal feedthroughs, as well as to make possible a design of anode wire frames which can be completely assembled with integral readout electronics off the detector site, transported and installed in the cryostat. The charge calibration scheme described in the previous section allows rapid testing at every step of this process. The main features of such a design are illustrated in figure 12.

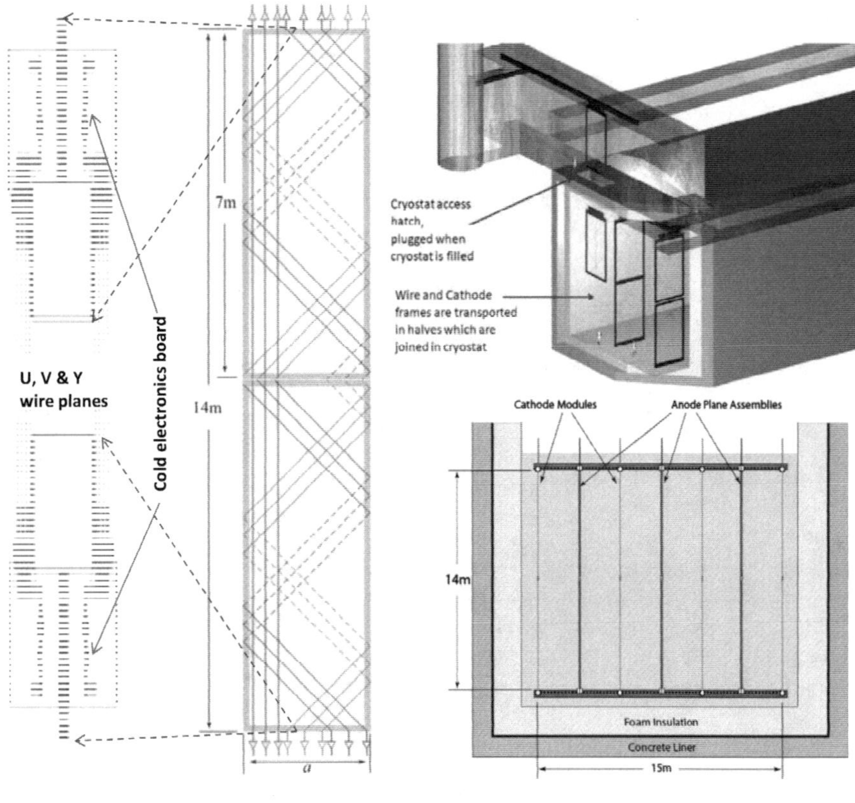

Figure 12. Key components of a possible ~20 kton LAr TPC: The cryostat (a cutout on upper right and a crossection on lower right); a wire frame (middle of the figure: two 7 m long wire frames on top of each other, with readout electronics at the top and at the bottom); horizontal structural parts of the wire frames showing electronics boards and sense wire attachments, on the left of the figure. The very few feedthroughs needed are at the near end of the cryostat. A feedthrough design proven in LHC/ATLAS is described in reference [9]. The "folded *u,v*, wires" concept is due to Hans Jostlein (FNAL).

Journal of Physics: Conference Series **308** (2011) 012021 doi:10.1088/1742-6596/308/1/012021

While the electronics in some large experiments (e.g., LHC) can hardly be considered accessible (an access usually requiring a long shutdown and disassembly of large equipment), the difficulties of access to a giant underground detector are much more apparent. While it is possible to conceive of a solution with a storage tank where the liquid argon could be transferred allowing access into the tank for a major overhaul, we concentrate in our development work to design a reliable detector system for long operation. In addition to understanding the dominant aging phenomenon, such as the one mentioned in section 5, as well as the design process to minimize it, a carefully designed testing procedure for all cold electronics components will be essential. Such a verification and testing procedure will have to include, at a minimum, the following steps:

- The CMOS devices will be designed and fully qualified in the Design and Development phase for operation at cryogenic temperatures. All circuit boards (fully assembled with ASICS) will be cycled several times between liquid nitrogen temperature (by immersion) and somewhat above room temperature (~50 C).
- At the TPC assembly stage, once each TPC module (wire frame with electronics) is installed in the cryostat, a calibration run of all electronics channels will be conducted … and then again, after cooling with Ar gas, prior to filling with LAr.

With the charge calibration built into the ASICS such testing of all channels will be straightforward.

In addition to the noise and performance advantages of the cold electronics location at the sense wires, all the advantages for the cryostat and TPC design are entirely due to a high degree of multiplexing (1/128, 1/512/, or 1/1024), and the resulting radical reduction in the number of signal feedthroughs and cables (with the attendant reduction in outgassing and probability of leaks). This raises the question of a single point failure in the output data links. *Redundant links and associated circuits will have to be used, with the redundancy increasing toward the top of the multiplexing pyramid.* Given the decreasing number of links with the degree of multiplexing, the cost of redundancy will be low.

{Discussion of costs is beyond the scope of this paper. Since this question comes up naturally, a glimpse will be given: Initial estimates indicate that the cost of the cold electronics readout expressed as a fraction of the cost of the detection medium (liquid argon) alone is less than one. Somewhat of a surprise to many would be that the cost of a warm electronics readout would come out higher (mainly due to the cost of feedthroughs and cables, the cost of the active electronics being nearly the same; the cost of added complexity in the cryostat design not taken into account).}

Note added in proof: The front-end ASIC containing preamplifier and shaper, outlined in figures 8. and 11, has been designed and successfully fabricated and tested. The results have been presented in Ref. [10].

Acknowledgements: The role of Bruce Baller, Bonnie Fleming, Hans Jostlein, David Lissauer, Steven Pordes, Jack Sondericker and Bill Willis in the development of concepts for the large TPC LAr detectors has been vital. Many useful discussions with them are gratefully acknowledged. Donald Makowiecki has performed numerous tests on circuit boards by immersion in liquid nitrogen to establish the integrity of various connections. Whatever progress we have made in understanding the properties of transistors at low temperatures would not have been possible without the input from John Cressler of Georgia Tech.

References

[1] Amerio S et al. 2004 Design, Construction and Tests of the ICARUS T600 Detector *Nucl. Instr. and Meth.* A **527**(2004) 329; and references therein.

[2] Radeka V 1988 Low-Noise Techniques in Detectors, Ann. Rev. Nucl. Part. Sci. **38** (1988) 217

[3] Rescia S and Radeka V 2009 Effect of sense wire resistance on readout noise of large liquid argon time projection chambers *Nuclear Science Symposium Conference Record (NSS/MIC)* IEEE Digital Object Identifier: 10.1109/NSSMIC.2009.5402155 (2009) 1932

[4] De Geronimo G, Fried J, Frost E, Phlips B F, Vernon E and Wulf E A 2008 Front-end ASIC for a Si Compton telescope, *IEEE Trans. Nucl. Sci.* **55** 2323

[5] Srinivasan J, Adve S V, Bose P and Rivers J A 2004 The impact of technology scaling on lifetime reliability *IEEE The International Conference on Dependable Systems and Networks (DSN-04)* June 2004.

[6] Cressler J D 2005 On the Potential of SiGe HBTs for Extreme Environment Electronics *Proc. IEEE* **93** 1559

[7] Chen T, Najafizadeh L, Zhu C, Ahmed A, Diestelhorst R, Espinel G and Cressler J D 2005 CMOS device reliability for emerging cryogenic space electronics applications *IEEE International Semiconductor Device Research Symposium*

[8] De Geronimo G, Fried J, Smith G C, Yu B, Vernon E, Britton C L, Bryan W L, Clonts L G and Frank S S 2007 ASIC for small angle Neutron scattering experiments at SNS, *IEEE Trans. Nucl. Sci.* **54** 541

[9] Axen D et al. 2005 Signal Feedthroughs for the ATLAS Barrel and Endcap Calorimeters 2005 *Rev. Sci. Instrum.* **76** 063306

[10] De Geronimo G et al 2010 Front-end ASIC for Liquid Argon TPCs, IEEE NSS/MIC Conference Record N56-2

Recent Results from Liquid Argon R&D at KEK

Masashi Tanaka

High Energy Accelerator Research Laboratory

E-mail: `Masashi.Tanaka@kek.jp`

Abstract. A Giant Liquid Argon TPC detector is considered as next generation neutrino and nucleon decay detector. In this article we report about recent liquid argon R&D activities at KEK. Starting in Summer 2008 we have built and successfully operated two different detectors: a 10L size double phase argon TPC and a 250L size single-phase TPC. So far we established liquid argon purities of (1) ~10 ppb in the 10L detector and (2) ~1 ppb in the 250L detector.

1. Overview of KEK Liquid Argon Test Stand

The Liquid Argon Time Projection Chamber (LAr TPC) is considered to be one of the most powerful charged particle imaging detectors. It's possible to cover a large volume ($\mathcal{O}(100)$ kt) with excellent position and energy resolution and real time readout. We are interested in applying this technique for the next generation neutrino and nucleon decay detector and have started R&D efforts at KEK in summer 2008. In January 2010 we have proposed a program to the J-PARC PAC: "*Towards a Long Baseline Neutrino and Nucleon Decay Experiment with a next-generation 100 kton LAr TPC detector at Okinoshima and an intensity upgraded J-PARC Neutrino beam*" (P32).

Although LAr is an excellent material for detecting charged particles it is known that the lifetime of drift electron strongly depends on the purity of the LAr (electronegative impurities such as oxygen capture drift electrons) and the attenuation length (λ) of the drift electrons is effectively parameterized in terms of the impurity concentration (σ)

$$\lambda(\mathrm{m}) = 0.5/\sigma(\mathrm{ppb}).$$

This means that even small detectors with a maximum drift length of 5 cm require a LAr purity of ~10 ppb in order to observe the ionization electron signal. For 5 m drift length we need to achieve ultra pure LAr with $\mathcal{O}(10)$ parts-per-trillion (ppt) of impurities. The technique to obtain ultra high purity LAr has been pioneered by ICARUS, a ~1 kt LAr-TPC with a drift distance of ~1.5 m[1]. Based on this achievement our goal is to study the possibility of a 100 kT scale detector with a $\mathcal{O}(10)$ m drift distance. Currently we are concentrating the effort to:

- establish basic techniques of LAr TPCs at KEK (cryogenic vessel, LAr purification, ionization charge and scintillation light readout, electronics, etc.) so that we can catch up with the current forefront.
- study the realization of large (~ 100kt) LAr TPC detectors (establish long drift, high voltage, etc).

- understand the physics performance of LAr TPC detectors using charged particle and neutrino beams.

We have built two test setups with different sizes: the so called "10L setup" and the "250L setup". The 10L setup consists of a small 20cm$\phi\times$25cmL vacuum chamber and its goal is (1) to establish 5 cm drift of ionization electrons which corresponds to 10 ppb of impurities in the liquid argon and (2) to establish the double phase argon TPC detector. The 250L setup consists of a 70cm$\phi\times$100cmL cryogenic vessel and its goal is (1) to establish the technique to build a double phase LAr-TPC with 150 kg active medium and 40 cm drift length, (2) to test the detector performance under a charged particle beam (e,mu,pi,K) of 100-1000 MeV/c and (3) finally to expose it to a neutrino beam.

In this article we report on the current status of the 10L and the 250L setup.

2. Results from the 10L Test Stand

2.1. 10L Test Chamber

We have designed and built the 10L setup based on previous studies done by other groups [2, 3]. Figure 1 shows schematics of the 10L setup. We have built two vessels with the sizes of 20 cm$\phi\times$25 cm and 30 cm$\phi\times$45 cm, respectively. The larger vessel is a dewar, used as LAr bath to refrigerate the smaller vessel which is our test chamber.

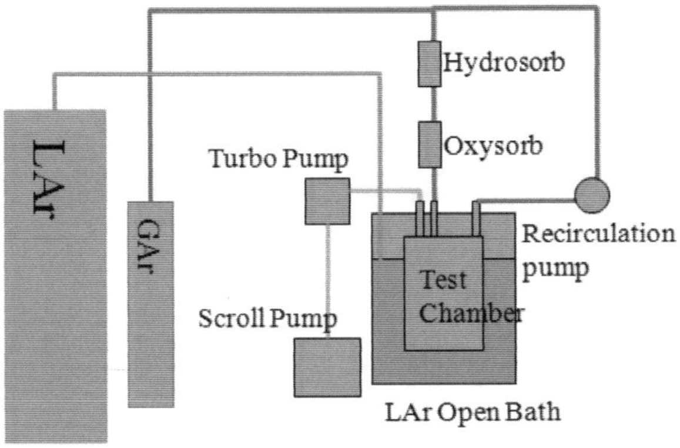

Figure 1. Schematic view (left) and picture (right) of 10L setup.

The argon gas we used for the test was "G3" grade which ensures the contamination of oxygen to be less than 0.2 ppm. Then argon gas is purified by using small cartridges of Hydrosorb® and Oxysorb® by Messer. These filters are specified to achieve better than 5 (20) ppb of oxygen (moisture) contamination if the initial contamination is less than 10 ppm. Since the test chamber is cooled by liquid argon boiling at atmospheric pressure, by applying certain overpressure (+0.5 bar typically) inside the chamber, the argon gas gets liquefied. Using this setup, we have achieved to liquefy 3 L/hour of high purity LAr inside the chamber.

Although the argon gas right after the output of the filter is pure, it can degrade because of several reasons. The following is a list of such sources and actions we took to solve the problems.

- Out-gassing of the material inside the chamber. We have baked-out the test chamber at 100 °C while we were evacuating the chamber with a molecular turbo pump (80 L/sec capacity) directly mounted on the top flange for few days. Figure 2 shows a profile of the baking process. The test chamber was kept at high temperature for 80 hours, then it cooled down to room temperature. The achieved vacuum level is $\sim 1\times10^{-4}$ Pa at room temperature and the out-gassing rate is measured to be 0.01 Pa/hour which is improved by a factor of 100 compared to before the baking.

- The contamination from the transfer line between the output of the filter and the test chamber. We shortened the transfer line as much as possible by directly mounting the filter cartridge to the test chamber.
- Leaks from flanges, feed-throughs and valves of the test chamber. The leak-rate at room temperature was checked by using a He leak detector which is able to detect leaks down to the level of 10^{9} Pa m^3s^{-1}. We have also observed that the CF flange leaks at liquid argon temperature because of the different thermal expansion coefficients of flange (SUS) and gasket (Cu). Thus we keep the LAr level of the bath inside the dewar lower than the flange of the test chamber.

Using the above procedure we have successfully filled the test chamber with less than $\sim$5 ppb of impurities in the liquid argon.

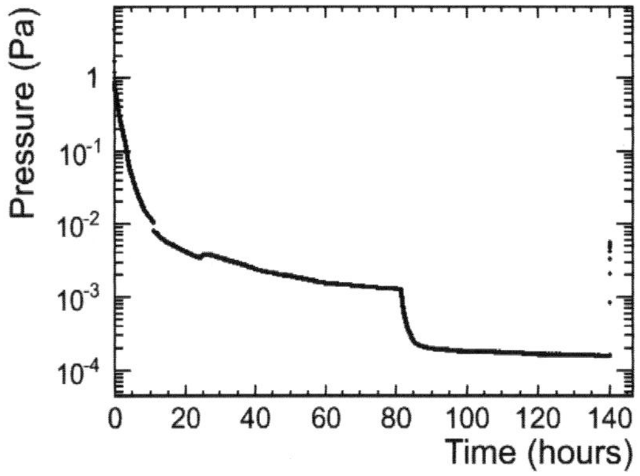

Figure 2. Vacuum level while baking process of the 10L system.

2.2. 10L Detector

Inside the test chamber we put a double-phase TPC detector with an active volume of 9cm×9cm×5cm (Figure 3). The detector consists of a cathode plate, field shaping electrodes, 2 extraction grids, a THGEM (thick Gas Electron Multiplier) and a readout anode.

The cathode plate is a 9 cm×9 cm× 1 mm copper plate and the field shaping electrodes are 10 cm×10 cm× 0.8mm stainless steel plates with a 9 cm×9 cm hole inside. The distance between each electrode is 1 cm. As extraction grids we soldered 100 $\mu m\phi$ stainless steel wires with a pitch of 5 mm on such a field shaping ring. The surface of the liquid argon is adjusted to be in the center of the two extraction grids within 1 mm accuracy. By applying an electric field of $\geq$ 3kV/cm, the drift electron is extracted from liquid to gas phase. The used THGEM was produced by REPIC (Fig. 3). Its base is made of glass epoxy (G10) with gold plated copper electrodes. The geometry is:

- thickness 400 μm
- hole diameter 300 μm
- hole distance 700 μm

The 9cm×9cm anode is segmented into 4 pads with a size of 9 cm×2.2 cm. The electric signal is extracted from the test chamber through a vacuum tight BNC connector (Kyocera). Finally a charge sensitive preamplifier (Amptek A250, 1 V/pC) and a post shaper amplifier (Hosin N012, shaping time 1 μs) were used to amplify and shape the signals. A typical voltage setting for operating the detector is listed below:

- E_{drift}: Cathode-Bottom extraction grid (≡Drift field) : 300 V/cm
- E_{ext}: Between extraction grids (≡Extraction field) : 5 kV/cm
- E_{ind1}: Top Extraction grid - THGEM : (≡Induction field 1): 700 kV/cm
- V_{THGEM}: Between THGEM electrodes: 400∼1750 V
- E_{ind2}: THGEM - Anode (≡Induction field 2): 1 kV/cm

Figure 4 shows a typical signal of cosmic muon track. To ensure to trigger on the cosmic muons passing through the TPC detector we have arranged two scintillation trigger counters with the TPC detector in between. The top plot shows the signal waveform for 4 TPC channels. Black, red, green, and blue lines correspond to channel 1,2,3, and 4, respectively. Figure 5-Left shows the gain of the THGEM as a function of the voltage between THGEM electrodes. The triangle (circle) points are with an induction 2 field equal to 1.0 kV/cm (3.5 kV/cm). Up to now the maximum achieved gain with this configuration is ∼ 2.5.

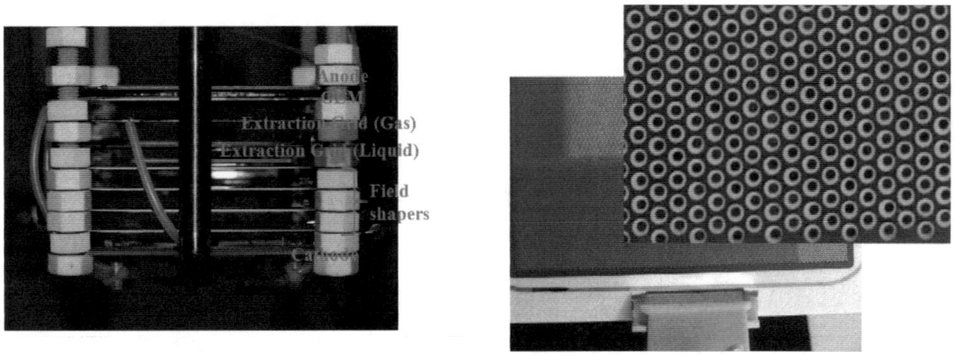

Figure 3. Pictures of 9x9x5cm^3 double phase TPC detector (left) and the THGEM (right).

2.3. THGEM Performance at Cryogenic Temperature

The mean free path of drifting electrons inside argon gas depends on the molecular density of the argon gas and is proportional to pressure (atm)/temperature (K). An electron needs to be accelerated to at least ∼25 eV within the mean free path in order to produce secondary electrons and initiate gas multiplication. This means that the breakdown voltage increases with increasing molecular density. At LAr temperature (90K) the molecular density is ∼ × 3 larger compared to room temperature. For this reason it is rather difficult to achieve high gas gains at low temperature.

Figure 5-right shows the measured gain of the THGEM as a function of P/T. These measurements were taken by varying the gas pressure between 1-2 atm at room temperature (300 K) with a constant THGEM voltage of 800 V. The data points show good agreement with an function $e^{\alpha\frac{T}{P}}$. The plot also shows the gain with liquid argon temperature equivalent density (P/T = 1/100 = 0.01), is about 50 lower than the gain at room temperature (P/T = 1/300 = 0.003), keeping the voltage constant.

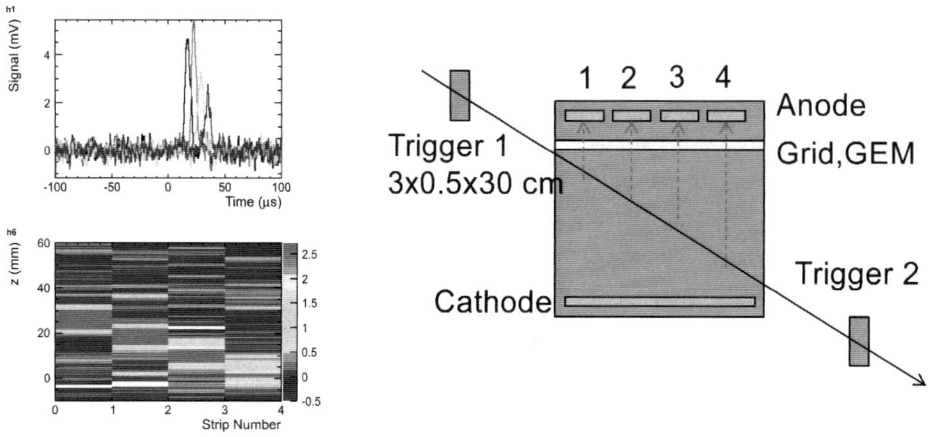

Figure 4. Cosmic track signal with 10L detector (left), and trigger counter configuration (right).

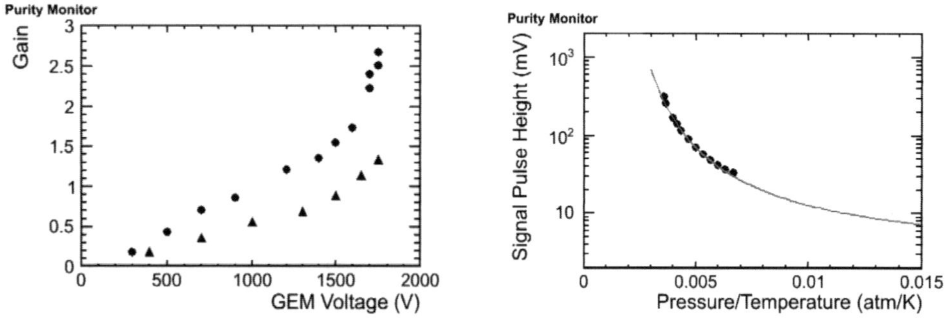

Figure 5. Left: Gain of the 10L Detector; Right: THGEM gain as a function of P/T

Most of the gas multiplication devices developed so far are designed to operate at room temperature and standard pressure. So there is still room for improvement to achieve high gain at cryogenic temperatures. Recently it has been successfully achieved a gain >10 using Large Electron Multipliers (or thick GEM/THGEM) produced at CERN, which is already sufficient for neutrino physics application [4].

3. LAr Purification for the 250L Test Stand

For a first test campaign in October 2010, of a 250L TPC in the K1.1BR beamline of the J-PARC slow extraction facility (see [5] for a detailed description of the setup), we built a prototype detector with coarse readout sampling, operated in liquid argon phase. The setup is housed in a cryostat originally built as a prototype liquid Xenon calorimeter for the MEG ($\mu \rightarrow e\gamma$) experiment [6]. The cryostat consists of two transverse cylindrical vessels (inner vessel and outer vessel). The inner vessel is filled with liquid argon and has $\sim$70 cm diameter and $\sim$100 cm length, thus the volume of the cryostat is about 400L. The maximum size of cuboid which fits inside the inner vessel is $\sim$50$\times$50$\times$100 cm^3 (250L). The volume between inner and outer vessels is for vacuum insulation and super insulator (20 layers of aluminum mylar films) is inserted to reduce thermal radiation. The total heat input of the cryostat is $\sim$30W at LAr temperature (90K).

The TPC detector has an active volume of $40 \times 40 \times 76$ cm^3. For this test, instead of a double-phase TPC, we use a single-phase grid chamber with the anode segmented into 76 strips (40 cm$\times$1 cm strip).

In the 250L setup the maximum drift distance is 40 cm, thus we need to achieve at least 1 ppb of LAr purity so that the attenuation of the signal does not cause significant charge attenuation. Figure 6 shows a schematic drawing of the cryogenic and purification system for the 250L test. The procedure to fill the chamber and to achieve and maintain high purity liquid argon in the vessel is as follows:

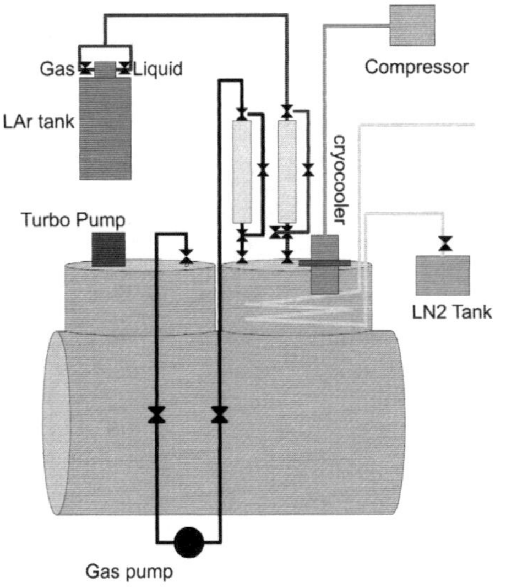

Figure 6. Cryogenic and purification system diagram of the 250L test (left), and purification cartridge (right).

- Evacuate the inner vessel with a molecular turbo pump (Edwards, 300L/s capacity) and a getter pump (SAES, 40 L/s capacity) directly mounted on the top flange for $\sim$ 1 week. The achieved vacuum level is 1×10^{-4} Pa and the out-gas/leak rate is $\sim$ 1 Pa/hour. Since the 250L cryostat is a thermal insulated vessel, it is difficult to heat the material inside the inner vessel while evacuation. Thus we separately bake-out some of the detector components in another vacuum chamber to reduce the out-gassing of the material efficiently.
- Fill the inner vessel with argon gas from LAr tank through the purification cartridge up to 80–100 kPa(Gauge). We have developed the purification cartridge based on Fermilab's study [7]. Figure 6 shows the purification cartridge which consists of activated copper (Angelhard Cu-0226, black granules) and molecular sieves (Union Showa type 4A, while granules).
- Start cooling down the vessel using a Gifford-McMahon (GM) cryocooler made by Sumitomo Heavy Industries and a LN2 heat exchange coil. The cooling power of the GM cryocooler is 160 W at LAr temperature (90K) and the LN2 heat exchange coils provide up to 600 W.
- The gas recirculation is turned on at the same time with a typical gas flow of 70L/min. For the gas purification, we use a commercial filter (SAES microtor).
- As temperature and pressure inside the vessel go down, we add argon gas and keep the pressure at 80 kPa (Gauge).

- After ~5 hours, the vessel is cooled down and LAr begins to be accumulated at the bottom of the vessel. At this stage, argon gas is still liquefied by cryocooler and LN2 coil. The typical liquefaction speed which is limited by the cooling power is 10 L/hour.
- At this point, we start filling the vessel with LAr from LAr tank. It takes another ~3 hours until the purification cartridge is completely cooled down and we are directly filling the vessel with LAr. After that the filling speed is ~ 50 L/hour which is limited by the impedance of the purification cartridge.
- We stop filling once the detector inside is completely immersed.
- Further purification to improve and/or keep the initial purity by recalculating the argon gas. Evaporate liquid argon inside the cryostat using cryogenic heater and purify the argon gas, then liquefy using the GM cryocooler and the LN2 coil. The system is designed to recirculate one full volume of the LAr inside the vessel in few days.

By using the procedure explained above we have successfully filled the 250L vessel at J-PARC K1.1Br beamline on October 2010, and exposed it to a charged particle (K, π, p, e) beam (see [5] for a detailed report). Figure 7 shows an event display of a cosmic muon track. The horizontal axis shows the TPC channel, the vertical axis shows drift time in μs and the color of the plot corresponds to the signal pulse height in ADC counts after pedestal subtraction. By an analysis of the released ionization (dE/dx) versus drift distance, for cosmic muon tracks, we measured an electron lifetime in LAr better than 300μs, corresponding to an upper limit of <1 ppb of electronegative impurities.

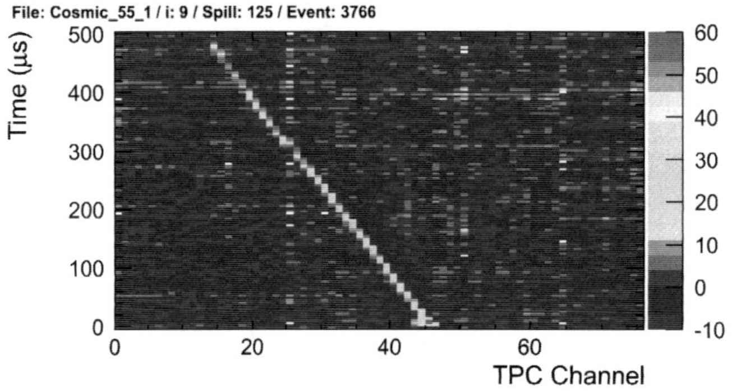

Figure 7. Typical cosmic track event recorded with the 250L detector

4. Summary

We started a LAr R&D program at KEK mainly concentrating on LAr purification techniques. We also tested a double-phase LAr TPC detector, where ionization electrons extracted from the liquid are amplified in the gas phase by means of Gas Electron Multipliers. We have built two different test stands for LAr TPC R&D, namely "10L setup" and "250L setup". In 10L setup, we have established better than 10 ppb LAr purity and 4 ch readout double phase readout TPC. In the 250L setup, we have achieved better than 1 ppb LAr purity and observed cosmic muon tracks as well as charged particles from an accelerator beam with a 76 channel readout single phase TPC.

References

[1] S. Amerio *et al.* [ICARUS Collaboration], Nucl. Instrum. Meth. A **527**, 329 (2004).
[2] E. Aprile, K. L. Giboni and C. Rubbia, Nucl. Instrum. Meth. A **241**, 62 (1985).
[3] A. Badertscher *et al.*, Nucl. Instrum. Meth. A **617**, 188 (2010) [arXiv:0907.2944 [physics.ins-det]].
[4] A. Badertscher *et al.*, Nucl. Instrum. Meth. A **641**, 48 (2011) [arXiv:1012.0483 [physics.ins-det]].
[5] T. Maruyama *et al.*, *A tagged low-momentum kaon test-beam exposure with a 250L LAr TPC (J-PARC T32)*, to appear in the Proc. of the 1st International Workshop towards the Giant Liquid Argon Charge Imaging Experiment (GLA2010).
[6] S. Mihara [MEG Collaboration], Nucl. Instrum. Meth. A **518**, 45 (2004).
[7] A. Curioni *et al.*, Nucl. Instrum. Meth. A **605**, 306 (2009) [arXiv:0903.2066 [physics.ins-det]].

Journal of Physics: Conference Series **308** (2011) 012023 doi:10.1088/1742-6596/308/1/012023

Results from the Fermilab Materials Test Stand and Status of the Liquid Argon Purity Demonstrator

B Rebel, M Adamowski, W Jaskierny, H Jostlein, C Kendziora, R Plunkett, S Pordes, R Schmitt, T Tope and T Yang

Fermilab - MS 220, PO Box 500, Batavia, IL 60540, USA

E-mail: brebel@fnal.gov

Abstract. The Fermilab Materials Test Stand was developed to test the suitability of materials for use in large liquid argon time projection chambers (TPCs). In addition to determining which materials are viable for use in TPCs, the test stand has also shown that water is especially detrimental to maintaining long electron lifetimes. The Liquid Argon Purity Demonstrator is currently under construction at Fermilab. Its goal is to show that long electron lifetimes can be achieved without evacuation of the cryostat, which is of particular interest in designing large liquid argon TPCs.

1. Introduction

Liquid argon time projection chambers (LArTPCs) offer the possibility of imaging charged particle interactions with incredible spatial resolution. A charged particle passing through liquid argon will ionize the argon liberating $\approx 55,000$ electrons per centimeter. The ionization electrons are drifted through an electric field towards readout wires where they produce current that is digitized by the readout electronics. A primary concern in the operation of an LArTPC is the ability of the electrons to reach the readout wires without being captured by electronegative contaminants. The large LArTPCs that are planned for future high energy physics experiments will require the electrons to be able to drift over distances of 2.5 m or more. The corresponding level of contamination in the liquid argon is at the level of 0.05 parts per billion (ppb) oxygen equivalent or less.

Two test stands at Fermilab are investigating aspects of LArTPC operation that relate directly to electronegative contamination in the liquid argon. The primary goal of the Materials Test Stand (MTS) is to qualify materials for use within LArTPCs and show that they do not increase the contamination in the liquid argon beyond acceptable levels. The primary goal of the Liquid Argon Purity Demonstrator (LAPD) is to show that the required level of purity can be achieved in a large volume of liquid argon without first evacuating the vessel containing the liquid and LArTPC. This test is motivated by the desire to save the costs associated with the construction of an evacuable cryostat for future multi-kiloton detectors.

2. The Materials Test Stand

The MTS is shown in Fig. 1 and is decribed in detail in [1]. The supply argon is sent through a molecular sieve to remove water and then through activated copper to remove oxygen before entering the evacuated cryostat. The cryostat has a capacity of 250 L and is vacuum insulated.

Figure 1. Photograph of the MTS at Fermilab showing the liquid argon supply, filters and cryostat.

Figure 2 shows a schematic of the cryostat and its instrumentation. The heart of the MTS is the airlock-sample cage assembly located in the center of the schematic. The airlock allows materials to be introduced into the cryostat without first having to empty it of liquid argon. The sample cage can be positioned anywhere between the airlock and the bottom of the cryostat to understand the temperature dependence of the impurities introduced by different materials.

In addition to the airlock, the cryostat also contains an ICARUS style purity monitor [2] to determine the electron lifetime and an internal filter to remove impurities introduced by the materials placed into the cryostat. The cryostat can be operated as a closed system when the liquid nitrogen cooled condenser is in operation. The argon gas that passes into the condenser is liquified and returned into the bulk liquid.

The condenser caused consternation in the performance of the MTS as the electron lifetime would rapidly decrease to zero whenever the condenser was used. The initial design of the condenser returned the liquid argon directly into the bulk liquid, and the method of return was hypothesized to cause the decrease in electron lifetime. Several new return paths were created in order to test the veracity of this hypothesis and it was found that increasing the surface area of cold metal encountered by the liquid before returning it to the bulk improved the electron lifetime. This result indicated that some impurity desorbed from warm surfaces, mixed with the gas and was then incorporated back into the liquid in the condenser. Further, that impurity absorbed to cold metal and could be removed from the liquid with an appropriate return path.

Water was a clear suspect for the impurity as it remains on surfaces in vacuum and has an affinity for cold surfaces. A moisture analyzer with 2 ppb detection capability was used to monitor the water concentration in the cryostat. Figure 3, taken from [1], shows the water concentration inside the cryostat as a material sample is placed at various depths in the cryostat.

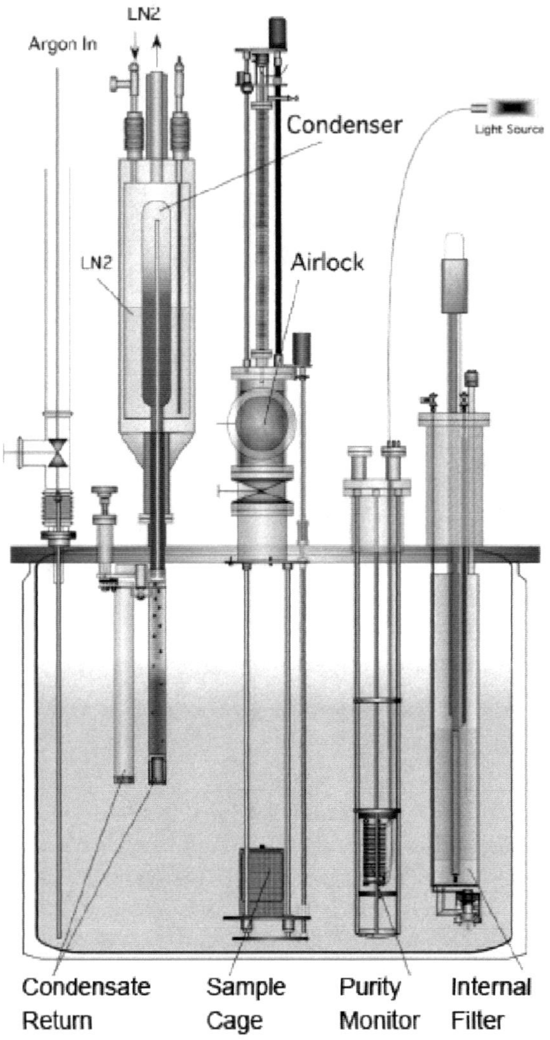

Figure 2. Schematic of the MTS cryostat and instrumentation.

The sample temperature and electron lifetime are also shown. As seen in the figure, the water concentration increases as the sample temperature increases, and the electron lifetime decreases. This figure demonstrates two major findings from the MTS. First, there is a direct relationship between electron lifetime and water concentration. Second, the water concentration does not change when materials are submerged in the liquid, but it does increase when materials are in the vapor space.

3. The Liquid Argon Purity Demonstrator

Currently operating LArTPC systems all rely on evacuation as the first step in obtaining a system capable of supporting long electron lifetimes. While that approach works well for small

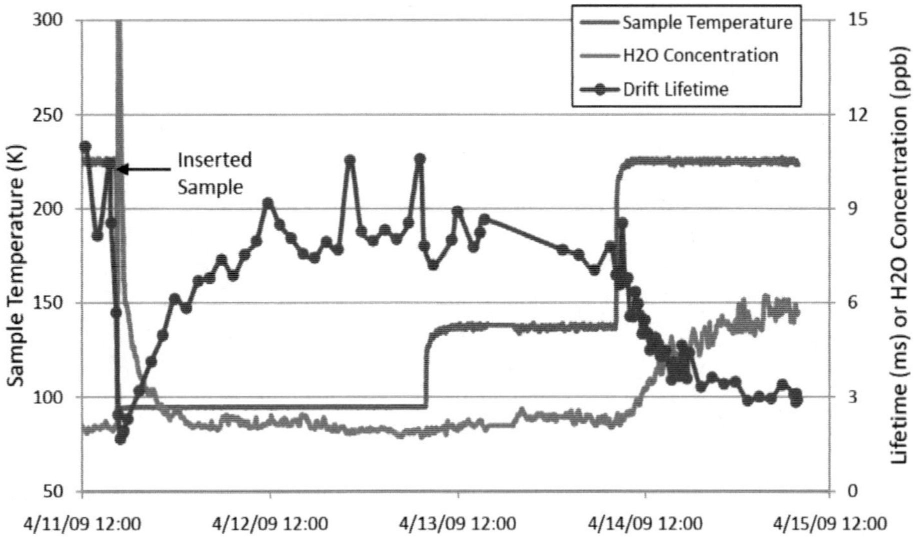

Figure 3. Material sample temperature, water concentration inside the cryostat, and electron lifetime as a function of time.

systems, it is not an ideal solution for large systems where the mass of the liquid argon is on the scale of kilotons. The cost of an evacuable cryostat necessary to contain that mass of liquid argon is at least a factor of two larger than a comparable non-evacuable vessel.

The LAPD is designed to test whether a large mass of liquid argon can be purified to the necessary level to allow for electron lifetimes on the order of several milliseconds without evacuating the vessel first. This test will be done with nothing in the vessel except for the liquid argon and monitoring devices. The LAPD vessel is shown in Fig. 4; it is 3 m in diameter and 4.6 m tall. The vessel will be filled with 30 tons of liquid argon. The LAPD relies heavily on the experience from the MTS in its design and operation plan. The purity monitors to be used are copies of the MTS monitors as is the data acquisition system.

In addition to showing that evacuation is not necessary for achieving long electron lifetimes, the LAPD will also monitor temperature gradients in the bulk liquid and concentrations of water and oxygen in order to check our models for the behavior of the liquid. After achieving the long electron lifetimes the LAPD will be emptied and typical materials used in an LArTPC will be placed in the volume and the process will be repeated to show that the technique works in the presence of material as well.

The purification of the volume will proceed in two stages. The first stage will purge the vessel with argon gas. It has been shown that the concentration of oxygen in a vessel purged with gaseous argon can be reduced to 100 parts per million (ppm) after 2.6 volume exchanges [3]. After the initial volume exchange, the gas will be heated slightly to help dry the surfaces of the vessel and monitors. Once the water and oxygen concentrations inside the vessel are at the level of ppm, the gas will be circulated through filter vessels similar to those used in the MTS. The liquid will be introduced into the vessel after a concentration less than 1 ppm has been achieved. The liquid will be continuously circulated through the filter vessels in order to achieve concentrations of water and oxygen on the order of 0.1 ppb.

The components of the cryogenic system for LAPD had to be redesigned from their counterparts in the MTS to account for the much larger size of the system. New filter vessels

Figure 4. The vessel for the LAPD. It is 3 m in diameter and 4.6 m tall.

were designed to accommodate the larger volumes of liquid argon as well as making significant investment in new high purity cryogenic valves. The condenser and phase separator to be used in LAPD were also designed specifically for the system. Attention was given to how one would scale the system for kiloton scale masses of liquid argon and the potential scaling of the cost for such large systems. Initial estimates indicate that the cost of larger systems does not scale linearly with volume.

The LAPD is an ideal system in which to study the operation of kiloton scale LArTPCs. To that end, it will perform several tests. The first is to understand the number of liquid argon volume exchanges in order to achieve drift distances on the scale of 2.5 m. Additionally the filter capacity as a function of flow rate will be closely monitored. After achieving the required lifetime, a known contamination will be introduced into the system to see how quickly it can recover the electron lifetime.

The LAPD is intended to provide information for the design of the LArTPC detector proposed for the Long Baseline Neutrino Experiment. As such, it must operate before decisions of whether to make the cryostat evacuable are made. Several components of the system have already been constructed, including the purity monitors, cryogenic pump and condenser. The cryogenic valves and filter vessels are currently under construction. The LAPD is expected to begin operations in September, 2010.

4. Conclusions

The MTS has demonstrated several aspects of purifying volumes of liquid argon to achieve long electron lifetimes, including the negative impact of water. The LAPD is building on the experience of the MTS and will begin operation in the fall of 2010. Its first results will provide key input to the design of the LArTPC detector proposed for the Long Baseline Neutrino Experiment.

5. References

[1] Andrews, R. *et al.* 2009 *Nucl. Instrum. Meth.* **A608** 251
[2] G. Carugno *et al.* 1990 *Nucl. Instr. Meth.* **A292** 580
[3] Jaskierny, W. *et al.* 2006 FERMILAB-TM-2384-E

Journal of Physics: Conference Series **308** (2011) 012024 doi:10.1088/1742-6596/308/1/012024

Towards a liquid Argon TPC without evacuation: filling of a 6 m^3 vessel with argon gas from air to ppm impurities concentration through flushing

A. Curioni[1], L. Epprecht[1], A. Gendotti[1], L. Knecht[1], D. Lussi[1], A. Marchionni[1], G. Natterer[1], F. Resnati[1], A. Rubbia[1], J. Coleman[2], M. Lewis[2], K. Mavrokoridis[2], K. McCormick[2], C. Touramanis[2]

[1] ETH Zurich, Institute for Particle Physics, CH-8093 Zurich, Switzerland
[2] University of Liverpool, Liverpool, United Kingdom

E-mail: alessandro.curioni@cern.ch

Abstract. In this paper we present a successful experimental test of filling a volume of 6 m^3 with argon gas, starting from normal ambient air and reducing the impurities content down to few parts per million (ppm) oxygen equivalent. This level of contamination was directly monitored measuring the slow component of the scintillation light of the Ar gas, which is sensitive to *all* sources of impurities affecting directly the argon scintillation.

1. Introduction

GLACIER (Giant Liquid Argon Charge Imaging ExpeRiment) is a proposed very large LAr TPC with a well defined conceptual design [1, 2]. The proposed cryostat is a single module cryo-tank based on industrial liquefied natural gas (LNG) technology, with a cylindrical shape giving an optimal surface-to-volume ratio. The detector design is meant to be highly scalable, up to a total mass of at least 100 kton.

One severe key technical challenge of the GLACIER design is to achieve long drift paths of free ionization electrons in liquid argon over a very long distance, possibly as large as 20 meters. The GLACIER concept therefore relies on a double phase operation with a readout with charge amplification to compensate for the S/N loss due to diffusion and remaining charge attenuation from impurity attachment during the long drift path, but in addition requires very special care to achieve and maintain very high purity in the liquid argon, at the level of $\lesssim$ 20 ppt oxygen equivalent [3], and a very high voltage to create the drift field to quickly drift the electron cloud across the liquid argon volume [4].

Argon purification systems based on oxygen reactants and molecular sieves, commercially available, have been studied since many years as part of the R&D for liquid Argon TPCs [5, 6], and are known to be very effective in reducing the oxygen contamination and also water, carbon oxydes, fluorates elements or hydrocarbons to very small levels. During the ICARUS 50L exposure to the WANF neutrino beam at CERN [7], a free electron lifetime of 15 ms, corresponding to 20 ppt oxygen equivalent, was achieved using such standard (gas) purification methods.

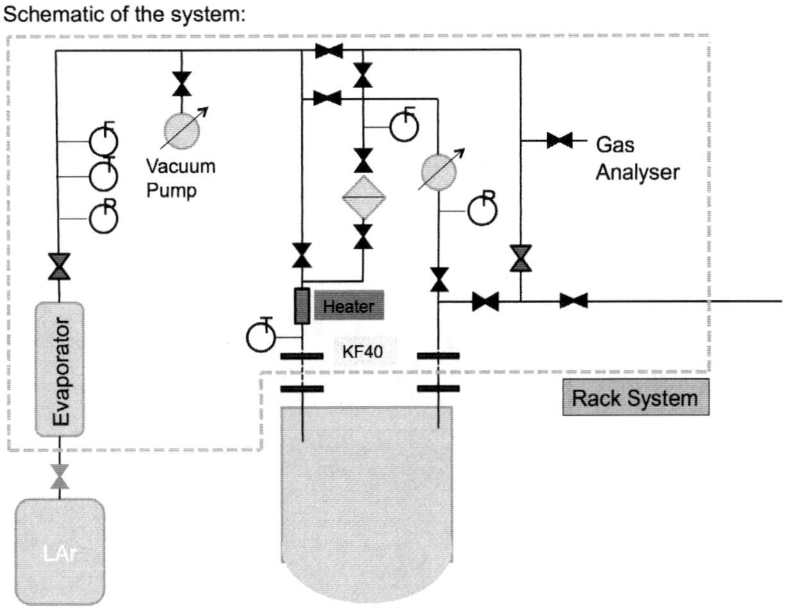

Figure 1. Schematics of the apparatus, completed with recirculation and gas analyzing systems.

However, like the ICARUS 50L, all LAr TPCs operated up to now were housed in a vacuum vessel, and ultra-high vacuum evacuation was always performed prior to cooling and filling with liquid argon. This will not be possible for a LNG tank, because this kind of vessel, although a time tested solution, cannot withstand under- or over-pressures above 50-100 mbar, so evacuation is impossible. As of today, the experimental demonstration that high liquid argon purity can be achieved and maintained over long periods of operation in an LNG tank, is lacking. As a first step towards this milestone, we started studying new methods and scalable techniques to manage non-evacuable vessels.

2. Avoiding the ultra-high vacuum phase

The vacuum phase covers at least three important separate functions in providing conditions for a low level of impurities in LAr:

(i) it removes air from the vessel prior to cooling and filling with LAr;

(ii) it removes outgassing (water and other elements) from the detector materials and the vessel walls (baking is also sometimes performed);

(iii) it verifies the integrity (tightness) of the system itself.

In the present work, we focused on the item (i). Items (ii) and (iii) will be studied later. Concerning (i), it should be pointed out that commercially available (high grade) bulk liquid argon, which is handled, transported and stored without using high vacuum vessels, has an oxygen content of about 1 ppm, which is an acceptable starting point for further purification in the liquid phase. Therefore it looks very reasonable to expect that at least the "remove air" function can be easily achieved without vacuum. Prior to our test, we therefore anticipated that the air in a vessel could be efficiently displaced by a flow of Ar gas, and we have experimentally tested this technique for reducing the air concentration using a 6 m^3 vessel.

In our test, we first proved that oxygen can be effectively reduced from its air concentration (20%) to less than 0.1% with oxygen detectors. However, in order to be sensitive to all potential

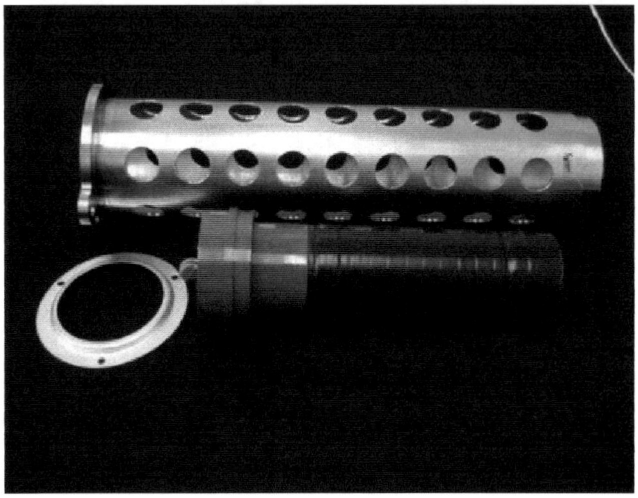

Figure 2. Photomultiplier tube (2" ETL 9831KB) and corresponding stainless steel holder.

contaminants that affect drift electron properties, we simultaneously instrumented our vessel with DUV photosensors (coated PMTs) to directly measure the argon scintillation light produced by embarked radioactive sources. Indeed, several components are observed in the scintillation of gaseous argon. The mean lifetime of the slowest (triplet) component (about 3 μs) is very sensitive to traces of impurities and can be very effectively used to monitor the impurities [8, 9] that affect the drift properties of electrons. See section 3.2.

To our knowledge, this is the first time this sort of experiment is conducted in a vessel of this size at such level of contamination. A test of argon gas flushing was previously reported in Ref. [10], however, the vessel was smaller than in our case and only the oxygen content was monitored. In our work, we therefore measured for the first time the level of *several* impurities by a direct observation of the argon scintillation light.

More experimental work is ongoing, in particular to address the outgassing and leak checking aspects, and results are expected in the near future.

3. Experimental setup

3.1. Ar vessel

A schematic of the apparatus, completed with a recirculation and gas analyzing systems, is shown in Fig.1. A 6 m^3 vessel, previously used as a liquid argon dewar for testing of LAr/U calorimeters, was refurbished and made available for this test. The vessel is a double walled dewar, to allow for vacuum insulation in case of cryogenic operation. The walls of the inner vessel are stainless steel, while the top flange, 7 cm thick, is made out of aluminum, as the outer vessel. The diameter of the inner vessel is 200 cm and the total height of the dewar is 235 cm. On the top flange there are 12 KF sealed nipples for feedthroughs. The top flange is sealed with an o-ring. The tightness of the vessel has been measured through pressure decay leak testing, which allowed to determine a leak rate of 0.15 mbar lt / sec (which compares to a leak rate of $\sim 10^{-6}$ mbar lt/sec characteristic of high vacuum vessels). This relatively large leak rate was deemed acceptable for our goals, since it corresponds - all things equal but a reversed pressure gradient - to a flow of 20 ppm of O_2 per hour; considering that the vessel is operated with an overpressure of $\sim$ 100 mbar, the actual inflow of O_2 from air is likely to be rather negligible. The Ar gas transfer line provided a maximum flow of about 200 lt/min. The gas line was connected

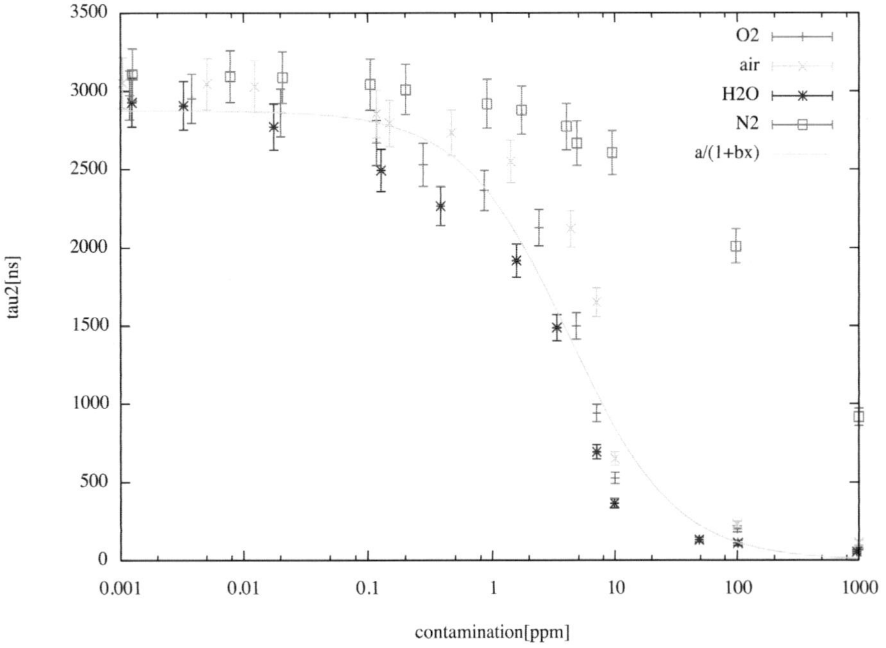

Figure 3. The dependence of τ_2 from the concentration of different impurities in Ar gas. Data from [9]. A fit with a Birks' law-type function to the O_2 data is superimposed.

to a 35 mm diameter stainless steel tube which went all the way to the bottom of the vessel. The Ar gas was exhausted through a flow meter, which could measure a maximum flow of 100 lt/min.

3.2. Detection of impurity traces with PMTs

During operation, the vessel was equipped with two percent level O_2 monitors[1], which are sensitive to a minimum concentration of 0.1% O_2. The oxygen sensors were left hanging from the top flange, one at a depth of 30 cm and the second one 90 cm below.

Three photomultiplier tubes (PMT) were deployed inside the vessel to monitor the DUV argon scintillation light, at a distance of 40, 110 and 180 cm from the top flange. The PMTs are 2" ETL 9831KB, each housed in a cylindrical stainless steel holder with a 40 kBq ^{241}Am (α emitter) source placed in front of the PMT, at a distance of 10 cm from the photocathode. There are large circular apertures on the sides of the holder, to allow for efficient gas circulation (Fig.2). Once inside the holder, a PMT sees a negligible amount of light from outside the holder itself, so that we could monitor the scintillation light of the argon at a given position. The inside of the PMT holder was lined with 3M foil specular reflector coated with 1 mg/cm^2 of tetraphenylbutadiene (TPB), rendering the reflection form the surface 90% diffuse [11]. In addition the PMT windows were coated with 0.05 mg/cm^2 TPB. TPB acts as a wavelength shifter for the DUV scintillation light of argon to approximately 430 nm (blue), within the high quantum efficiency range of the PMT [11]. The signal from each PMT was digitized at a sampling rate of 1 GS/s, using an Acqiris DP1400 digitizer.

[1] OXY-SEN model, from Alpha Omega Instruments.

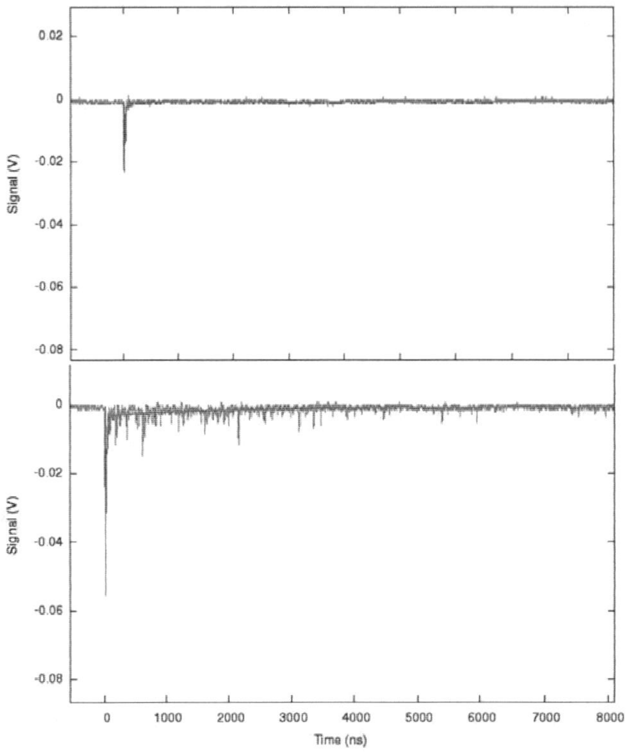

Figure 4. PMT traces for alpha particles from an Am source. *Top:* an event at the beginning of the run, τ_2 consistent with zero. *Bottom:* an event taken near the end of the run, with τ_2 exceeding 1200 ns.

The scintillation light of pure argon gas has a fast and a slow component, with decay times τ_1 shorter than 10 ns and τ_2 of about 3 μs [12, 13]. The fast component is rather independent of the purity of the Ar, while the decay time of the slow component is sensitive to the concentration of impurities. In particular, very clean Ar shows a decay time exceeding 3 μs, while for O_2 concentration larger than 0.1% the decay time is 100 ns or less. See for example Fig. 3 (data from [9]), which is directly comparable to the measurements presented here since it was obtained with the same hardware and analysis procedure (in a smaller cell).

The quenching effect of impurities on the slow component of the scintillation light can be roughly described by

$$\tau_2[\text{ns}] = \frac{\tau'}{1 + k\rho[\text{ppm}]}$$

where τ' is the asymptotic value of τ_2 for zero impurities, ρ[ppm] is the concentration of impurities in ppm and k is a constant; this function is reminiscent of the Birks' law that describes the quenched light yield in scintillators for an increased dE/dx. For O_2, $\tau' = 2878$ ns and $k=$ 0.24. This method is more sensitive for τ_2 in the 500 to 2500 ns range (i.e. 0.5 to 10 ppm of O_2), while for smaller τ_2 the variation is rather modest over a couple of orders of magnitude change in O_2 concentration.

For each measurement of the τ_2 10,000 events are acquired, a procedure which typically takes few seconds. Each waveform is fitted offline, to determine the value of τ_2. The average value for

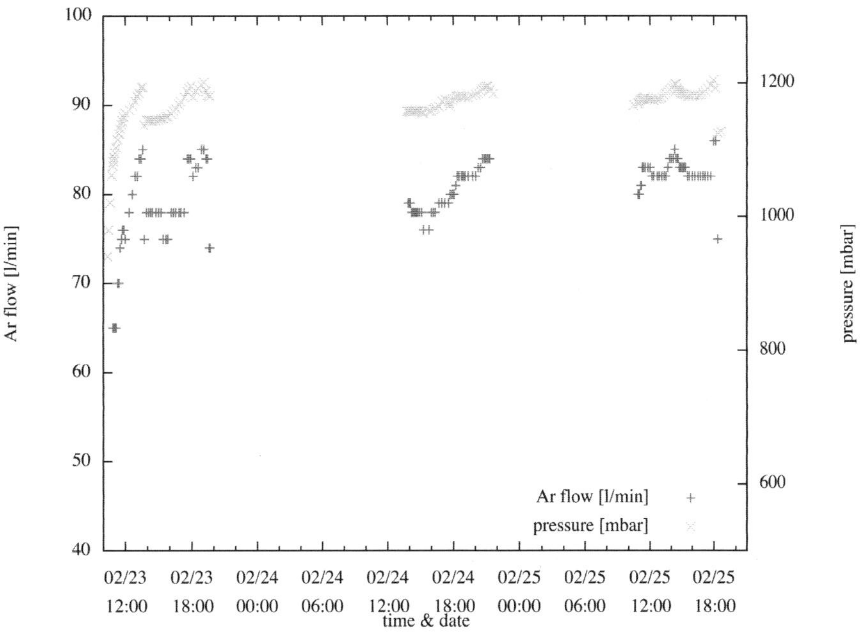

Figure 5. Ar gas flow and pressure in the vessel during Ar flushing.

the 10,000 events is taken as the measured τ_2 at a given time. Traces taken at the beginning and at the end of the run are shown in Fig.4.

4. Results

The 6 m^3 vessel was flushed with argon gas, not continuously, over more than a week. We started using a bottle (10 m^3 stp) of Ar57 (impurities less than 3 ppm) for a preliminary flushing which reduced the O_2 concentration from 20% to 0.5%, measured with one OXY-SEN sensor placed on the vessel exhaust. After 48 hrs without flushing any argon and the vessel with 150 mbar overpressure, the O_2 measured on the exhaust line actually improved, since purer argon had time to diffuse from the bottom to the top of the vessel. A flange at the top was then opened to insert the two OXY-SEN sensors inside the "clean" argon volume. Flushing with Ar57 was then restarted - as can be seen in Fig.5. Part of the data from the OXY-SEN readings are collected in Tab.1; the two sensors at different depths show very clearly the pure Ar gas acting as a piston, removing the residual air starting from the bottom to the top. The Ar57 flow was approximately constant at 80 lt/min, limited by the flowmeter on the exhaust line; this is equivalent to one volume change per 75 min. The vessel was always flushed with Ar57, except for the last 1.7 volume changes (starting at [02/25 14:00]) when Ar60 (impurities less than 1 ppm) was used, because the purification rate seemed to decrease, possibly due to ~3 ppm of O_2 in the input gas.

Fig.6 and Fig.7 summarize the main results versus acquisition time. When the sensitivity limit of the OXY-SEN sensors was reached (0.1%), data have been taken continuously using the three PMTs. The decay time of the slow component of the Ar scintillation light (τ_2) improves continuously due to the removal of impurities. The PMT near the bottom of the vessel (and the Ar inlet) consistently shows the largest τ_2, and shows the reduction in of contamination earlier than the other PMTs, as expected from the Ar gas acting as a piston. The PMTs at the top and in the middle show a similar τ_2, and the top one shows larger variations, correlated with sudden changes in the flow rate and pressure, which is measured at the top flange. The Ar flow near the top flange is expected to be quite turbulent and non-uniform, since the outlet is - in practice - point-like when compared to the size of the flange itself. When the flushing stops, τ_2 decreases by an amount roughly equivalent to the introduction of 2ppm of O_2, possibly coming from outgassing and/or leaks to the outside at a rate of 0.15 ppm per hour. The O_2 equivalent concentration at the end of the flushing procedure at the location of the bottom PMT is lower than 4 ppm, while the average of the 3 PMTs is slightly larger than 4 ppm. The O_2 equivalent concentration [ppm] *vs* time is shown in Fig.7 for $\tau_2 > 500$ ns, i.e. the optimal range for this technique. The O_2 concentration drops of about a factor of two after 5 hrs of flushing, i.e. 24,000 lt of Ar gas stp or 4 volume changes.

Table 1. O_2 concentrations as measured with OXY-SEN monitors (see text).

Time&Date	Exhaust O_2 [%]	Top O_2 [%]	Bottom O_2 [%]
2010-02-19 15:30	19.1	-	-
2010-02-19 16:18	9.1	-	-
2010-02-19 16:28	6.9	-	-
2010-02-19 16:39	5.2	-	-
2010-02-19 16:48	4.2	-	-
2010-02-19 16:57	3.5	-	-
2010-02-19 17:06	3.3	-	-
2010-02-19 17:12	2.9	-	-
2010-02-20 10:29	1.4	-	-
2010-02-23 10:24	-	0.3	0.3
2010-02-23 10:30	-	0.3	0.3
2010-02-23 10:39	-	0.3	0.3
2010-02-23 10:55	-	0.3	0.3
2010-02-23 11:05	-	0.25	0.3
2010-02-23 11:20	-	0.25	0.3
2010-02-23 11:27	-	0.2	0.3
2010-02-23 12:01	-	0.2	0.3
2010-02-23 12:21	-	0.2	0.25
2010-02-23 12:38	-	0.2	0.25
2010-02-23 12:54	-	0.2	0.2
2010-02-23 13:35	-	0.15	0.2
2010-02-23 13:44	-	0.1	0.2
2010-02-23 16:09	-	0.1	0.15
2010-02-23 16:21	-	0.1	0.1
2010-02-23 18:34	-	0.1	0.1

5. Conclusions

The main results of this test is the purging of a 6 m^3 vessel from air to a residual contamination of few ppm of O_2 equivalent in pure Ar gas. The impurity concentration in argon has been

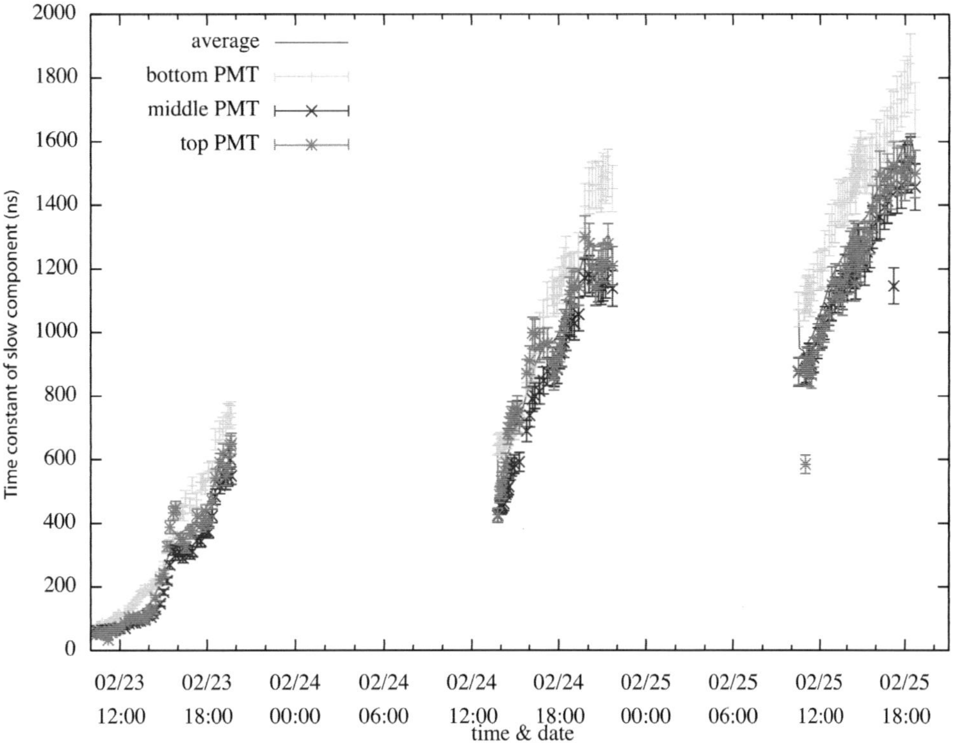

Figure 6. Fitted time constant of the slow component of the Ar scintillation light *vs.*time for three PMTs at different depths. Data taken during Ar flushing.

measured down to the few ppm range using a simple system, based on the quenching of the Ar scintillation due to impurities. The piston effect of argon displacing impurities has been seen both for percent level contamination and ppm level contamination. The final O_2 equivalent concentration, after ten volume changes, was $\sim$4 ppm, which corresponds to a reduction of the O_2 of about five orders of magnitude, faster than a simple exponential trend. The experimental setup was remarkably simple, and the reduction in concentration of impurities in the vessel proved to be rather insensitive to interruptions in the flushing procedure. This is an experimental demonstration that impurities in argon can be reduced easily and efficiently to ppm level without vacuum pumping.

Acknowledgments
We thank the organizers of GLA2010 for inviting us to this excellent and very fruitful workshop. This work was supported by ETH Zürich, the Swiss National Science Foundation (SNF) and the University of Liverpool. We are grateful to CERN for their hospitality where these tests were performed.

References
[1] A. Rubbia, "Experiments for CP-violation: A giant liquid argon scintillation, Cerenkov and charge imaging experiment?," arXiv:hep-ph/0402110.

[2] A. Rubbia, "Underground Neutrino Detectors for Particle and Astroparticle Science: the Giant Liquid Argon Charge Imaging ExpeRiment (GLACIER)," J. Phys. Conf. Ser. **171** (2009) 012020 [arXiv:0908.1286 [hep-ph]].

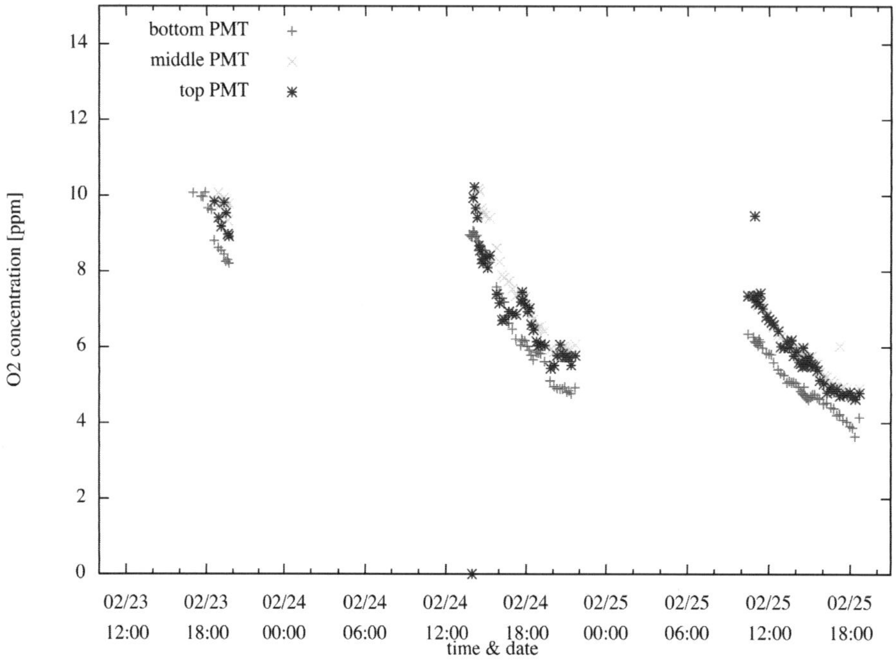

Figure 7. Same as Fig.6, converted in ppm of O_2 equivalent in argon.

[3] A. Rubbia, these proceedings.
[4] S. Horikawa *et al.*, these proceedings.
[5] A. Bettini *et al.*, "A Study Of The Factors Affecting The Electron Lifetime In Ultrapure Liquid Argon," Nucl. Instrum. Meth. A **305**, 177 (1991).
[6] P. Benetti *et al.*, "Argon purification in the liquid phase," Nucl. Instrum. Meth. A **333**, 567 (1993).
[7] F. Arneodo *et al.* [ICARUS-Milano Collaboration], "Performance of a liquid argon time projection chamber exposed to the WANF neutrino beam," Phys. Rev. D **74**, 112001 (2006) [arXiv:physics/0609205].
[8] C. Amsler et al., *Luminescence quenching of the triplet excimer state by air traces in gaseous argon*, JINST **3** (2008) P02001.
[9] K. Mavrokoridis. Characterisation of Liquid Xenon and Argon as Targets for Direct Dark Matter Detection. PhD thesis, Department of Physics & Astronomy, University of Sheffield, 2009.
[10] W. Jaskierny, H. Jostlein, S. H. Pordes, P. A. Rapidis and T. Tope, "Test of purging a small tank with argon," FERMILAB-TM-2384-E
[11] V. Boccone *et al.* [The ArDM Collaboration], "Development of wavelength shifter coated reflectors for the ArDM argon dark matter detector," JINST **4**, P06001 (2009) [arXiv:0904.0246 [physics.ins-det]].
[12] Colli L 1954, *Phys. Rev.* **95** 4
[13] Keto J W, Gleason Jr. R E and Walters G K 1974, *Phys. Rev. Letters* **33** 23

Journal of Physics: Conference Series **308** (2011) 012025 doi:10.1088/1742-6596/308/1/012025

Monitoring the parameters of a large size liquid Argon Time Projection Chamber using UV laser beams

Biagio Rossi
on behalf of LHEP liquid Argon group
Albert Einstein Center for Fundamental Physics, Laboratory for High Energy Physics (LHEP), University of Bern, Switzerland
E-mail: biagio.rossi@lhep.unibe.ch

Abstract. This paper reports on multiphoton ionization at 266 nm of liquid Argon in a Time Projection Chamber (LAr TPC) detector. The electron signal produced by the laser beam is a formidable tool for the monitoring of the next generation large mass LAr TPCs [1]. The detector we have designed and tested allowed us also to measure the two-photon absorption cross-section (σ_{ex}) of LAr with unprecedented accuracy and precision, appropriate for using the UV laser technique also for a quantitative calibration of the detector.

1. Introduction

The liquid Argon Time Projection Chamber (LAr TPC) detector technology allows for uniform and high resolution imaging of massive detector volumes. The operating principle of the LAr TPC is based on the undistorted transport of tracks of ionizing electrons in highly purified LAr by a uniform electric field over distances up to a few metres. Imaging is provided by wire planes or other read-out devices placed at the end of the drift path. The drifting electrons are collected by the outermost wire plane which gives position and calorimetric information. Additional planes with different orientation can be positioned in front of the collection plane to record the signal induced by the passage of the drifting electrons. This provides different projective views of the same event thus allowing space point reconstruction. The third coordinate is given by the measurements of the drift time given by the time interval between the passage of the ionizing particle in the active volume (t_0) and the arrival of the drifting electrons on the wire planes. The t_0 can come from the detection of the scintillation light of liquid Argon by means of photomultipliers, or from an external source. The particle momentum of an incoming particle is inferred by the measurement of its multiple scattering [2], while the detection of the local energy deposition provides particle identification. For more information, the reader is referred to [1] and references therein quoted.

A very large LAr TPC with a mass ranging from 10 kton to 100 kton would produce an important physics output thanks to the excellent event reconstruction performance [3, 4, 5, 6, 7]. The purity of the liquid Argon is a key ingredient to achieve imaging over long drift distances (>5 m), needed for building such a large masses. In particular, a purity corresponding to less than 0.1 ppb of electronegative elements such as Oxygen, has to be achieved in order to allow for electron lifetimes of milliseconds and, hence, drift distances of meters. Moreover, the purity of LAr has to be monitored continuously while TPC is running, thus calling for practical and efficient purity monitoring techniques.

To this end, laser based methods are very interesting as they provide a reliable route to tackle the problem. In addition, a laser beam is an ideal tool for performig the energy calibration of the detector, for measuring the drift velocity, the space and time resolution of the detector, the capability of double track separation, the diffusion of the drifting electrons, and also for verifing the homogeneity of the drift field. Furthermore, the laser technique can be also used in the *double phase* LAr TPCs [3, 8] for the monitoring of the stability of the read-out amplification stages.

In this paper, the UV laser multiphoton ionization of liquid Argon with the 4th harmonic of a Nd-YAG laser is illustrated. In the first part, the monitoring of the parameters of a LAr TPC with laser beams is pointed out. In the second part, a theoretical introduction to the multiphoton ionization with an accurate measurement of the two-photon absorption cross section of LAr (σ_{ex}), which is the atomic intermediate step to ionization, fundamental for reliable calibration of LAr TPCs, is presented.

The interaction of laser light with liquid matter is comparatively less studied with respect to gas and solid phase. Nevertheless, some important features are well understood [9]. Laser ionization can occur as a multiphoton process through the simultaneous absorption of two or more photons via virtual states in a medium. This process requires high photons flux, namely, pulsed lasers. In a multiphoton process, bound electrons absorb several laser photons simultaneously, thus exciting the atom to high energy levels or even to ionization.

The measurements described in this paper are done with a prototype LAr TPC described in [1] that is part of a R&D project we are conducing in Bern envisioning the realization of LAr TPC detectors of increasing size. First results have been published also on the study of a TPC we have built and operated employing a mixture of LAr and Nitrogen [10].

2. Experimental setup

The detector (figure 1) used for the measurements consists of a TPC (figure 2) housed in a tube filled with purified liquid Argon, in thermal contact with a bath of liquid Argon (figure 3). The TPC is complemented by an electronic read-out and data acquisition system (DAQ), a liquid Argon recirculation and purification system, a photomultiplier (PMT), a series of ancillary equipments for monitoring and control, and a UV-laser with the related optics to allow the study of multiphoton ionization of liquid Argon.

The TPC detector is contained in a stainless-steel cylindrical tube. The latter measures 50 cm in diametre and 110 cm in height, for an inner volume of about 200 litres. The tube is placed in thermal contact with a bath of liquid Argon and is closed by a stainless-steel flange. A series of additional smaller flanges are positioned on the top flange. They include feed-throughs needed to evacuate the inner volume of the vessel, to fill it with liquid Argon, to read-out the detector signals, to supply the high-voltage, to house the pressure monitors, level metre and safety valves, and to provide the transport of the UV laser beams to the inner part of the TPC. The various components are described in detail in [1].

Figure 4 shows the arrangement and the path of the laser beam used for the measurements. The laser beam passes through a system of prisms to be *cleaned* from the residuals of the 1st and 2nd harmonic components. After a deflection on a 266 nm coated mirror, it is then fed in a variable, motorized attenuator. After the attenuator two parallel beams of variable intensity are present. One is dumped, while the other is used for the measurements. The use of the attenuator is important, as it allows to vary the energy of the laser beam without changing the laser generation parameters. Consequently, the laser shows better stability performance. Both energy and pulse length measurements are performed before the beam is finally steered into the

Journal of Physics: Conference Series **308** (2011) 012025 doi:10.1088/1742-6596/308/1/012025

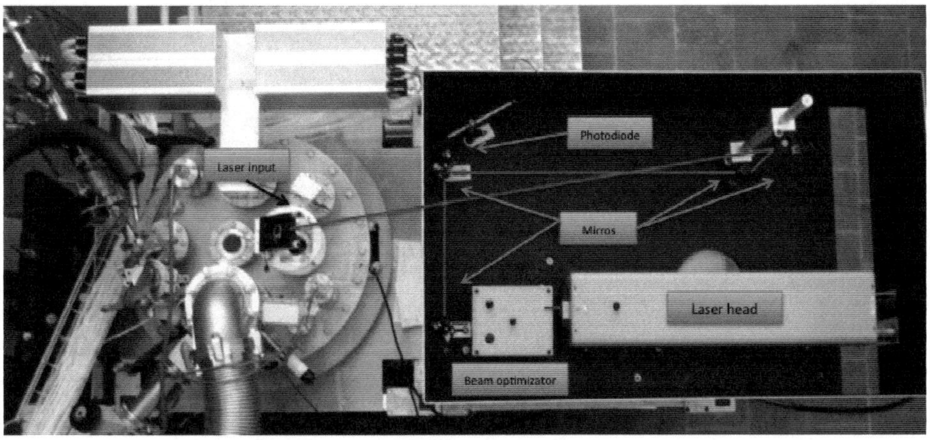

Figure 1. Photograph of the experimental setup (top view) including the laser beam. Red lines show the beam path before being steered into the detector.

Figure 2. Photograph of the experimental setup. The TPC, the PMT and a mirror installed on a stainless steel support are clearly visible. The drift direction is horizontal with induction (collection) wires running at an angle of 25 degrees relative to the vertical (horizontal) direction.

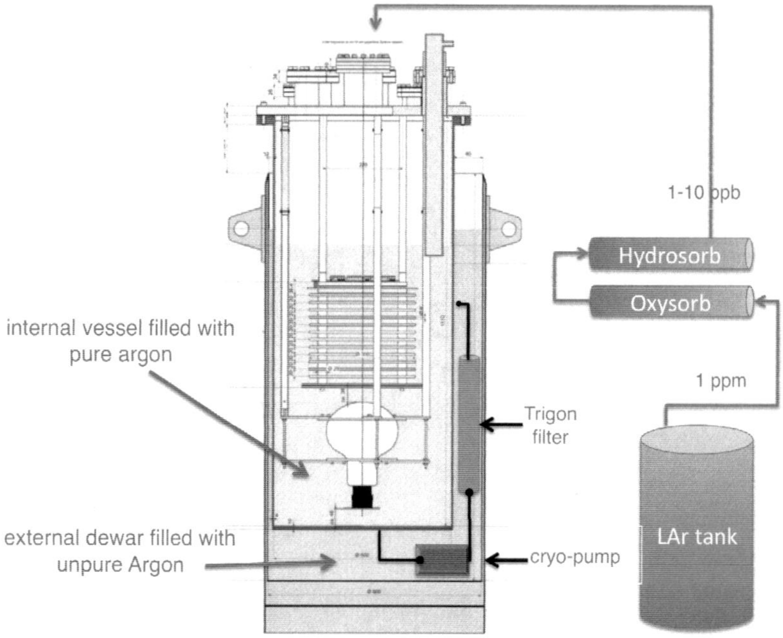

Figure 3. Schematic drawing of the detector set-up. The TPC, mounted with the drift coordinate vertically. The PMT and the relative supports are housed in a cylindrical tube filled with pure Argon. The external vessel that contains un-purified Argon hosts the liquid purification/recirculation system.

chamber.

In theoretical calculations (section 5.2), we assume for simplicity that the transverse distribution of the laser intensity has a square shape. In the experiment, the laser beam passes after the attenuator through a diaphragm of 4 mm diameter selecting the more uniform, central part of the beam. In this way one obtains a definite beam cross section also preventing any effect of the laser beam tails (halo) on the ionization process. In fact, the contribution to ionization of photons in the halo is negligible due to the nonlinear character of the process.

After the diaphragm, the light is fed into the TPC inner volume by passing through a system of mirrors (figure 1) and a custom made optical feedthrough [1].

As described in figure 4, the beam passes two times in the active volume, thus producing a vertical track orthogonal to the drift coordinate and a second track *inclined* with respect to the drift coordinate. After the second passage, the beam is steered out of the dewar by means of another optical feedthrough. The extraction of the laser beam from the TPC allows for the monitoring of its energy, after ionization of LAr has occurred. Cross check of the energy measurements at the entrance and exit optical feedthroughs indicates no depletion of the ionizing laser beam.

The measurement of the pulse width is performed by means of a fast photodiode[1], calibrated in energy with an energy meter[2] and readout by a CAEN V1729 ADC. The board, a Switched-Capacitor Digitizer, performs the coding of 4 analog channels of bandwidth up to 300 MHz over 12 bits at a sampling frequency of up to 2 GS/s and over a depth of 2520 usable points. It is suited for acquisition of fast analog signals.

[1] Thorlabs DET 10 A.
[2] GEANTEC - SOLO 2.

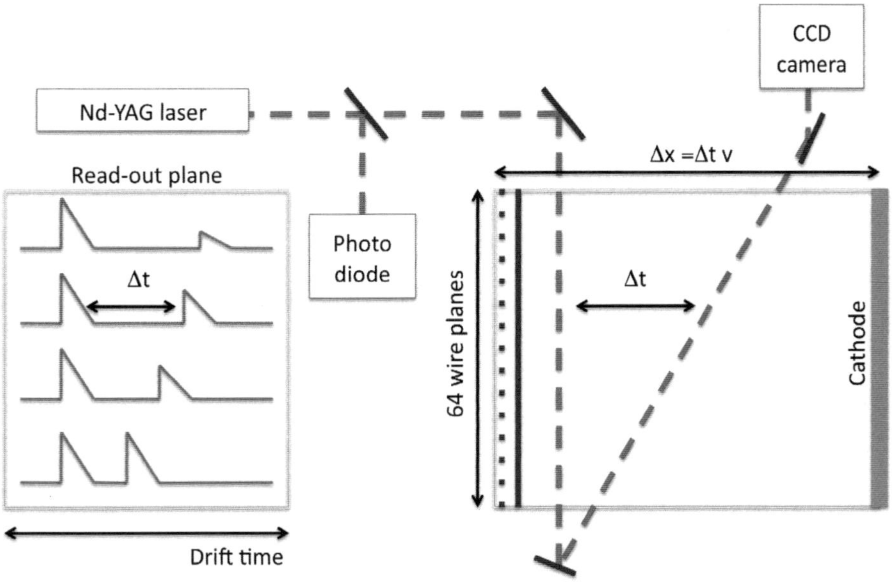

Figure 4. Schematic view of the laser beam path and the readout response.

3. Detector operation and data taking

Before filling with liquid Argon, the detector vessel is evacuated to a residual pressure of 10^{-5} mbar. The detector is then filled with ultra pure liquid Argon flown through the purification cartridges in liquid phase, analogous to the purification unit inserted in the recirculation system. The purity of the liquid Argon before passing through the OXYSORB [11] cartridge has a nominal concentration of water and Oxygen of the order of a ppm. The procedure is described in details in [1]

Several measurements were performed with the TPC installed with the electrons drifting along the horizontal direction. The detection of the first UV laser induced ionization tracks have been published in [1].

In figure 5 the display of a typical UV laser induced event is shown. Two tracks along all the wires at different drift time are well visible. The beam first generates the same signal over all the wires of the TPC collection plane at the same time (drift coordinate), and then, after the reflection on a mirror the beam passes through the active volume again inducing a second track *inclined* with respect to the drift coordinate. In figure 6 typical waveform peaks induced by a single UV laser pulse on a collection wire are shown. Two peaks at different drift time are clearly visible.

4. Monitoring drift velocity and LAr purity

The drift velocity is determined knowing *a priori* the distance between the two tracks, and measuring the difference between the drift time of the peaks (wire by wire) induced by the tracks.

In figure 7 the velocity of the drifting electrons as a function of the drift field intensity is shown. We performed two campaigns of measurements, collecting for each 1000 UV laser-induced events, for different values of the drift field. Results, for different LAr temperatures, are listed in the column of table 1. For the same field intensity, the higher the temperature, the lower

Journal of Physics: Conference Series **308** (2011) 012025 doi:10.1088/1742-6596/308/1/012025

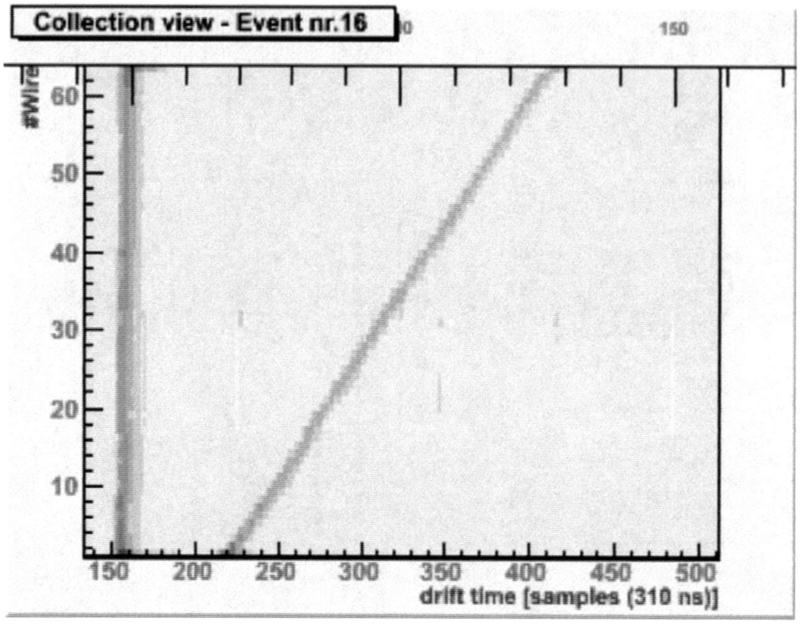

Figure 5. Event display of a UV laser ionization event in the setup described above. Two tracks along all the wires at different drift time are visible. The beam first generates the same signal over all the wires of the collection plane at the same time (drift coordinate) and then after the reflection on a mirror passes through the active volume again indcucing another track *inclined* with respect to the drift coordinate.

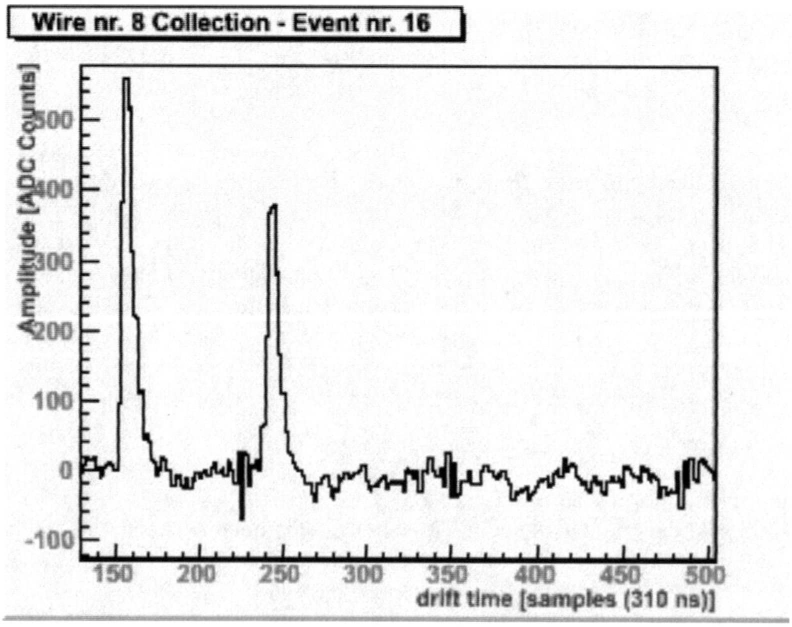

Figure 6. Typical signals induced on a wire of the TPC collection plane by the two UV laser induced tracks. Two peaks at different drift time are clearly visible. The area under each peak provides the electron yield.

Journal of Physics: Conference Series **308** (2011) 012025 doi:10.1088/1742-6596/308/1/012025

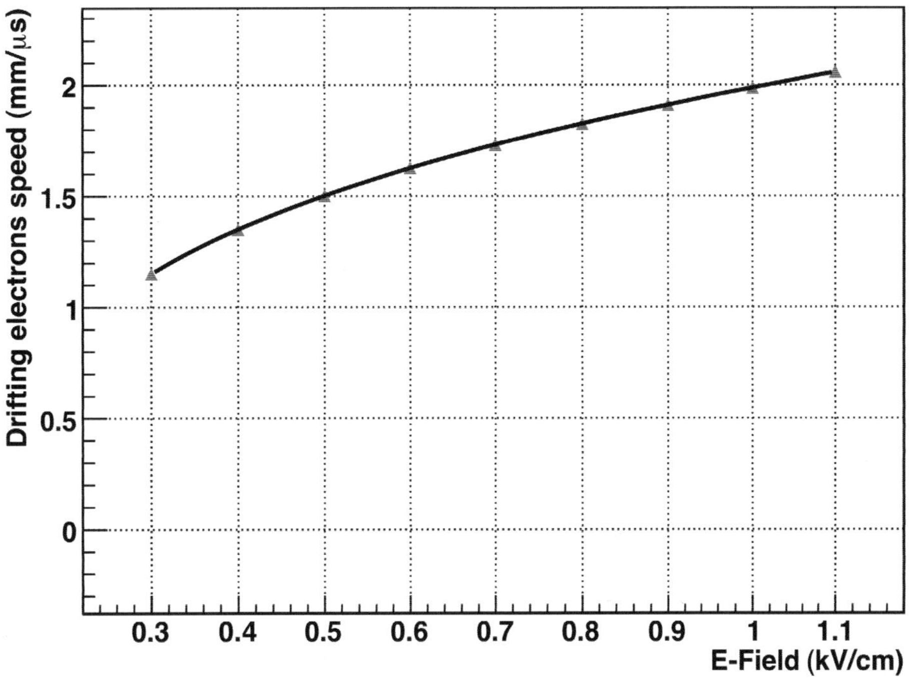

Figure 7. Drifting electrons lifetime (μs) as a function of the electric field intensity (T=89.1 K). The error bars are smaller than the points. The fit was performed with an empirical formula taken from [12, 13].

is the drift velocity. Data well agree with the theoretical predictions and with previous results [12, 13, 14].

Table 1. Velocity of drifting electrons for different electric fields.

E (kV/cm)	v (mm/μs)@88.9 K	v (mm/μs)@89.1 K
1.20	2.138 ± 0.006	
1.10	2.089 ± 0.006	2.059 ± 0.006
1.00	2.021 ± 0.006	1.987 ± 0.006
0.90	1.943 ± 0.006	1.910 ± 0.005
0.80	1.857 ± 0.005	1.826 ± 0.005
0.70	1.759 ± 0.005	1.738 ± 0.005
0.60	1.656 ± 0.004	1.627 ± 0.008
0.50	1.530 ± 0.004	1.503 ± 0.004
0.40	1.371 ± 0.003	1.350 ± 0.004
0.30	1.158 ± 0.004	1.150 ± 0.005
0.20	0.883 ± 0.004	

The electrons attachment to the electronegative impurities present in the LAr during the drift diminish the charge collected by the wire planes. The measurement of the charge attenuation along the drift time provides an estimate of the LAr purity. In figure 8 the profile histogram of the charge produced by 1000 *inclined* laser tracks as a function of the drift field is shown.

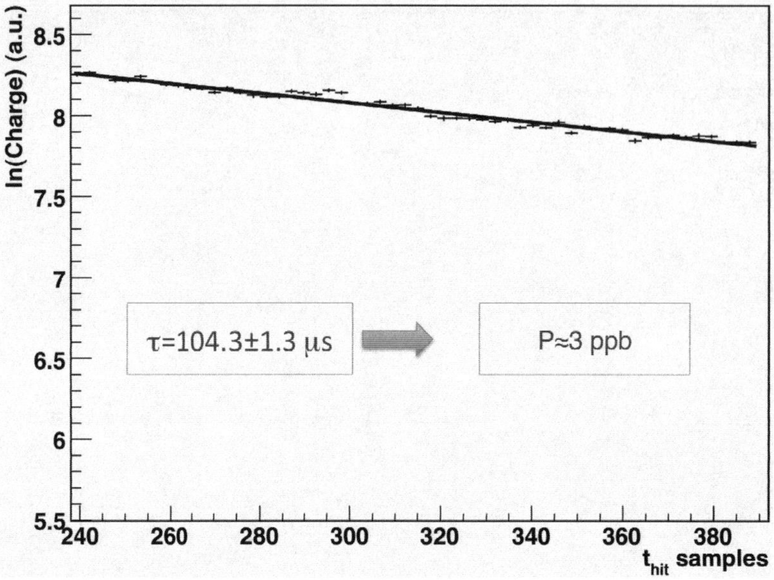

Figure 8. Profile histogram of the charge induced by 1000 laser tracks as a function of the drift time. The points are fit with line $f(t) = q_0 - (t_{hit} - t_0)/\tau$, where q_0 is the charge produced by the laser track, t_0 is the position (along the abscissa) of the wire planes, t_{hit} is the drift coordinate, and τ is the drifting electrons lifetime in LAr. For this sample of data $\tau = (104.3 \pm 1.3)$ μs.

The points on the figures were fit with a line $f(t) = q_0 - (t_{hit} - t_0)/\tau$, where q_0 is the charge produced by the laser track, t_0 is the position (along the abscissa) of the wire planes, t_{hit} is the drift coordinate, and τ is the drifting electrons lifetime in LAr. The τ is related to the purity via the following formula:

$$\mathrm{P[ppb]} \sim \frac{300}{\tau[\mu\mathrm{s}]} \tag{1}$$

The fit yields a lifetime of $\tau = 104.3 \pm 1.3$ μs and, hence, a purity of about 3 ppb of O_2 eq.

5. Multiphoton absorption of LAr

An atom with ionization energy E_i can be ionized by irradiation with photons with energy $E = h\nu$ much lower than E_i if the photon flux is intense enough to promote multiphoton absorption.

One of the features of multiphoton absorption is that it can occur via laser-induced virtual states, which are not eigenstates of the atom. Thus, it does not require any intermediate atomic state. These laser induced virtual states are characterized by a lifetime $\tau^{virtual}$ of the order of one optical cycle. For the laser used in the present experiment we have:

$$\tau^{\mathrm{virtual}}_{266\ \mathrm{nm}} \approx \frac{\lambda}{2\pi c} \approx 1.4 \cdot 10^{-16}\ \mathrm{s} \tag{2}$$

while the lifetime of an intermediate (excited) *real* state is of the order of $\tau \sim 10^{-8}$ s [15]. When an atomic state is located not too far from a laser-induced virtual state (quasi-resonant ionization), the multiphoton absorption process is stronlgy enhanced, thus requiring a laser intensity much lower than in the non-resonant case [15].

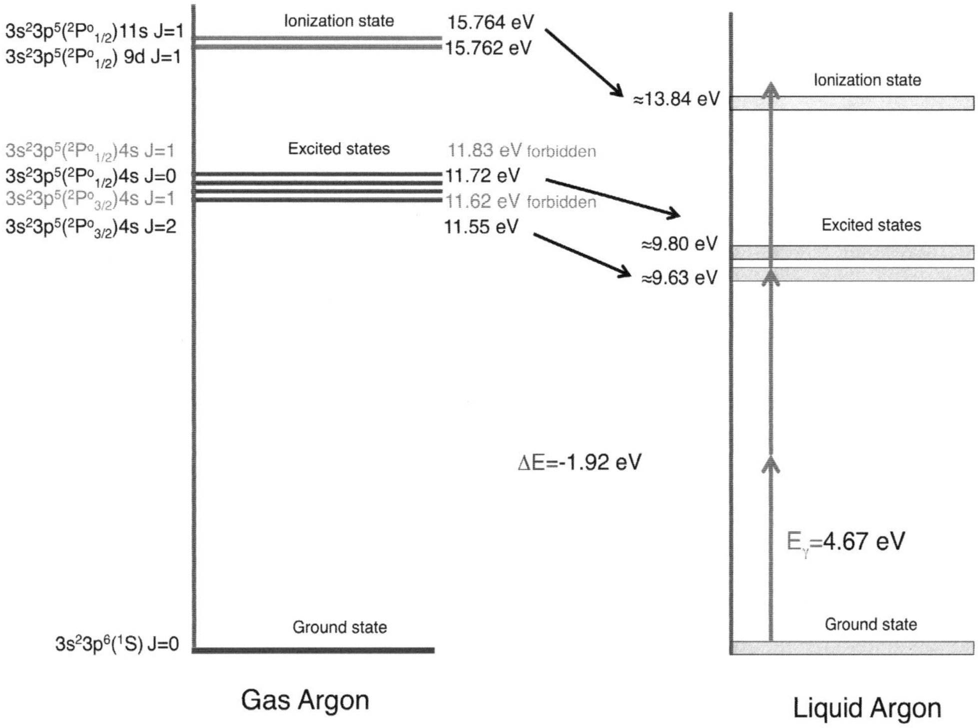

Figure 9. Schematics of the gas/liquid Argon energetic levels. In the passage from gas to liquid phase, the absolute values of the energetic levels of the atom are reduced [9] and the lines broaden, becoming energy bands as in the condensed matter.

5.1. Energy levels in liquid Argon

The interaction of laser light with matter in the liquid phase has been comparatively less studied with respect to gas and solid phase. Nevertheless, some important features are well understood [9]. In the passage from gas to liquid phase, the energy separation between the energy levels reduces. This is sketched in figure 9 where the energy levels of gas and liquid Argon are compared. The ionization potential for Argon gas (GAr) is $I_p = 15.76$ eV [16], and four excited states potentially involved in a multiphoton process lie at 11.83 eV, 11.72 eV, 11.62 eV and 11.55 eV, respectively. For two-photon absorption $\Delta J = 0, 2$ and, hence, only the transitions to levels with J=0 and J=2 are allowed by conservation rules. The effects leading to the energy shift can be summarized in the following formula [17]:

$$I_p^{\text{liquid}} = I_p^{\text{gas}} + P_+ + V_0 \tag{3}$$

where I_p^{gas} (I_p^{liquid}) is the ionization potential in gas (liquid) phase and V_0 is the energy of the quasi-free electron at the bottom of the conduction band of the liquid. The polarization energy P_+ of the positive ion is given by the Born equation [18]:

$$P_+ = -\frac{e^2}{2r_+}(1 - \epsilon^{-1}) \tag{4}$$

In the classical approximation, a liquid is treated as a uniform continuum with the optical dielectric constant ϵ, while a positive ion is regarded as a sphere with radius r_+ and charge e.

For Argon the values experimentally measured are ϵ=1.5 at 550 nm and ϵ=1.9 at 128 nm, but an extrapolation from a plot published in [19] leads to ϵ=1.59 at 266 nm. Since $V_0 = -0.17$ eV [20], $r_+ = 0.088$ nm, we have $P_+ = -1.75$ eV thereby the ionization state in liquid turns out to be $I_p^{\text{liquid}} = I_p^{\text{gas}} - 1.92$ eV. Consequently, the ionization and the lowest excited state energies become, respectively:

$$I_p^{\text{liquid}} = 13.84 \text{ eV} \quad , \quad I_e^{\text{liquid}} = 9.63 \text{ eV} \tag{5}$$

in our experiment we have used the 4th harmonic (λ=266 nm, corresponding to a photon energy of 4.67 eV) of a Nd-YAG laser source, the main features of which are reported in Table 2.

Table 2. 4th harmonic UV laser specifications (Continuum, model Surelite I-10)

wavelength (nm)	max rep. rate (Hz)	max energy (mJ)	energy stability
266	10	82	5%
pulse width (ns)	rod diameter (mm)	divergence (mrad)	FWHM (ns)
5-7	6	0.6	5.7

At this wavelength at least three photons are needed to produce ionization in LAr, although the intermediate *real* state can make the actual multiphoton transition a 2+1 transition. In fact, the lowest allowed excited state lies at 9.63 eV, which is only 0.3 eV higher than the energy of two photons at 266 nm. However, it is very plausible that this energy gap disappears in the liquid phase due to band formation and the consequent level broadening of the atomic energy levels [9].

5.2. Rate equations

In a multiphoton ionization process the N-photon ionization rate W is given by $W = \sigma_N F^N$, where σ_N is the generalized N-photon ionization cross-section. The rate W is expressed in reciprocal seconds, σ_N is expressed in cm^{2N} s^{N-1} and the laser flux F is in photon cm^{-2} s^{-1}.

For gases the one-photon absorption cross-section σ_1 is typically of the order of 10^{-17}cm^2[21], whereas the two-photon and the three-photon cross-sections are of the order of $\sigma_2 = W/F^2 \sim 10^{-50}$cm^{4}s and $\sigma_3 = W/F^3 \sim 10^{-83}$cm^{6}s^2, respectively.

From the energy levels depicted in figure 9 we can write the following rate equation to retrieve the expected rate of excited electrons produced:

$$\frac{dN_{ex}}{dt} = (N_0 - N_{ex})F^2\sigma_{ex} - N_{ex}\sigma_i F - N_{ex}/\tau \tag{6}$$

where N_0 is the initial density of atoms in the ground state, N_{ex} the density of atoms in the excited state, σ_{ex} the two-photon transition cross-section from ground to the intermediate excited state, σ_i is the single photon transition cross-section from the excited state to the continuum, and τ the lifetime of the excited state, respectively. The rate of ionization is given by:

$$\frac{dN_i}{dt} = N_{ex}\sigma_i F \tag{7}$$

where N_i is the density of ionized atoms (or number of extracted electrons). Here we consider the ion/electron recombination negligible for UV laser induced events at our operational drift field [22]. Defining $h = \sigma_{ex}F^2 + \sigma_i F + \frac{1}{\tau}$, replacing equation (6) in equation (7) and time-integrating ($dt = s$) one obtains:

$$N_i = \frac{N_0\sigma_i\sigma_{ex}F^3}{h}\Big(\frac{e^{-hs} - 1}{h} + s\Big) \tag{8}$$

where s is the FWHM of the laser pulse (in our case s=$(5.7 \pm 5\%) \cdot 10^{-9}$ s).

Three different regions can be defined:

- low photon flux for $hs << 1$;
- high photon flux for $hs >> 1$, $\sigma_{ex}F^2 << \sigma_i F$ and $1/\tau << \sigma_i F$;
- very high photon flux for $hs >> 1$, $\sigma_{ex}F^2 >> \sigma_i F$ and $1/\tau << \sigma_{ex}F^2$.

In the region of low photon flux equation (8) can be written as:

$$N_i = \frac{1}{2} N_0 \sigma_{ex} \sigma_i s^2 F^3 \tag{9}$$

showing a cubic dependence of the density of the electrons in the ionized state with respect to the flux of photons. In the high photon flux regime equation (8) becomes:

$$N_i = N_0 \sigma_{ex} s F^2 \tag{10}$$

exibiting a quadratic dependence. Finally, for very high photon flux equation (8) can be approximated as:

$$N_i = N_0 \sigma_i s F \tag{11}$$

with a linear dependence due to the saturation of the two-photon transition.

6. Measurement of the two photon absorption cross section

In order to estimate the cross-sections σ_{ex} and σ_i of the rate equations presented above, and the lifetime of the excited state of LAr, we measured the rate of electrons produced by the laser beam as a function of the photon flux. For each laser track we determined the total electron yield produced by the laser (for each wire) simultaneously, monitoring pulse-by-pulse the energy, and FWHM of the beam. For the data set shown below, the beam energy was varied in the range 0.1÷2.0 mJ.

Figure 10 reports the rate of electrons (measured in electron atom^{-1} s^{-1}) as a function of the photon flux (measured in photon cm^{-2} s^{-1}), in a log-log plot. In the investigated intensity range the data are well described by a straight line, and the experimental points are very well fitted with a power law dependence on photon flux F (see equation 9-11). In particular, the best fit is obtained by using equation 10 in the following form:

$$\frac{N_{ex}}{N_0 s} = \sigma_{ex} \cdot F^n \tag{12}$$

with n and σ_{ex} as free parameters. The corresponding best fitted values are:

$$n = 2.02 \pm 0.05 \tag{13}$$

$$\sigma_{ex} = (1.24 \pm 0.10_{\text{stat}} \pm 0.30_{\text{syst}}) \cdot 10^{-56}\ \text{cm}^4\text{s} \tag{14}$$

The value of n is consistent with 2, indicating that at our laser intensities the limiting step of the three-photon ionization process is the two-photon absorption to the intermediate state.

The value of σ_{ex} presented here provides a reliable and accurate measurements of the two-photon absorption cross section of LAr [22]. It is within the range of the two-photon absorption cross-section for liquids [23, 24]. The assumption made ($hs >> 1$, $\frac{1}{\tau} << \sigma_i F$ and $\sigma_i F >> \sigma_{ex} F^2$) to approximate equation (8) to equation (10) are justified by the experimental data. Moreover, the results are well consistent with the values $\sigma_i = 10^{-16}$-10^{-17} cm^2 and $\tau = 10^{-7}$-10^{-8} s provided by [21, 25, 26].

Systematic uncertainties due to the absolute energy calibration of the TPC and to the measurement of the photon flux have been evaluated. We assume a 25% uncertainty on the absolute number of electrons collected, and 10% uncertainty on the pulse-by-pulse measurement of the beam energy.

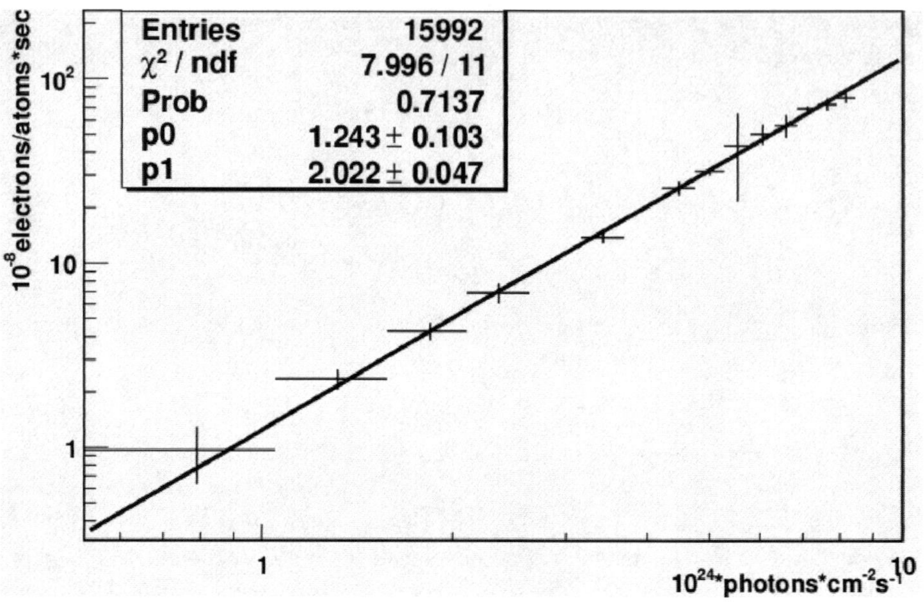

Figure 10. Profile histogram of the rate of electrons as a function of the photon flux of the laser beam.

7. Conclusions and outlook

We reported on the measurement of UV laser-induced multiphoton ionization of LAr performed with a liquid Argon Time Projection Chamber detector. The possibility of using such a technique for the general monitoring of the parameters (i.e. measurements of LAr purity) of LAr TPC detectors of large size has been shown. Furthermore, for the first time, the value of the two-photon absorption cross-section $\sigma_{ex} = (1.24 \pm 0.10_{\text{stat}} \pm 0.30_{\text{syst}}) \cdot 10^{-56}$ cm^{4}s has been determined with a good accuracy appropriate for quantitative calibration of LAr TPC detectors.

The results obtained and described in this paper make us confident on the practical use of UV laser beams technique for the general monitoring and for the calibration of LAr TPC detectors, as requested for the envisioned large scale applications of the technique.

Acknowledgements

The work presented in this paper was conducted thanks to grants from the Swiss National Science Foundation (SNF) and from the University of Bern. We warmly acknowledge both institutions. We wish to warmly thank our technical collaborators R. Haenni, P. Lutz and F. Nydegger, for their skillful help in building and operating the detector and the related infrastructure.

References

[1] B. Rossi *et al.*, A prototype liquid Argon Time Projection Chamber for the study of UV laser multi-photonic ionization, JINST **4** (2009) P07011.

[2] A. Ankowski *et al.*, Measurement of through-going particle momentum by means of multiple scattering with the ICARUS T600 TPC, Eur. Phys. J. C 48 (2006) 667.

[3] D. Autiero *et al.*, Large underground, liquid based detectors for astro-particle physics in Europe: scientific case and prospects, JCAP 0711:011 (2007).

[4] A. Ereditato and A. Rubbia, Ideas for future liquid Argon detectors, Nucl. Phys. Proc. Suppl. 139 (2005) 301.

[5] A. Ereditato and A. Rubbia, The liquid Argon TPC: a powerful detector for future neutrino experiments and proton decay searches, Nucl. Phys. Proc. Suppl. 154 (2006) 163.

[6] A. Ereditato and A. Rubbia, Conceptual design of a scalable multi-kton superconducting magnetized liquid Argon TPC, Nucl. Phys. Proc. Suppl. 155 (2006) 233.
[7] B. Baibussinov *et al.*, A new, very massive modular Liquid Argon Imaging Chamber to detect low energy off-axis neutrinos from the CNGS beam. (Project MODULAr), arXiv:0704.1422.
[8] A. Badertscher *et al.*, Construction and operation of a Double Phase LAr Large Electron Multiplier Time Projection Chamber, arXiv:0811.3384, (2008).
[9] W.F. Schmidt, Electronic Conduction Processes in Dielectric Liquids, IEEE Trans. Electrical Insulation EI-19 (1984) 389.
[10] A. Ereditato *et al.*, Study of ionization signals in a TPC filled with a mixture of liquid Argon and Nitrogen, JINST **3** (2008) P10002.
[11] Messer Schweiz AG, http://www.messer.ch
[12] A.M.Kalinin *et al.*, ATLAS Internal Note, ATLAS-LARG-NO-058, (1996) CERN.
[13] W. Walkowiak, Drift velocity of free electrons in liquid argon, Nucl. Instr. Meth. in Phys. Res. **A449** (2000) 288.
[14] S. Amoruso *et al.*, Analysis of the liquid argon purity in the ICARUS T600 TPC, Nucl. Instr. Meth. in Phys. Res. **A516** (2004) 68.
[15] S.L Chin and P.Lambropoulos, Multiphoton ionization of Atoms, Academic Press Canada (1984).
[16] NIST - National Institute of Standards and Technology, NIST Chemstry WebBook, NIST Standard Reference Databse N. 69.
[17] B. Raz and J. Jortner, Energy of the quasi-free electron state in liquid and solid rare gases, Chem. Phys. Lett. **4** (1969) 155.
[18] M. Born, Volumen und Hydratationswaerme der Ionen, Z. Phys. **1** (1920) 45.
[19] M. Antonello *et al.*, Detection of Cherenkov light emission in liquid argon, Nucl. Instr. Meth. in Phys. Res., A 516 (2004) 348.
[20] R. Reininger, U. Asaf, I. T. Steinberger, and S. Basak, Relationship between the energy V0 of the quasi-free-electron and its mobility in fluid argon, krypton, and xenon, Phys. Rev. **B** 28 (1983) 4426.
[21] H. J. Hilke, Detector calibration with lasers - A review, Nucl. Instr. Meth. in Phys. Res. A 252 (1986) 169.
[22] I. Badhrees *et al.*, Measurement of the two-photon absorption cross-section of liquid argon with a time projection chamber, New J. Phys. **12** (2010) 113024, [arXiv:1011.6001 [physics.ins-det]].
[23] A. C. Tam and C. K. N. Patel, Two-photons absorption spectra and cross-section measurements in liquids, Nature Vol. 280 (1979).
[24] C. G. Elles *et al.*, Electronic structure of liquid water from polarization-dependent two-photon absorption spectroscopy, The journal of chemical physics 130 (2009) 084501.
[25] J. Bourotte and B. Sadoulet, Ionization of multiwire proportional chamber gas by double photon absorptio, Nucl. Instr. and Meth. 173 (1980) 463.
[26] G.S. Hurst *et al.*, Resonance ionization spectroscopy and one-atom detection, Rev. Mod. Phy. 51 (4) (1979) 767.

1st International Workshop towards the Giant Liquid Argon Charge Imaging Experiment IOP Publishing
Journal of Physics: Conference Series **308** (2011) 012026 doi:10.1088/1742-6596/308/1/012026

Test of a Liquid Argon TPC in a magnetic field and investigation of high temperature superconductors in liquid argon and nitrogen

A. Badertscher, L. Knecht, M. Laffranchi, G. Natterer, A. Rubbia, Th. Strauss

ETH Zurich, Institute for Particle Physics, CH-8093 Zurich, Switzerland

E-mail: `badertscher@phys.ethz.ch`

Abstract. Tests with cosmic ray muons of a small liquid argon time projection chamber (LAr TPC) in a magnetic field of 0.55 T are described. No effect of the magnetic field on the imaging properties were observed. In view of a future large, magnetized LAr TPC, we investigated the possibility to operate a high temperature superconducting (HTS) solenoid directly in the LAr of the detector. The critical current I_c of HTS cables in an external magnetic field was measured at liquid nitrogen and liquid argon temperatures and a small prototype HTS solenoid was built and tested.

1. Test of a LAr TPC in a magnetic field

The liquid argon time projection chamber (LAr TPC) [1] is a homogeneous 3D tracking device for charged particles with excellent - bubble chamber like - imaging properties and at the same time it is a fine grain calorimeter for fully contained particles due to the measurement of the energy loss dE/dx. The ICARUS collaboration has demonstrated the feasibility of this novel technology for large mass detectors. A 600 ton (T600) detector consisting of two identical 300 ton half-modules was built and successfully tested [2]. The detector is now installed at the Gran Sasso underground laboratory in Italy. The possibility to operate the LAr TPC in a magnetic field would add the very interesting features of determining the electric charge of particles and the momentum, also for particles leaving the chamber [3, 4, 5, 6]. The measurement of the charge, e.g., is a must for future experiments trying to measure CP violation in the leptonic sector at a neutrino factory.

An R&D program to investigate a small LAr TPC in a magnetic field was performed. The goal was to study the drift properties of free electrons in LAr in the presence of a magnetic field and to prove that the imaging capabilities are not affected. A detailed description of the experiment is given in [7] and results were published in [8, 9]. Fig. 1 is a 3D CAD drawing showing a cut through the setup with the essential components of the experiment. The scintillators on top of the magnet, in the bore hole on top of the cryostat, and at the bottom of the magnet were used to trigger on cosmic ray muons. The LAr cryostat was inserted into the recycled SINDRUM I magnet from PSI[1] which allowed to test the chamber in a maximal field of 0.55 T.

[1] Paul Scherrer Institute, CH-5232 Villigen, Switzerland

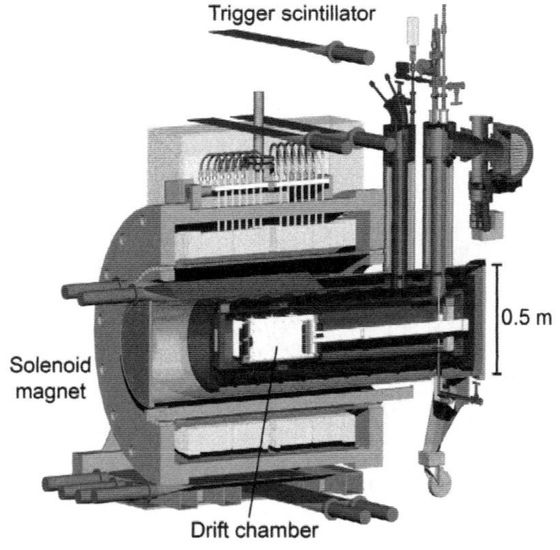

Figure 1. Global view of the experiment.

The cryostat consists of three concentrical stainless steel cylinders: the innermost cylinder contains the purified LAr with the drift chamber, the second cylinder is a LN_2 bath kept at a pressure of 2.7 bar in order not to freeze out the LAr at about 1 bar, and the outermost cylinder is for the insulation vacuum.

The chamber has a length (along the B-field direction) of 300 mm, a height of 150 mm and a maximal drift length of 150 mm (horizontal drift field perpendicular to the B-field). The left side of Fig. 2 shows a CAD drawing into the open chamber, and the right side is a picture of the chamber ready to slide into the cryostat. After pumping the LAr cryostat, it was filled through a purification cartridge containing activated Cu powder to remove impurities, mainly O_2, the LAr was not recirculated anymore through the cartridge after the filling.

The chamber consists of a stainless steel cathode, 27 field shaping electrodes to produce a

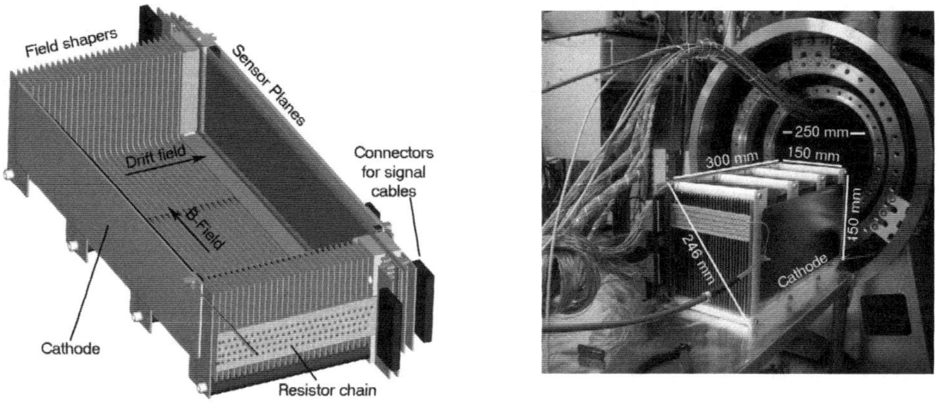

Figure 2. Left: CAD drawing of the open TPC. Right: Picture of the TPC ready to slide into the cryostat.

Journal of Physics: Conference Series **308** (2011) 012026 doi:10.1088/1742-6596/308/1/012026

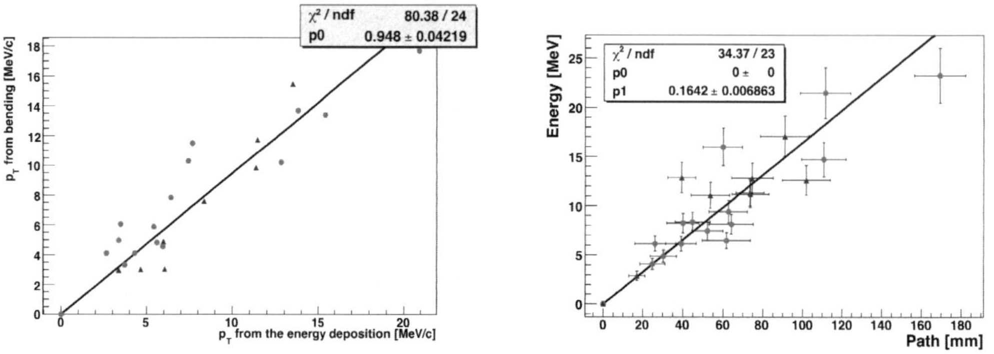

Figure 3. left: Correlation between the two momentum measurement methods for δ-electrons (circles) and decay-positrons (triangles); right: Measured energy as a function of the reconstructed path length of the δ-electrons (circles) and the decay-positrons (triangles).

homogeneous drift field and 3 sensor planes. The first two detector planes (induction planes) are wire chambers with 100 μm wires oriented at $\pm 60^0$ to the vertical and a pitch of 2 mm, and the third plane (collection plane) is a PCB with horizontal strips with a width of 1 mm and a pitch of 2 mm; the chamber has a total of 329 channels. In the beginning of the run a maximal drift field of 1.5 kV/cm was applied and had to be reduced later to 0.3 kV/cm because of HV breakdowns. The Lorentz angle for a drift field of 0.5 kV/cm and a B-field of 0.5 T was estimated to be $\approx 1.7°$.
For initial tests without magnetic field, a coincidence of the scintillators was used to trigger on through-going muons (trigger rate 0.55 Hz). To trigger on stopping muons, the scintillators on top of the cryostat in the bore hole were used, yielding the time t_0 of the event needed to determine the drift time, together with the analog sum of 32 channels of the TPC; the rate was about 1/min. About 30'400 events were collected and visually scanned, and a small sample of 15 δ-electrons and 8 decay positrons from stopped muons were selected for a first analysis [10]. This represents a highly biased sample of well measurable events fully contained in the chamber. The tracks were reconstructed in 3D and their momentum and kinetic energy were calculated from the magnetic bending and the summed energy loss along the track. Fig. 3 shows on the left side the comparison of the p_t obtained from the magnetic bending and from the energy measurement; on the right side is the measured kinetic energy plotted versus the total track length.

2. Investigation of high temperature superconductors in liquid argon and nitrogen

A very interesting option to magnetize a LAr TPC would be to build a high temperature superconducting (HTS) solenoid directly in the LAr cryostat of the detector. Since the manufacturers deliver data on the temperature dependence of the properties of their commercial HTS cables only up to LN_2 temperature (77 K), we performed a small R&D program to investigate the performance of different HTS cables also at LAr temperature (87 K) [11]. We tested BSCCO (first generation HTS cable) and YBCO (second generation) from American Superconductors AMSC[2] and SuperPower Inc.[3]. Table 1 summarizes some properties of the tested cables.

We measured the critical current I_c in LN_2 and LAr as a function of an external magnetic field applied parallel (in the direction of the width of the cable) and perpendicular to the HTS cable. Here, the critical current I_c is defined as the current, for which the voltage drop in the

[2] www.amsc.com
[3] www.superpower-inc.com

AMSC	Width	Thickness	Critical temp.
BSCCO	4 mm	0.4 mm	110 K
YBCO	4.3 mm	0.25 mm	90 K
SuperPower			
YBCO	4 mm	0.1 mm	90 K

Table 1. Properties of the tested HTS cables.

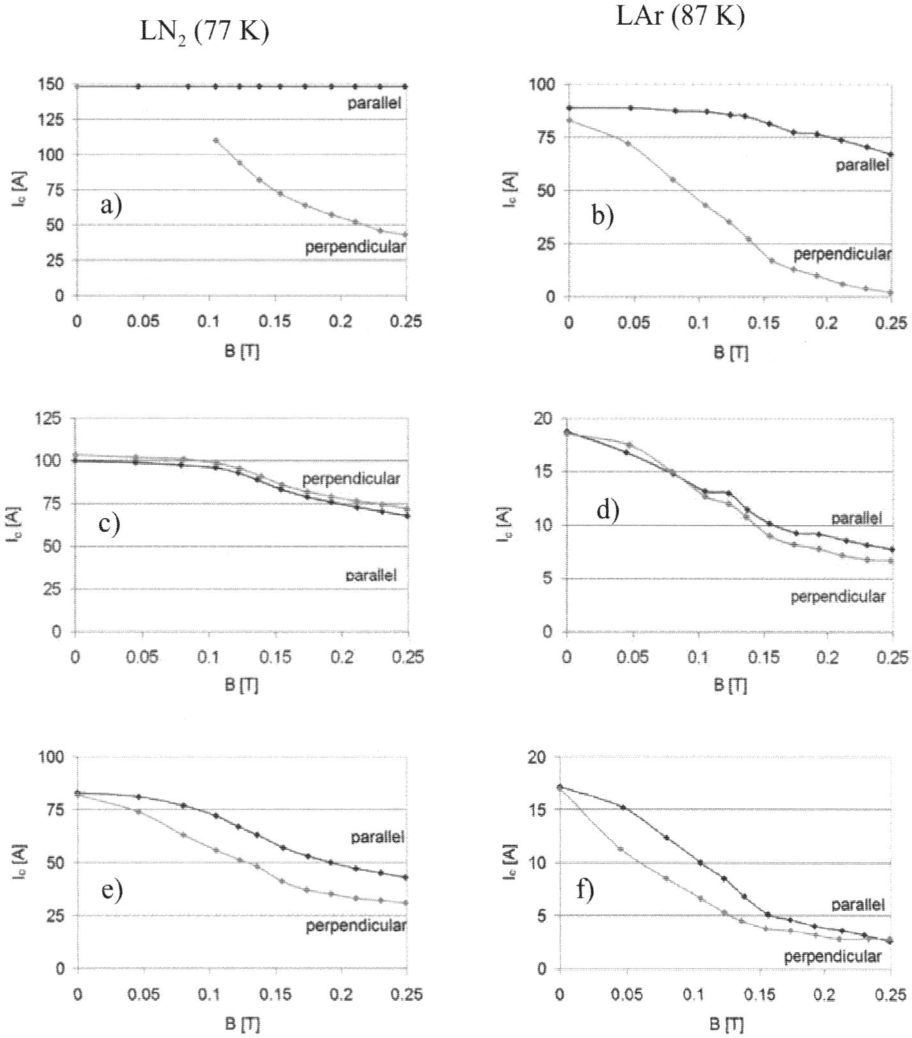

Figure 4. Measured critical current I_c for the three tested HTS cables as a function of the applied parallel and perpendicular magnetic fields. Left column: measurements in LN_2 at 77 K, right column: measurements in LAr at 87 K. a) and b) BSCCO cable from AMSC, c) and d) YBCO cable from AMSC, e) and f) YBCO cable from SuperPower Inc.

cable reaches $1\,\mu\mathrm{V/cm}$. Fig. 4 shows, that in LAr the critical current is already substantially reduced compared to the value measured in LN_2, in particular for the second generation YBCO cables.

After testing the cables, we built a small prototype HTS solenoid with about 100 m of the AMSC BSCCO cable [12]. The solenoid shown in Fig. 5 consists of four so-called pancakes with 25 (on average) windings. It has a total length of 130 mm, including the iron end plates and copper spacers and iron shielding rings between the pancakes. The average diameter of the windings was 230 mm and the bore hole was 210 mm. The iron shielding rings was inserted between the pancakes to suppress the perpendicular B-field component at the innermost coil windings. In LN_2 a maximal B-field of 0.2 T was achieved at I = 145 A and in LAr it was 0.11 T at I = 80 A.

Figure 5. Left: Connections of the pancakes of the HTS solenoid. Right: The complete solenoid.

References

[1] C. Rubbia, *The Liquid-Argon Time projection Chamber: a new concept for Neutrino Detector*, CERN-EP/77-08, (1977).
[2] S. Amerio et al., Nucl. Instr. and Meth. A 527 (2004) 329.
[3] M. Campanelli, A. Bueno and A. Rubbia, *Optimization Studies for CP- and T-violation*, Nucl. Instr. and Meth. A 503 (2003) 133.
[4] A. Bueno, M. Campanelli, S. Navas-Concha and A. Rubbia, *On the energy and baseline optimization to study effects related to the delta-phase (CP-/T-violation) in neutrino oscillations at a neutrino factory*, Nucl. Phys. B 631 (2002) 239.
[5] A. Rubbia, *Neutrino factories: Detector concepts for studies of CP and T violation effects in neutrino oscillations*, arXiv:hep-ph/0106088. Appeared in *Proceedings of the 9th International Symposium on Neutrino Telescopes, Venice, Italy, March 2001.*
[6] A. Bueno, M. Campanelli and A. Rubbia, *Physics potential at a neutrino factory: Can we benefit from more than just detecting muons?*, Nucl. Phys. B 589 (2000) 577, arXiv:hep-ph/0005007.
[7] M. Laffranchi, PhD thesis No. 16002, ETH Zurich, 2005.
[8] A. Badertscher et al., New J. of Phys. 7 (2005) 63.
[9] A. Badertscher et al., Nucl. Instr. and Meth. A 555 (2005) 294.
[10] A. Müller, Diploma thesis, ETH Zurich, 2005.
[11] Th. Strauss, PhD thesis, ETH Zurich, 2010.
[12] Th. Strauss, Diploma thesis, ETH Zurich and Humboldt-Universität Berlin, March 2006.

Feasibility of high-voltage systems for a very long drift in liquid argon TPCs

S. Horikawa, A. Badertscher, L. Kaufmann, M. Laffranchi, A. Marchionni, M. Messina[1], G. Natterer and A. Rubbia

ETH Zurich, Institute for Particle Physics, CH-8093 Zürich, Switzerland

E-mail: `Sosuke.Horikawa@cern.ch`

Abstract. Designs of high-voltage (HV) systems for creating a drift electric field in liquid argon TPCs are reviewed. In ongoing experiments systems capable of ∼100 kV are realised for a drift field of 0.5–1 kV/cm over a length of up to 1.5 m. Two of them having different approaches are presented: the ICARUS-T600 detector having a system consisting of an external power supply, HV feedthroughs and resistive voltage degraders and the ArDM-1t detector having a cryogenic Greinacher HV multiplier inside the liquid argon volume. For a giant scale liquid argon TPC, a system providing 2 MV may be required to attain a drift length of ∼20 m. Feasibility of such a system is evaluated by extrapolating the existing designs.

1. Introduction

The intensity of the drift electric field is one of the important design parameters for liquid argon time projection chambers (LAr-TPCs). A better collection of ionisation charges is attained by increasing the field intensity because: (1) more electron-ion recombination is prevented [1] and (2) attenuation of the drift electrons due to their attachment to residual electronegative impurities such as oxygen decreases. The dependence of the attachment cross section on the electric field is known to be weak in the practical range of the intensity (0.05–1 kV/cm) [2]. The attenuation then is described well by an exponential decrease with drift time, characterised by the drift electron lifetime τ which is determined dominantly by the impurity concentration. The mean drift velocity of the electrons increases with increasing electric field, leading to the shorter collection time and consequently to the less attenuation. The drift velocity increases by 30% by doubling the electric field intensity from 0.5 to 1 kV/cm, and again 30% from 1 to 2 kV/cm [3]. From a technical point of view, however, difficulties increase with increasing field intensity which requires higher voltages. Therefore, it should be determined by a right compromise between the detector performance and the practicality of the high voltage. A field intensity of 1 kV/cm is a reasonable compromise for very long drifts.

A drift field of 1 kV/cm requires a potential difference of 100 kV over 1 m. A drift length of 20 m which can be the case in giant scale detectors [4] thus requires 2 MV. There are two different approaches to realise such a high voltage (HV) system for LAr-TPCs. The first type uses an external HV power supply and feed it into the detector volume using feedthroughs as in the ICARUS-T600 detector [5]. This type has the advantage that the power supply is able to be

[1] Now at Albert Einstein Center for Fundamental Physics, Laboratory for High Energy Physics, University of Bern, CH-3012 Bern, Switzerland.

repaired in case of problems without emptying the LAr volume. In addition, the output voltage can be monitored precisely and continuously during the data taking from the current flowing through a resistive voltage degrader. The second type has an internal HV generator directly inside the LAr volume as the Greinacher HV multiplier of the ArDM-1t detector [6], which we have developed at CERN. Its advantages are: (1) all the HV parts are immersed in LAr which has a large dielectric strength (~1 MV/cm), (2) thus feedthroughs for very HV are not needed, (3) the circuit itself can be used as a voltage divider, so the system needs no resistive load, (4) thus the power dissipation is virtually zero and (5) this allows a low frequency (e.g. 50 Hz) of the AC input signal which is fully outside of the bandwidth of the charge amplifiers used for this type of detectors.

2. ICARUS-T600 system with an external HV power supply and feedthroughs

The ICARUS-T600 detector [5] is based essentially on four ionisation chambers occupying two parallelepiped volumes each having a size of roughly 20 m long, 3.6 m wide and 3.9 m high containing approximately 240 tons of LAr. In each volume a central vertical cathode plane is facing a wire chamber on each side consisting of three planes. The nominal cathode potential is −75 kV creating a drift field of 0.5 kV/cm over the maximum drift length of 1.5 m, while the system has been tested up to 1 kV/cm with −150 kV without problems. Each of the four drift volumes is surrounded by a set of 29 regularly spaced field shaping electrodes (Figure 1). Linearly decreasing potentials are distributed to the electrodes with the aid of a resistive voltage degrader consisting of 30 times 25-MΩ series resistors. The nominal current is 0.1 mA dissipating 7.5 W per volume. For redundancy each resistor is made of four 100-MΩ resistors inserted in parallel between adjacent electrodes. Mechanical details of the resistors connecting adjacent field shaping electrodes are shown in Figure 2. Calculations show that the field distortion is limited in the region close to the electrodes in case that two or even three resistors break out of four in the same group. The probability of such a case is anyhow negligibly small.

Figure 1. Field shaping electrodes of the ICARUS-T600 detector.

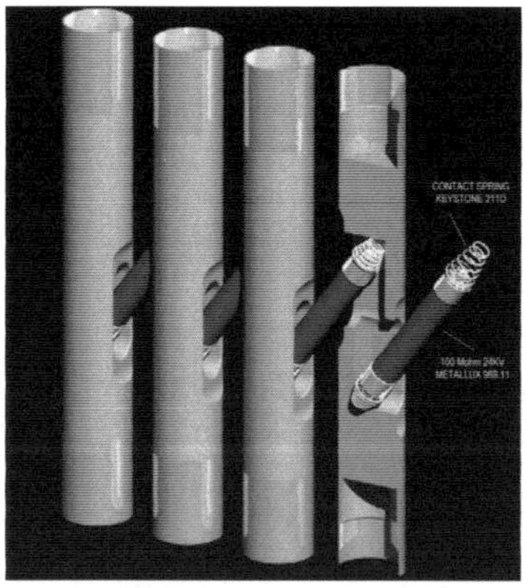

Figure 2. 100-MΩ resistors inserted between adjacent field shaping electrodes.

Figure 3 shows the ICARUS HV feedthrough having a coaxial geometry using UHMW PE (ultra high molecular weight polyethylene) as an insulator. The inner conductor is made of a thin wall stainless-steel tube to minimise heat input and bubbles around the spring contact at the bottom end, which is immersed in LAr. The connection to the cathode plane is made via a cup-shaped receptacle which can tolerate a small displacement due to different thermal contractions between the LAr vessel and the detector structure (Figure 4). The outer conductor,

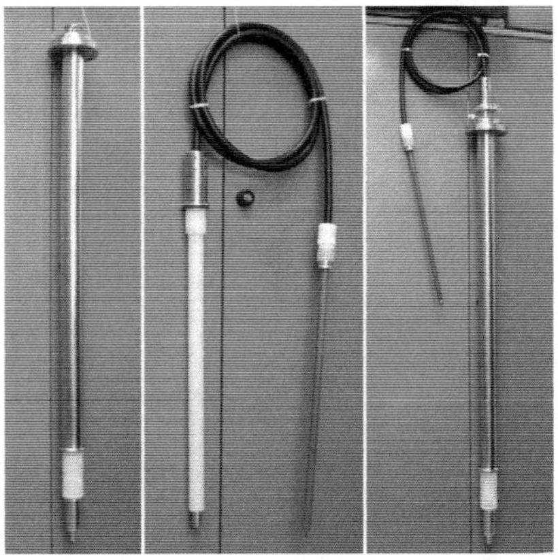

Figure 3. Left; HV feedthrough. Center; HV cable. Right; the feedthrough with the cable inserted.

Figure 4. Spring contact at the bottom end of the feedthrough and the cup-shaped receptacle mounted on the cathode.

made of a stainless-steel tube, surrounds the PE insulator extending inside the cryostat down to the LAr level. During the first tests of the feedthrough occasional micro discharges were observed, due to the air presence in the interstices in the cable-to-plug and in the plug-to-socket connections. The micro discharges immediately disappear by filling these interstices with dry nitrogen gas.

A 150-kV/1-mA power supply from Heinzinger electronic GmbH[2] is used for a the operation of T600. It gives a peak-to-peak voltage ripple less than 10^{-5} at maximum load, which is reduced by a factor five under the operating conditions of T300 (half of T600). The ripple mainly consists of two components, i.e. 50 Hz and 37.45 kHz. Using a proper filter the high frequency components can be eliminated and the low frequency component can be reduced by more than a factor of three.

3. ArDM-1t system with an internal Greinacher HV multiplier

ArDM-1t is a 1-ton LAr-TPC built for a direct detection of WIMPs [6]. Its HV system is based on an internal Greinacher [7] (also known as Cockcroft-Walton [8]) HV multiplier (Figure 5).

Figure 6 illustrates the ArDM-1t detector. The cylindrical drift volume is 80 cm in diameter and 124 cm high. A vertical electric field is generated with the aid of 30 ring-shaped electrodes called "field shapers" spaced every 40 mm. The cathode grid which is at the highest potential in the system is located on the bottom. Ionisation electrons drift upwards and are collected by the charge readout system (near ground potential) on the top. The DC high voltages generated by the Greinacher circuit are linearly distributed from the cathode to the first (top) field shaper. The system is capable to generate about −400 kV resulting in an electric field of above 3 kV/cm. For the first phase of the experiment it will be operated at −124 kV to create a drift electric field of 1 kV/cm.

The ArDM Greinacher circuit has 210 stages (Figure 5). It outputs DC voltages on the bottom row of the diagram as denoted with V_i. The potential at the ith stage is ideally equal to the peak-to-peak value of the input AC voltage multiplied by the stage number, i.e. $V_i = i \cdot V_{\rm pp}^{\rm in}$. Each capacitor symbol corresponds to four (two) 82-nF polypropylene capacitors[3] in parallel for

[2] Heinzinger electronic GmbH, D-83026 Rosenheim, Germany. `http://www.heinzinger.com/`
[3] Evox Rifa PHE450; KEMET Electronics S.A., CH-1211 Geneva 20, Switzerland. `http://www.kemet.com/`

Journal of Physics: Conference Series **308** (2011) 012027 doi:10.1088/1742-6596/308/1/012027

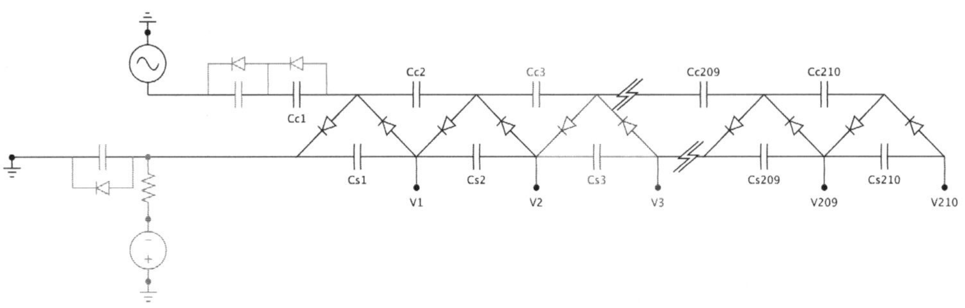

Figure 5. Diagram of the ArDM Greinacher circuit. Additional capacitors, diodes and the DC voltage source indicated in red are for adjusting the potential of the lowest stage.

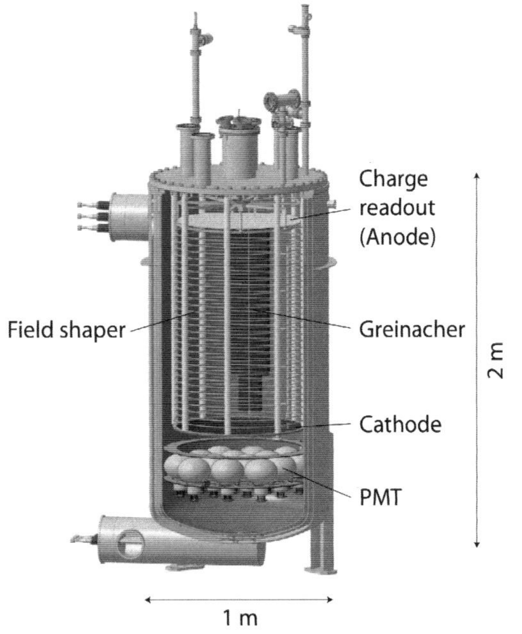

Figure 6. ArDM-1t detector.

Figure 7. ArDM Greinacher circuit.

stages 1 to 170 (171 to 210) and each diode symbol to three avalanche diodes[4] in series. The components have been chosen for their reliability in LAr. Thanks to the redundancy failures of a single component do not cause a critical loss of the functionality of the whole circuit. A picture of the ArDM Greinacher circuit mounted on the detector is shown in Figure 7.

Before the ArDM system was realised, substantial R&D efforts were undertaken with several prototype Greinacher circuits having a smaller number of stages. At the last stage of the R&D a 10-stage prototype was built with exactly the same design and components (i.e. the polypropylene capacitors and the avalanche diodes) as in ArDM. An extensive study was done with it in a range of different circumstances, i.e. in air and in LAr [9], where we obtained a good understanding of the fundamental characteristics of the circuit and of the components. In LAr the maximum voltage of 24 kV was reached.

[4] BY505, High-voltage soft-recovery rectifier; Philips Semiconductors. No longer in production.

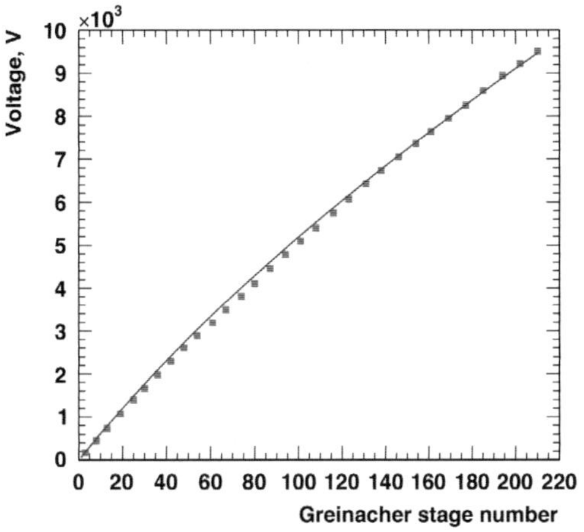

Figure 8. Potentials measured at each field shaper plotted as a function of Greinacher stage number.

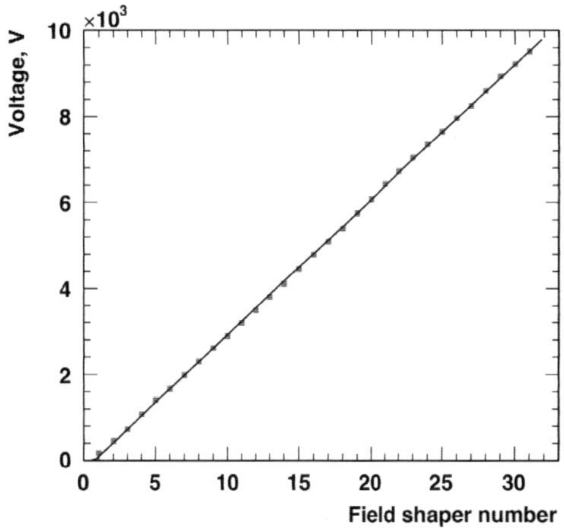

Figure 9. Potentials measured at each field shaper as a function of the field shaper number.

The full 210-stage system was first tested in air up to −10 kV cathode voltage. Figure 8 shows the potentials measured at each field shaper plotted as a function of the corresponding Greinacher stage number. Under real conditions the output voltage does not depend linearly on the stage number. This non-linearity is explained by the "shunt" capacitance between the top and bottom rows of the circuit as in Figure 5 (see also Section 4), which is essentially due to the diode capacitance and the stray capacitance determined by the geometry of the circuit. Fitting the measured curve with a model [10] as shown in blue in Figure 8 the value of the shunt capacitance was evaluated to be 2.35 pF. Note that the influence of the resistive load of the circuit (mainly the reverse resistance of the diode chain) to this effect is negligibly small in this system compared to that of the shunt capacitance. The non-linearity however can be compensated simply by choosing an appropriate Greinacher stage connecting to each field shaper when the number of Greinacher stages N_{Gr} is significantly larger than that of field shapers N_{fs}. This is the case in ArDM ($N_{\mathrm{Gr}} = 7N_{\mathrm{fs}}$) and the potentials are distributed linearly to the field shapers as shown in Figure 9, where the measured potentials are plotted now as a function of the field shaper number (No. 31 corresponding to the cathode).

The ArDM Greinacher circuit is designed so that the resistive load is minimised in order to allow operation at low frequency (50 Hz). Hence, once it is charged up, it keeps the high voltage even if the input AC voltage source is switched off. Its natural discharging is dominated by the leakage current through the diode chain and was measured to be consistent with an exponential decay with a lifetime of 19 hours in air at room temperature. It can be even slower in LAr because the leakage current is much smaller at 87 K. The total energy stored in the system is 22 J at 124 kV output voltage and can be ∼1 kJ at ∼400 kV. For safety and also for protecting the apparatus including the circuit itself in case of any unexpected events, it is therefore very important to have a system capable of actively and rapidly discharging the Greinacher circuit. Figure 10 illustrates the schematic diagram of the discharging system in ArDM. The whole circuit can be discharged by touching the cathode with an oval-shaped contact grounded through the resistor chain having a total resistance of 1.8 GΩ, which consists of nine 200-MΩ HV resistors. The discharge current and consequently the cathode potential before discharging can be measured through the voltage across the 220-kΩ readout resistor connected in series to the chain. Figure 11 illustrates the mechanical realisation of the system. The discharging contact and the chain can rotate around the polyethylene axis rod steered by a rotary motion vacuum feedthrough.

This technique was tested using a 10-stage prototype Greinacher circuit in liquid nitrogen. A 100-MΩ resistor of the same type as in ArDM was used as the discharging resistor with a 200-kΩ readout resistor. Figure 12 shows the voltage across the readout resistor at discharging from -10 kV. The measured curve is very consistent with an exponential decay. The peak voltage and the time constant $\tau = RC$ agreed with naive calculations from the resistance and the capacitance of the system, verifying the small temperature coefficients of the components. A simulation using Ngspice[5] shown in red in the same plot agreed with the measurement as well.

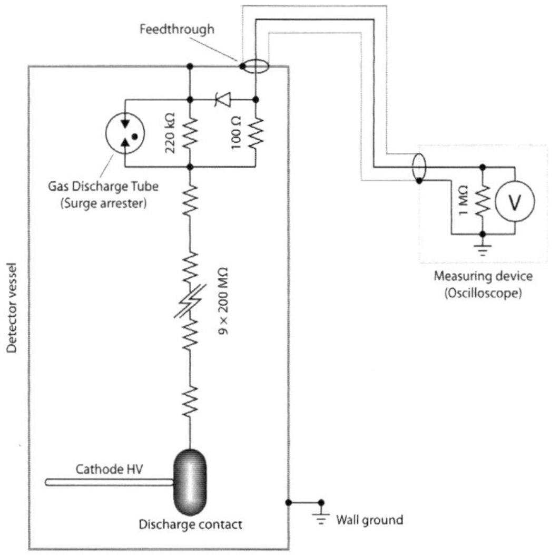

Figure 10. Schematic diagram of the ArDM Greinacher discharging system.

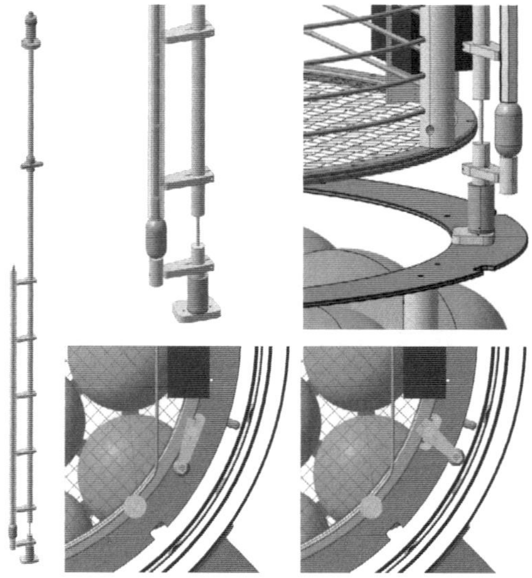

Figure 11. The ArDM Greinacher discharging system mechanics.

4. Extrapolation to a megavolt system

Keeping 1 kV/cm a very long drift of electrons over 20 m requires 2 MV. In order to realise such a HV system with an external power supply, a feedthrough which withstands megavolts is needed. Such a feedthrough can in principle be built by scaling the existing ICARUS design. One may also exploit expertise from industries, where devices for above 1 MV are in development e.g. for HVDC (high voltage direct current) power transmission systems [11]. Nevertheless, realising a HV generator for 2 MV is very challenging. Use of LAr can be very profitable in this context since it has a dielectric strength of larger than 1 MV/cm and is known as one of the best insulator materials. A Greinacher HV multiplier in pure LAr can therefore be an interesting option for such a HV generator in particular for a giant scale LAr-TPC. HV generators are often the most crucial source of high frequency noise which readout electronics suffer from. As mentioned above the ArDM-type system having an internal Greinacher circuit does not require any resistive load and thus can be operated at very low frequencies. This feature can be a remarkable advantage particularly for megavolt systems since a low pass filter for such high voltages can potentially be a critical issue.

[5] Ngspice circuit simulator. `http://ngspice.sourceforge.net/`

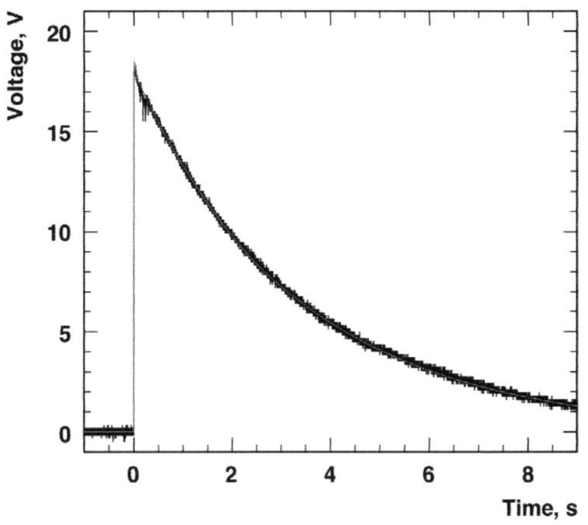

Figure 12. Discharging curve of the 10-stage Greinacher circuit measured in liquid nitrogen over the readout resistor (see text).

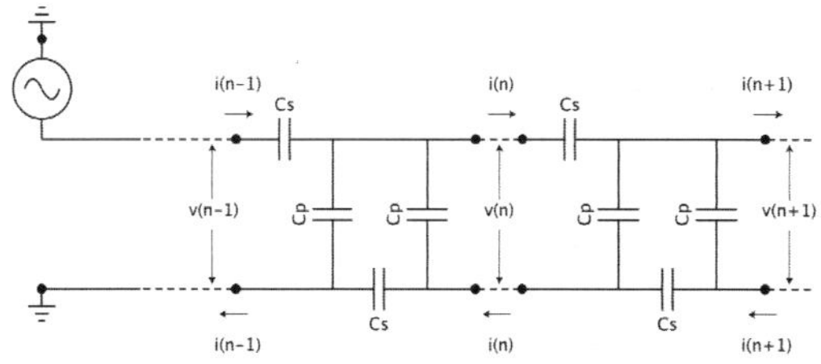

Figure 13. Transmission line equivalent to a Greinacher circuit.

The feasibility of extrapolating the ArDM Greinacher system to a megavolt generator has been studied based on the test results as described above. Theoretical studies based on a transmission-line model [10] suggest that the maximum output voltage that can be attained by a Greinacher circuit occurs for an infinite number of stages and is given by

$$V_{\rm max} = \frac{E}{\gamma}, \ \gamma \approx \sqrt{\frac{C_{\rm p}}{C_{\rm s}}}, \tag{1}$$

where E $(= V_{\rm pp}^{\rm in}/2)$ is the input voltage, $C_{\rm s}$ the capacitor capacitance and $C_{\rm p}$ $(\ll C_{\rm s})$ the shunt capacitance (Figure 13). To attain larger $V_{\rm max}$ one has to go for larger E, or/and smaller γ by increasing $C_{\rm s}$ or decreasing $C_{\rm p}$. The total output voltage V of an N-stage circuit can then be obtained by $V = V_{\rm max}\tanh(2\gamma N)$. To attain a desired output voltage V for given E and $C_{\rm p}$ it is favourable to choose N and $C_{\rm s}$ so that the total capacitor capacitance $N \cdot C_{\rm s}$ can be minimised. This corresponds to $2\gamma N \approx 1.42$ or $V \approx 0.89 \cdot V_{\rm max}$.

The results of the extrapolations are summarised in Table 1 with the actual ArDM parameters as reference. For the extrapolations to 10-m drift use of the same polypropylene capacitors as in ArDM is assumed and the input voltage thus is limited to their maximum operation voltage $2E = 2.5$ kV. For the shunt capacitance the ArDM value $C_{\rm p} = 2.35$ pF is assumed. The capacitor capacitance $C_{\rm s}$ is then increased until $V_{\rm max} = V/0.89$ is reached for the desired output voltage V. A 1-MV system for 10-m drift looks very feasible but the total capacitance becomes about 20 times larger than ArDM. The extrapolation to a 2-MV system for 20-m drift was done assuming E can be increased by a factor of $\sqrt{2}$ and the shunt capacitance $C_{\rm p}$ can be reduced by a factor of 2, leading to the twice larger $V_{\rm max}$ with the same $C_{\rm s}$. The former can be realised with a different capacitor type or by developing similar polypropylene capacitors having a higher operating voltage. The latter can be achieved by optimising the geometry of the system, which is possible since the space density of the circuit components becomes significantly lower for a giant-scale detector. These calculations demonstrate that it is feasible with engineering optimisation to build an ArDM-type Greinacher circuit which can generate 2 MV.

Table 1. Extrapolation of the ArDM Greinacher circuit to a long drift up to 20 m.

	Unit	ArDM		Extrapolation	
		first phase	nominal		
Drift length	m	1.24	1.24	10	20
Electric field	V/cm	1k	3.05k	1k	1k
Total output voltage	V	124k	378k	1M	2M
Input voltage $V_{\rm pp-in} = 2E$	V	820	2.5k	2.5k	3.5k
Shunt capacitance, $C_{\rm p}$	F	2.35p	2.35p	2.35p	1.18p
Capacitor, $C_{\rm s}$	F	328n/164n	328n/164n	1.90μ	1.90μ
Number of stages, N	–	210	210	638	903
N per 10 cm	–	16.9	16.9	6.38	4.51
Total capacitance	F	125μ	125μ	2.43m	3.43m
Capacitance per 10 cm	F	10.4μ	10.4μ	24.3μ	17.2μ
Total charge	C	73.6m	312m	6.06	12.1
Total energy	J	21.7	948	7.58k	21.5k

5. Conclusions and outlook

ICARUS-T600 has a well established HV system with an external power supply, feedthroughs and resistive voltage degraders for a nominal operation voltage of 75 kV with a possibility for 150 kV. The ArDM internal Greinacher HV multiplier was developed for ~400 kV operating voltage. Its fundamental characteristics has been well understood in the first tests in air and the system will be tested at 124 kV in LAr in 2010. A very long drift in giant scale LAr-TPCs requires megavolt systems. A feedthrough withstanding megavolts can in principle be realised by scaling up the existing ICARUS design. Calculations showed that a Greinacher HV multiplier of the ArDM type can be extrapolated to 1–2 MV with some engineering optimisations.

Acknowledgement

This work was supported by ETH Zürich and the Swiss National Science Foundation (SNF). We are grateful to F. Sergiampietri of INFN for the thorough and detailed information on the ICARUS-T600 HV system and also for the stimulating discussions about various possibilities of the future HV systems for a long drift in LAr-TPCs.

References

[1] S. Amoruso *et al.* [ICARUS Collaboration], Nucl. Instrum. Meth. A **523** (2004) 275.
[2] S. Amoruso *et al.*, Nucl. Instrum. Meth. A **516** (2004) 68.
[3] W. Walkowiak, Nucl. Instrum. Meth. A **449** (2000) 288.
[4] A. Rubbia, J. Phys. Conf. Ser. **171** (2009) 012020 [arXiv:0908.1286 [hep-ph]].
[5] S. Amerio *et al.* [ICARUS Collaboration], Nucl. Instrum. Meth. A **527** (2004) 329.
[6] A. Rubbia, J. Phys. Conf. Ser. **39** (2006) 129 [arXiv:hep-ph/0510320].
[7] H. Greinacher, Z. Physik **4** (1921) 195.
[8] J. D. Cockcroft and E. T. S. Walton, Proc. Roy. Soc. (London) **131** (1932) 619.
[9] L. Kaufmann, "Detector Performance and Background Studies for the ArDM Experiment," Diss. ETH No. 17806 (2008).
[10] M. M. Weiner, Rev. Sci. Instrum. **40** (1969) 330; E. Everhart and P. Lorrain, Rev. Sci. Instrum. **24** (1953) 221.
[11] For example: `http://www.abb.com/hvdc`

Journal of Physics: Conference Series **308** (2011) 012028 doi:10.1088/1742-6596/308/1/012028

Okinoshima site study

Masakazu Yoshioka[1], Takuya Hasegawa[1], Osamu Hirabayashi[2], Tokiyoshi Kaneta[2], Kiyokazu Kawakami[2], Nobuhiro Kimura[1], Takasumi Maruyama[1], Koichiro Nishikawa[1], Tetsunori Oshimo[2], André Rubbia[3], Satoshi Taguchi[2] and Masashi Tanaka[1]

[1]High Energy Accelerator Research Organization (KEK), Tsukuba, Ibaraki 305-0801, Japan
[2]PENTA-OCEAN construction co., Ltd., 2-2-8 Koraku, Bunkyo-ku, Tokyo, 112-8576, Japan
[3]ETH Zurich, Institute for Particle Physics, CH-8093 Zurich, Switzerland

E-mail: masakazu.yoshioka@kek.jp

Abstract. The Okinoshima Islands are located on an off-axis of 0.76 degree (almost on-axis) with the same neutrino beam setting as T2K (2.5 degree off- axis to Kamioka). The distance of Okinoshima from the neutrino source, KEK/J-PARC, is 658 km. This configuration enables to conduct precision measurement of the 1st and 2nd neutrino oscillation maxima of the appearance electron neutrino energy spectrum. We have made studies of the Okinoshima Island site from the geological, geographic and infrastructure points of view to investigate the possibility of building a large underground cavern there to install a 100 kton class Giant Liquid Argon Time Projection Chamber (TPC) detector for a neutrino oscillation and proton decay experiment. We have found that the Okinoshima Island site is a good candidate site for such an observatory.

1. General outline of Okinoshima

The Okinoshima Islands are located on an off-axis of 0.76 degree (almost on-axis) with the same neutrino beam setting as T2K (2.5 degree off- axis to Kamioka) and the distance of Okinoshima from the neutrino source, KEK/J-PARC, is 658 km as is shown in Figure.1. This configuration enables to conduct precision measurement of the 1st and 2nd neutrino oscillation maxima of the appearance electron neutrino energy spectrum [1]. Okinoshima mainly consists of Dogo Island and the Dozen group of islands. Of all the Okinoshima islands, the biggest is Dogo. The islands in the group were born as volcanic islands about 5~6 million years ago, and Dogo still has some of the original stable homogeneous bedrock, Oki-gneiss, in which a large underground cavern could be excavated.

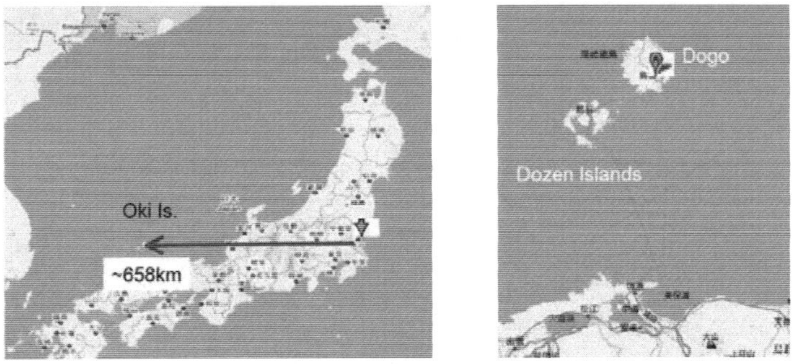

Figure1. Location of the Okinoshima Islands (http://maps.google.co.jp/maps).

Dogo is almost circular with a diameter of about 16 km and the center is mountainous with a 500 m altitude, this is where we would look for a candidate site for the underground neutrino observatory. The circumference of Dogo is 151 km and the area is 243 km^2. The distance between the main island of Honshu and Okinoshima is about 80 km and all of the Okinoshima islands belong to the Shimane prefecture. The seashore has been designated the Daisen-Oki National Park. Therefore, any developments there would be covered by various laws and regulations. The central mountainous region is, however, excluded from these restrictions. These islands are historically important as a place of exile for various unfortunate nobles between the 9th to 14th centuries. At present about 16000 people live there and the economy mainly depends on the fishery and tourist business.

2. Site selection: Geology and geography

The bedrock under Okinoshima was originally formed from metamorphic rocks, which were extruded when the Pacific and Eurasian plates collided. This rock is called Oki-Gneiss and is one of the oldest rocks in Japan. Following that, during volcanic activities between 5~6 million years ago, Igneous rock intruded into the most of the area of these Islands. But a circular wall of Oki-Gneiss remained at Dogo as shown in Figure 2. Geographical features along sections A-B and C-D lines are also shown in the same figure. The Igneous rock areas include not only lava but also pyroclastic material and underground water. Therefore, they would not be suitable for making an underground cavern. On the other hand, the remaining Oki-Gneiss is very stable and in general there are only a few cracks, dislocations or active faults. Typical specific gravity and axial strength of the Oki-Gneiss is 27 kN/m^3 and 79 MPa, respectively.

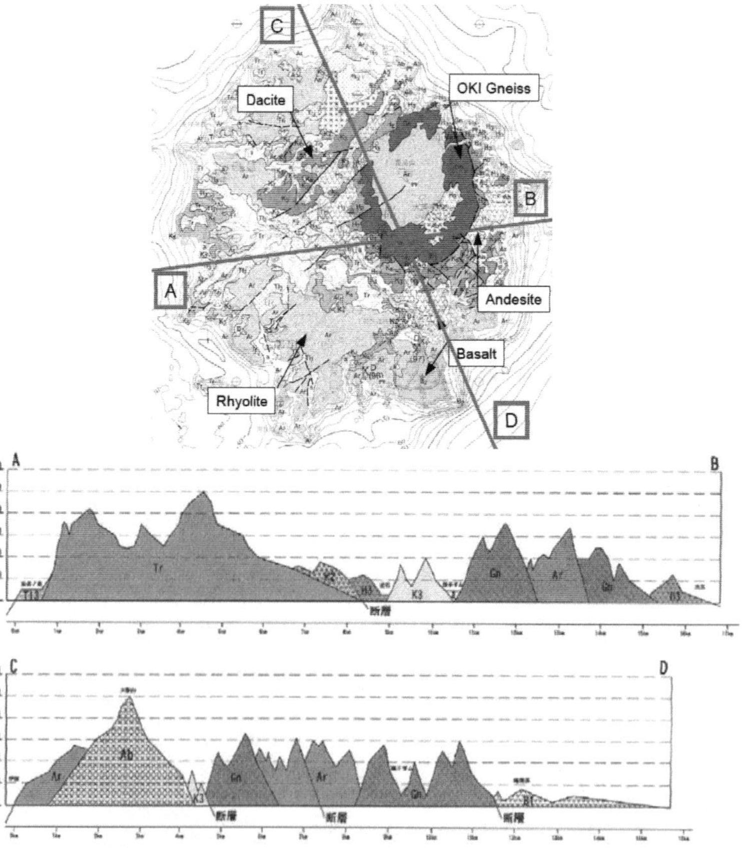

Figure2. The geology and geography of the biggest Island, Dogo.

Since the characteristics of the Oki-Gneiss are so good several quarries on Dogo provided crushed rock for the foundations of the Shimane Nuclear Power Plant. One of such quarry is shown in Figure 3.

Figure3. One of the quarries in Dogo.

A flood-control dam, the Choshi Dam (see Figure 4) was constructed in 2000, adjacent to the quarry in Figure 3 in the Oki-Gneiss area. The coefficient of water permeability was carefully examined and confirmed to be smaller than 10^{-7} m/s (Lugeon coefficient) in the Oki-Gneiss area.

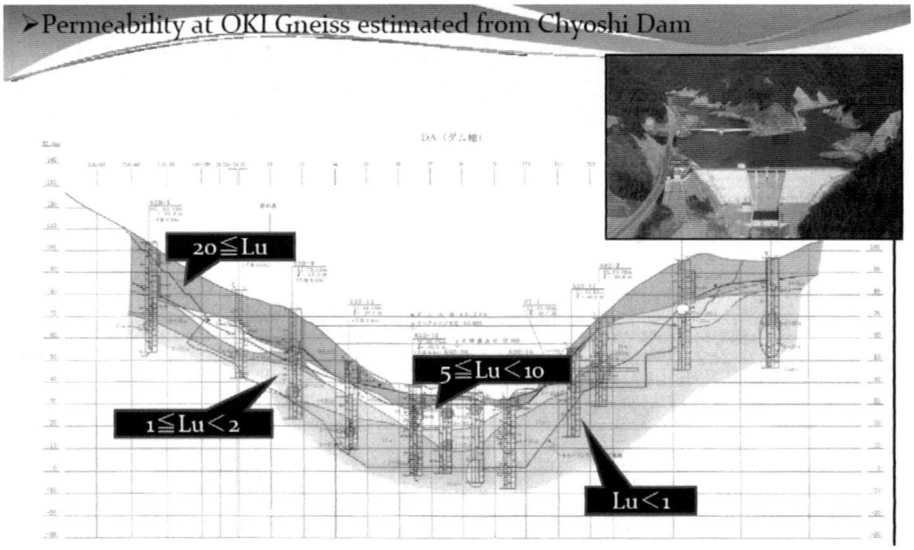

Figure4. Lugeon coefficient of Choshi Dam.

From the conservative civil engineering and safety point of view, the best location for the underground cavern is the quarries as shown in Figure 3. This is based on the following considerations:

- The location of the cavern should be in an Oki-Gneiss area and have enough distance from the igneous rocks to avoid rock brake down and degradation by the thermal effect of hot igneous rock intrusions during volcanic episodes.
- Fault zones should be avoided.
- Existing infrastructure, e.g., access to roads, sea port, etc., should be utilized as much as possible.
- Minimization of damage to the natural environment.
- The site should not be inside of the national park area.

Journal of Physics: Conference Series **308** (2011) 012028 doi:10.1088/1742-6596/308/1/012028

3. Conceptual design of the cavern

Figure 5 shows a cross-sectional drawing of the potential location of the cavern in the Oki-Gneiss area, where the nearest distance from the Rhyolite area is larger than 500 m. The elevation at the bottom of the cavern would be 63 m and the earth covering between the top of the cavern and the mountain top is 252 m. The bottom of the cavern could be lowered, for instance by another 100 m if necessary. In such a case, the access tunnel from the ground surface would have to be inclined down to the cavern floor.

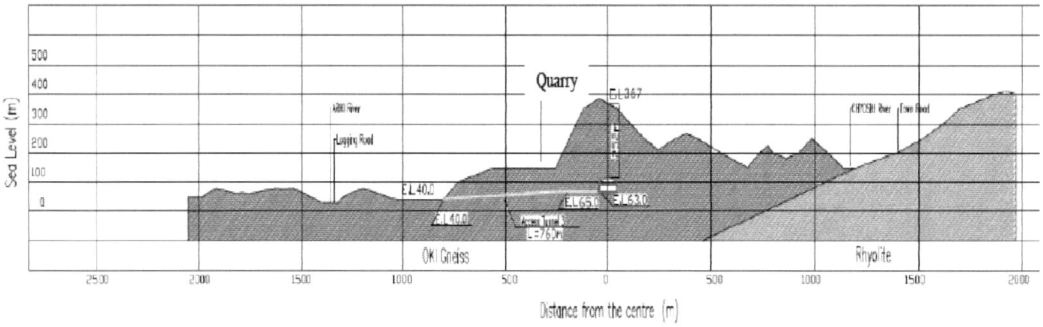

Figure5. Geological profile at the location of the cavern.

A conceptual design of the cavern has been carried out and is shown in Figure 6. In order to contain a cylindrical 100 kton Liquid Ar TPC detector with a base diameter of 80 m and a height of 20 m, the inner dimensions of the cylindrical cavern should provide a base diameter of 91 m, a height of 20 m and a spherical cap of 20 m in height.

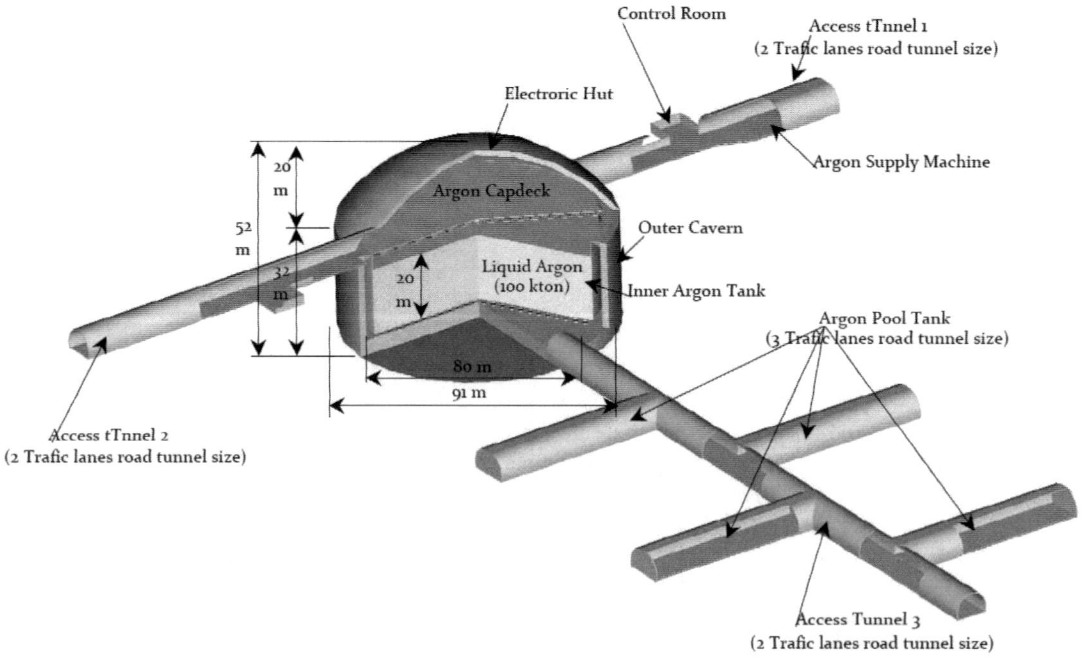

Figure6. A conceptual design of the cavern.

It should be mentioned that the biggest deep underground cavern existing in Japan now is the Kan-na-gawa underground power plant of Tokyo Electric Power Company shown in Figure 7. The present proposed cavern would have a similar volume.

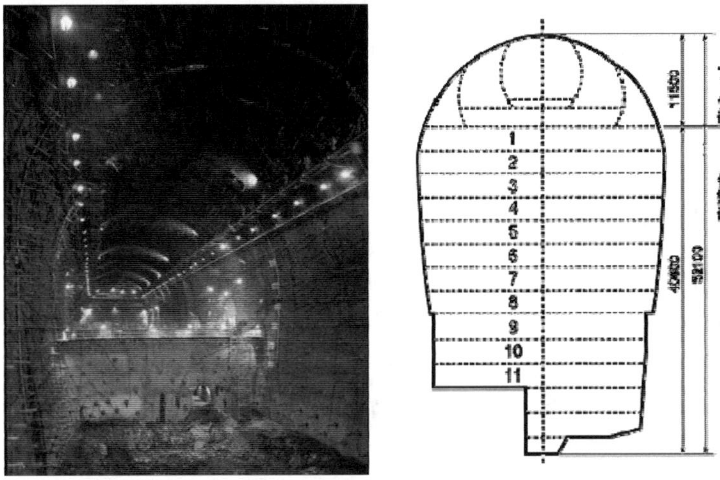

Figure7. The cavern of the Kan-na-gawa underground power plant.
(http://www.kajima.co.jp/news/digest/aug_2000/tokushu/toku03.htm)

As is shown in Figure 7, the cavern has an egg shaped cross-section provide for good stability against the strong underground pressures. Many rock bolts and pre-stressed rock anchors were used to achieve enough inner surface stability. For the neutrino observatory, detailed design of the cavern shape and the arrangement of rock bolts and rock anchors should be carried out based on multiple pilot surveys before and during the excavation work. The excavation volume of the cavern for the Liquid Ar TPC detector (Figure 6) is slightly less than that of Kan-na-gawa underground power plant (Figure 7). The shape of this cavern would be, however, different from the underground power plant case. As a consequence, careful studies of the construction methods and supporting arrangement of the rock bolts and anchors are necessary. Figure 8 shows one of the examples from our preliminary studies.

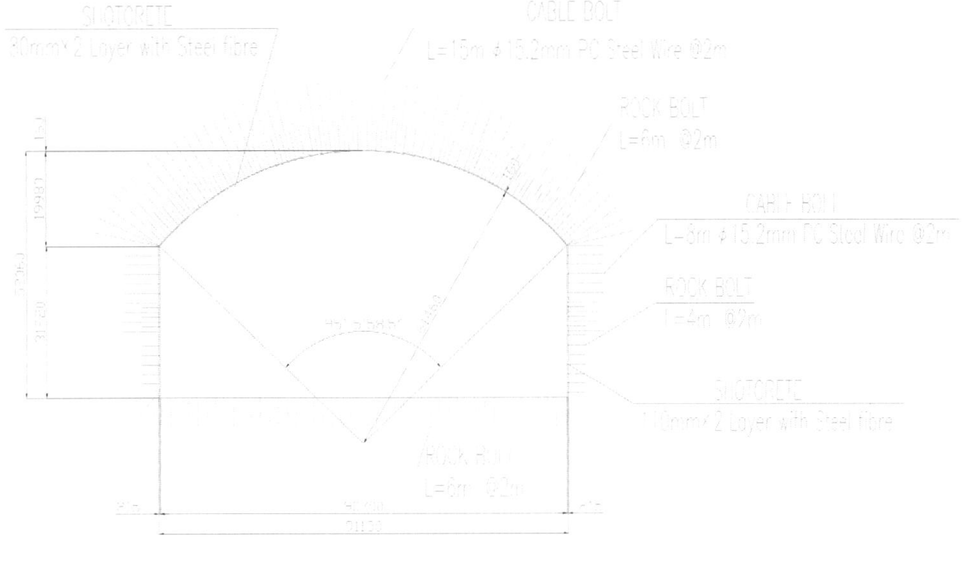

Figure8. An example of the cavern shape reinforcement study.

In order to excavate the cavern, access tunnel-1 should be constructed first. Access tunnel-2 is necessary for emergency escape, but would also be useful for the observatory operation, maintenance and so on. Access tunnels-1 and -2 come in at the top of the cavern. The main excavation starts from the top (the arch) and then proceeds to dig down to the bottom. Access tunnel-3 is mainly for the Liquid Ar pool and the construction of the main body and detector vessel, it should be constructed as early as possible. Four Liquid Ar tanks are shown in Figure 6, but the necessary number of tanks and their volumes would be determined by the means of procuring 100 kton of Liquid Ar . This will be discussed later. The size of these access tunnels should be about twice the standard single-lane road traffic tunnel size. The overall construction schedule is shown in Figure 9.

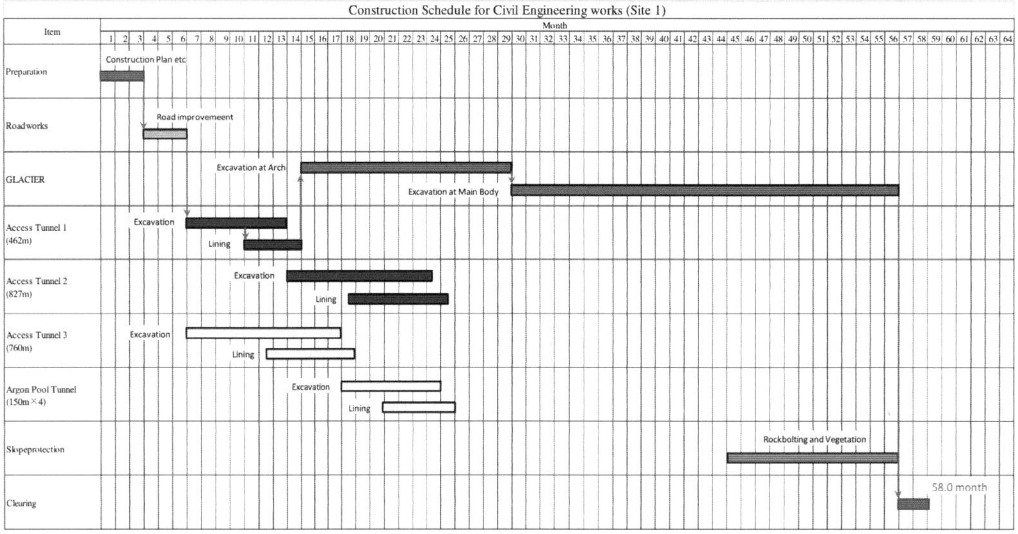

Figure9. Overall construction schedule.

In the present study, 58 months are necessary to complete the reference design construction in total. We would need more studies and necessary pilot surveys, including the construction of the pilot tunnels, boring studies, and other geotechnical investigations.

4. Infrastructure

Since there are regular daily commercial connecting flights and ferry services between Honshu and Dogo, we would have no transportation problem. The most important harbour is the Saigou port, which is close to the location of the candidate site. The existing access traffic situation from Saigou port to the site is almost adequate for carrying the heavy equipment needed for the civil engineering work, the large amount of Liquid Ar, and for the construction of the detector system. Only a short access road would need to be constructed for the final approach to the site.

The Chugoku Electric Power Company provides electricity for the Okinoshima Islands. The existing total electricity capacity is 32 MW and may be enough for the construction and operation of the observatory, but the adequacy depends on the status of industry and population of these islands.

5. Procurement of Liquid Ar

We have considered various case studies as to how to procure 100 kton of Liquid Ar over a 5 year period considering the need to minimize the cost. A solution could be as simple as the following;

- Buy Liquid Ar from several large-scale manufacturing plants having large production capacities. Many such plants are located in Japan. There are 5 candidate plants at this moment in the west side of Japan near Okinoshima, and we would contract to buy 10~15 % of their annual production capacity.

- Hire 4 tanker trucks dedicated for the Liquid Ar ground transportation, both in Honshu and Okinoshima for 5 years.
- Hire 1 barge dedicated for the marine transportation between Honshu and Okinoshima for 5 years.

We have studied other cases. For example:

- Construct an on-site production plant: But the cost for the initial construction would be very expensive and enough electricity for its operation would not be available in Dogo.
- Use an LNG carrier: We would have to solve the technical and legal problems in converting an LNG carrier to Liquid Ar. Further, the cost of hiring such a ship for 5 years would be expensive.
- Use regular commercial ferry transport: Currently they are not legally allowed to carry tanker trucks.

Therefore the present proposed solution is very simple and we would not need to solve any technical and legal problems. The cost of contracting for trucks and a barge for 5 years is not the major part of the total cost for the Liquid Ar.

6. Acknowledgement

Mr. Seishiro Yoshioka of OSAKA GAS CO., LTD. is the key person who provided us valuable information about the means of procuring 100 kton of Liquid Ar. He also introduced us Mr. Masataka Ito of NIPPON STEEL ENGINEERING CO., LTD. and Mr. Hiroshi Terai of AIR WATER INC.. We could not have arrived at the present solution without their help. The authors are very grateful for their assistance.

Reference

[1] A.Badertscher, T.Hasegawa, T.Kobayashi, A.Marchionni, A.Meregaglia, T.Maruyama, K.Nishikawa and A.Rubbia, "A Possible Future Long Baseline Neutrino and Nucleon Decay Experiment with a 100 kton Liquid Argon TPC at Okinoshima using J-PARC Neutrino Facility," arXiv:0804.2111

LAGUNA DESIGN STUDY, Underground infrastructures and engineering

Guido Alexander Nuijten

ROCKPLAN LTD, Asemamiehenkatu 2, 00520 Helsinki, FINLAND

E-mail: guido.nuijten@rockplan.fi

Abstract. The European Commission has awarded the LAGUNA project a grant of 1.7 million euro for a Design Study from the seventh framework program of research and technology development (FP7-INFRASTRUCTURES – 2007-1) in 2008. The purpose of this two year work is to study the feasibility of the considered experiments and prepare a conceptual design of the required underground infrastructure. It is due to deliver a report that allows the funding agencies to decide on the realization of the experiment and to select the site and the technology. The result of this work is the first step towards fulfilling the goals of LAGUNA. The work will continue with EU funding to study the possibilities more thoroughly. The LAGUNA project is included in the future plans prepared by European funding organizations. (Astroparticle physics in Europe). It is recommended that a new large European infrastructure is put forward, as a future international multi-purpose facility for improved studies on proton decay and low-energy neutrinos from astrophysical origin. The three detection techniques being studied for such large detectors in Europe, Water-Cherenkov (like MEMPHYS), liquid scintillator (like LENA) and liquid argon (like GLACIER), are evaluated in the context of a common design study which should also address the underground infrastructure and the possibility of an eventual detection of future accelerator neutrino beams. The design study is also to take into account worldwide efforts and converge, on a time scale of 2010, to a common proposal.

1. Physics goals

Large underground neutrino detectors, for instance Super-Kamiokande (Japan), BOREXINO (Italy) or SNO (Canada), have achieved fundamental results in particle and astroparticle physics. A next-generation very large multipurpose neutrino observatory of a total mass in the range of 100’000 to 1’000’000 tons will provide new and unique scientific opportunities in this field, very likely leading to fundamental discoveries. It will aim at a significant improvement in the sensitivity to search for proton decays, pursuing the only possible path to directly test physics at the GUT (Grand Unification Theory) scale, extending the proton lifetime sensitivities up to 10^{35} years, a range compatible with several theoretical models; it will measure with unprecedented sensitivity the last unknown mixing θ_{13} and unveil the existence of CP violation in the leptonic sector, which in turn could provide an explanation of the matter-antimatter asymmetry in the Universe; moreover it will detect neutrinos as messengers from astrophysical objects as well as from the Early Universe to give us information on processes happening in the Universe, which cannot be studied otherwise. In particular, it will sense a large number of neutrinos emitted by exploding galactic and extragalactic type-II supernovae, allowing an accurate study of the mechanisms driving the explosion. The neutrino observatory will also allow precision studies of other astrophysical or terrestrial sources of neutrinos like solar and atmospheric ones, and search for new sources of astrophysical neutrinos, like for example the diffuse neutrino background from relic supernovae or those produced in Dark Matter (WIMP) annihilation in the centre of the Sun or the Earth [1]

2. General

The acronym LAGUNA stands for Design of a pan-European Infrastructure for Large Apparatus studying Grand Unification and Neutrino Astrophysics. The LAGUNA Project is coordinated by the Swiss Federal Institute of Technology Zurich, led by Prof. A. Rubbia, and there are 30 parties (Academic Partners, Industrial Partners and Affiliated Academic Partners) taking part in the project. The purpose is to design three very large underground laboratories in Europe and to find a suitable location for them. The coordination of the technical design considering underground infrastructures and engineering is led by Prof. F. von Feilitzsch. On behalf of Finland the universities of Oulu & Jyväskylä as well their industrial partner ROCKPLAN are taking part in the project [2].

2.1. Possible experiments

Three detection techniques being studied for such large detectors in Europe, Water-Cherenkov (like MEMPHYS), liquid scintillator (like LENA) and liquid argon (like GLACIER), are evaluated in the context of a common design study which should also address the underground infrastructure and the possibility of an eventual detection of future accelerator neutrino beams.

2.1.1. GLACIER

GLACIER stands for Giant Liquid Argon Charge Imaging ExpeRiment. The basic design features of the GLACIER detector can be summarized as follows:

Single 100 kton "boiling" cryogenic tanker at atmospheric pressure for a stable and safe equilibrium condition (temperature is constant while Argon is boiling). The evaporation rate is small (less than 10^{-3} of the total volume per day given by the very favorable area to volume ratio) and is compensated by corresponding refilling of the evaporated Argon volume.

Charge imaging, scintillation and Cherenkov light readout for a complete (redundant) event reconstruction. This represents a clear advantage over large mass, alternative detectors operating with only one of these readout modes. Scintillation and Cherenkov light can be readout essentially independently for improved physics performance.

Charge amplification to allow for very long drift paths. The detector is running in bi-phase mode. In order to allow for drift lengths as long as 20 m, which provides an economical way to increase the volume of the detector with a constant number of channels, charge attenuation will occur along the drift due to attachment to the remnant impurities present in the liquid argon. This effect can be compensated with charge amplification near the anodes located in the gas phase.

2.1.2. MEMPHYS

MEMPHYS stands for MEgaton Mass PHYSics. From a practical point of view, the most straightforward liquid is water, where the detection is based on the Cherenkov light emission by the final state particles. This faint light is detected by a very large number of photomultipliers positioned on the surface of the container. The technology was pioneered by the IMB and Kamiokande experiments (USA and Japan, respectively) and successfully extended to Super-Kamiokande during many years of operation. Super-Kamiokande is the largest Water Cherenkov detector ever built so far. The possibility of building a water Cherenkov detector with a fiducial mass of about 500 kton observed by about 200 000 photomultipliers is currently being investigated by different groups around the world, and for different underground sites. While water is a cheap medium, the size of such detectors is limited by the cost of excavation and of the photomultipliers.

The MEMPHYS project is being discussed for deployment in an extended Frejus laboratory (South East France). In the US, the UNO detector is being proposed for a future underground facility in North America. In Japan Hyper-Kamiokande will provide an extension of Super-Kamiokande. All these investigations are based on Super-Kamiokande, where its size is increased by a factor 20, using 3 to 5 new caverns to be excavated.

2.1.3. LENA

LENA stands for Low Energy Neutrino Astronomy. Detecting a neutrino requires a very sensitive device in a low background environment. Due to a low interaction probability of a neutrino such a device must be very large, to get a reasonable amount of data in a reasonable time.

The core of LENA consists of a huge tank filled with a special organic liquid. The liquid may be composed of three components: solvent, scintillator and wavelength shifter. The solvent might be an aromatic hydrocarbon coded PXE (phenyl-o-xylylethane). It has a high light yield and large attenuation length. Particularly, it is considered a non-hazardous liquid (flash-point of 145°C, HMIS ratings <1). Laboratory experiments have shown that attenuation lengths of ~10 m (@ 430 nm) and a good photoelectron yield can be achieved in both pure PXE and a mixture containing 80% of dodecane. Another alternative may be LAB, mixed with small amounts of other chemicals. Its optical properties are probably better but need to be studied more.

The liquid target will have a very large volume, to minimize the presence of any radioactivity. The maximum width is given by the optical properties of the liquid. For PXE the maximum, and also optimal diameter is about 13 m, for LAB probably 15 m.

A neutrino hitting the scintillator target may induce a reaction resulting in a light pulse. The light pulses will be observed by light sensors at the outer walls of the tank. There will be about 12 000 light sensors, probably 50 cm photo multiplying tubes, and respective amount of wires to data acquisition system. The phototubes require also a high-voltage (> kV) line.

2.2. Possible sites

Seven different sites being studied for such large detectors in Europe are evaluated in the context of a common design study which should also address the underground infrastructure and the possibility of an eventual detection of future accelerator neutrino beams.

2.2.1. Pyhäsalmi, Finland

Pyhäsalmi Mine Oy is owned by Inmet Mining Corporation. Pyhäsalmi is an underground copper and zinc mine located in central Finland. The mine is at present the deepest mine in Europe, reaching a depth of 1440 meter below surface. The Mine produces three types of concentrates: copper, zinc and pyrite.

In 1962 the Mine began as an open pit operation. This phase lasted until 1967, the year when Mine operations commenced underground. In 1975 the open pit was completely worked out. As mining progressed the Mine was gradually deepened. The latest phase of deepening with a view to exploiting the ore lens below 1050 level was carried out between 1998 and 2001. The resulting new Mine started operation in 2001 and mining is carried out via the new 1440 meter deep Timo Shaft.

The Pyhäsalmi Mine is a very modern and safe operation and one of the best performing Mines worldwide with an annual production of 1.4 million tones. To secure extend mining into the future the Mine invests significantly in exploration both in the Mine site and regionally. Pyhäsalmi's safety and health management system was certified under the Occupational Health and Safety Assessment System (OHSAS) 18001 standard.

Present at location are all main needed infrastructure facilities as 110 kV power supply, railway connections and a good road infrastructure. The railway station in Pyhäsalmi is the second biggest freight station of Finland as provides a very good connection with the harbor in Kokkola, about 160 km distance. The harbor of Kokkola is situated on the west coast. The town also has an airfield. The main corridor from Helsinki to Oulu passes at only 6 km from the site. [2]

2.2.2. Frejus, France

The Fréjus site of LAGUNA is located near to the Italian-French border next to the Fréjus Highway tunnel connecting the villages of Modane in France and Bardonecchia in Italy. The distance from Fréjus to CERN is only 135 km.

Regarding to transport services the Fréjus tunnel is one of the main connections through the French Alps between the main cities of Lyon and Torino. The connection consists both the road tunnel as the railway tunnel. Connected to the road tunnel Fréjus is known for its present laboratory, LSM, standing for Laboratoire Souterrain de Modane, created in 1982. [3]

2.2.3. Boulby, United Kingdom

Boulby mine is located on the coast of North East England, 15km North of the town of Whitby in Yorkshire. The site is rather exceptional in the UK being within a major national park of outstanding beauty (the North York Moors National Park) yet close to one of the largest industrial zones in Europe, Middlesbrough and Tees valley (30 km away), with excellent transport links.

There are thus outstanding recreational facilities for the visitor, with an excellent living environment and some of the most spectacular coasts and beaches in England, but also easy access to all the industry and materials needed for LAGUNA, including liquid argon and mineral oils manufactured in the huge petro-chemical plants of the Tees valley, connected directly to the mine by a dedicated rail link. [4]

2.2.4. Umbria, Italy

The Italian site, unlike other sites analyzed in this feasibility study, is not corresponding to any existing infrastructure, such as a mine or a tunnel. It has been located referring to a hypothesis of an off-axis experiment, using existing neutrino beam from CERN in Geneve to LNGS, the Gran Sasso National Laboratory.

After research a site has been located in the region of Umbira, that answers to all scientific and technical requests. The location is in the mountains in the area of Valnerina. The location is about 650 km from CERN. Valneria is a narrow and tortuous valley, sparsely populated with a rural and high environmental value. [5]

2.2.5. Polkowice-Sieroszowice, Poland

The Polkowice-Zieroszowice mine has been under continuous operation for almost 30 years. Including the holding of KGHM there are 4000 employees in the mine and 18000 in the holding. The mine operates in three mining regions, covering a total surface area of about 175 km^2. In 2008 about 11 millions tons of ore were extracted.

The location of the mine is in the south west corner of Poland. The biggest nearby city is Wroclaw, about 90 km from the mine, with over 600000 inhabitants. Wroclaw offers a railway station, an international airport, universities, research institutions and a rich cultural life. The road infrastructure also offers good connections to Germany. [6]

2.2.6. Canfranc, Spain

The Canfranc site of LAGUNA is located near to the Spanish-French border next to the Somport Highway tunnel. The Canfranc Underground Laboratory (LSC) is a new facility for Underground Science. It is conceived as a Consortium of the Spanish Ministry of Science and Innovation, the Aragon Regional Government and the University of Zaragoza.

Canfranc is a small village located in Northern Spain. The Somport international mountain pass connects Spain and France in this region, and has provided communication since ancient times in the Pyrenees. The tunnel links the Jaca area in Spain with the city Pau. [7]

2.2.7. Slanic, Romania
Slanic town is situated in the sub Carpathian hills about 100 km north from Bucharest. Slanic hosts a salt mine, Unirea, which is administrated by SALROM, SA. The salt exploitation ended in 1971. For the time being the mine is used for tourism and medical purposes. The access into the mine is assured by elevator able to carry up 15 tons. The network of galleries in the Unirea salt mine are very large. The current area is 70000 m^2 and 2.9 million m^3 were excavated.

The temperature in the mine is very modest, around 12°C. The Unirea salt mine is connected to electricity, rads, railway, phone, internet and GSM networks[8]

3. General and scientific considerations for Finland

Within Finland the Pyhäsalmi Mine offers the best location for this purpose, as it is the deepest present location in Finland 1400 meters below surface.

Pyhäsalmi Mine has expressed its interest to be chosen as one of the locations for the underground laboratory as stated in the memorandum of understanding between the Mine and LAGUNA parties in Finland.

Pyhäsalmi is located in the municipality of Pyhäjärvi in the middle of Finland, 450 km north of Helsinki and 150 km south of Oulu.

LAGUNA has received a very positive support from the municipality of Pyhäjärvi as well from the local region Northern Ostrobothnia. A part from the scientific involvement by the Universities of Jyväskylä and Oulu, also the University of Helsinki, the Helsinki Institute of Physics and the Finnish Academy have shown to be interested in the LAGUNA project. The national government of Finland and members of the House of Parliament have been informed and their response, although unofficial, is genuinely positive.

The clearance from nuclear plants to avoid reactor neutrino events is good as there are no nuclear reactors nearby. The nearest nuclear plant is located in Olkiluoto, about 425 km from Pyhäsalmi.

The high level of the existing infrastructure as well as the excellent rock conditions offers the possibility to host all Laguna alternatives simultaneously.

For all options Pyhäsalmi offers the required depth to reduce sufficiently the muon flux at the laboratory level.

Concerning CERN's possible future long distance neutrino beam experiment Pyhäsalmi offers the ideal distance required, i.e. 2300 km from CERN. This offers a very interesting possibility to connect LAGUNA in Finland with experiments in CERN.

4. Technical considerations for Finland

Finland is known for its Precambrian tectonically stable hard bedrock. Earthquake zones are absent at the vicinity, therefore earthquakes occur extremely rarely and their magnitude may be neglected. The hard and very old bedrock of Finland provides by far one of the best locations in Europe to locate any of the LAGUNA laboratories deep under the ground.

Finland has high expertise and good experience in designing and construction of large, complex underground excavations and constructions.

Pyhäsalmi Mine has a suitable low temperature of surrounding rock. The temperature is only a moderate 16°C at 900m and about 23°C at 1400m depth. Pyhäsalmi Mine has very suitable hydrological conditions. The deep rock is nearly completely dry and there are no rock formations, that show time related deformations (like creep of salt or anhydrite).

The Mine has one decline, one main transport hoist shaft and one ventilation shaft. The present underground infrastructure only needs to be enlarged with a ventilation shaft and it provides a perfect layout to conduct excavation as well construction and filling of the tank.

On surface there is an excellent rail yard and with good rail and road connections to the harbor in Kokkola. Construction of the tunnels can be started directly at the depth desired due to excellent existing infrastructure as well on surface as below surface.

Pyhäsalmi has no challenges how to dispose excavated rock as all excavated rock can be handed over to the Mine and will be disposed underground.

5. Estimated costs and realisation time for Finland

Construction costs of the laboratory and observatory infrastructure can be divided in two parts: underground infrastructure and the observatory including detector, sensors and liquids. In the Design Study the main focus is towards the costs for the underground infrastructure. These costs content of preparation costs of the site and all design and legal questions, of excavation costs of the main detector cavern including all auxiliary rooms and needed access routes to the underground site and finally these costs content also all reinforcement works needed for the safety and stability of the cavern. Operation and maintenance costs are not included in this study. The total costs are estimated to range from 45M€ to 125M€.

The preparation time (design, decision taking etc.) is assumed to be finished at the end of 2012. Technically is can be started at any time. External factors may influence the possible date of starting of the project. The construction time, starting directly after the preparation phase, of the underground infrastructure is estimated to range from 3 to 7 years, depending on the size of the detector.

References

[1] The LAGUNA design study– towards giant liquid based underground detectors for neutrino physics and astrophysics and proton decay searches (LAGUNA CERN yellow report)

[2] PYHÄSALMI, LAGUNA Design Study, Feasibility Study for LAGUNA at PYHÄSALMI: Underground infrastructure and engineering, (EU, FP 7: Work Package 2: Deliverable 2.1) ROCKPLAN LTD, G.A. Nuijten, 12/04/2010

[3] LAGUNA, feasibility study for the Fréjus site, final report (EU, FP 7: Work Package 2: Deliverable 2.2), LOMBARDI, 14/05/2010

[4] BOULBY, LAGUNA Design Study – Interim Report, Geo-technical, Underground Infrastructure and Engineering (EU, FP7: Work Package 2: Deliverable 2.3) – University of Sheffield, CPL et. al., 16/11/2009

[5] LAGUNA, Design Study, Underground infrastructures and engineering for LAGUNA at Italian Site (EU, FP7: Work Package 2: Deliverable 2.4) Regione Umbira Site (Valnerina), (EU, FP7: Work Package 2: Deliverable 2.4) AGT Ingegneria - 04/2010

[6] Sieroszowice (SUNLAB), LAGUNA Design Study, Underground infrastructures and engineering Interim Report (EU, FP7: Work Package 2: Deliverable 2.5), KGHM Cuprum CBR, Wroclaw – 01/12/2009

[7] Feasibility Study for Large Underground Caverns and Auxiliary Infrastructure Facilities of the LAGUNA project at the LSC (Canfranc, Huesca, Spain), (EU, FP7: Work Package 2: Deliverable 2.6) Universidad Autonomia de Madrid, STMR & Iberinsa, 12/05/2010

[8] LAGUNA, Design Study SLANIC, Underground infrastructures and engineering (EU, FP7: Work Package 2: Deliverable 2.7) Universitatea din Bucuresti, IFIN-HH – 14/05/2010

Towards a 100 kton LAr experiment

André Rubbia

ETH Zurich, Institute for Particle Physics, CH-8093 Zürich, Switzerland

High Energy Accelerator Research Organization (KEK), Tsukuba, Ibaraki, Japan

E-mail: andre.rubbia@cern.ch

Abstract. The realization of a 100 kton LAr TPC demands concrete R&D in several areas. Although rightly relying on pioneering efforts performed up to now, in particular within the ICARUS programme, it cannot be simply “linearly” extrapolated from the current state-of-art. The GLA2010 workshop is intended to be the first of a series of workshops dedicated to these issues, which was initiated to bring together researchers having common interest in realizing a giant neutrino observatory based on the liquid Argon time projection chamber technology combining next-generation searches for proton decay and neutrino physics with natural and artificial sources. In this talk, I would like to present my personal collection of some results presented at this workshop and give an outlook on future options.

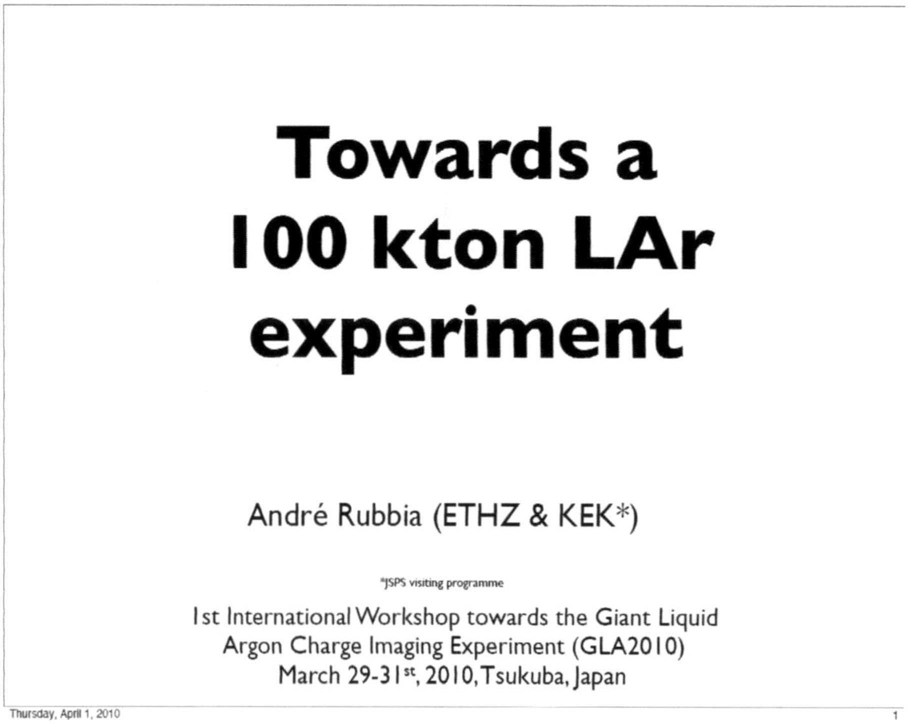

Journal of Physics: Conference Series **308** (2011) 012030 doi:10.1088/1742-6596/308/1/012030

ICARUS T600(*0.5) is the largest detector built and operated to date

Since its successful operation in Pavia in 2001 there have been tremendous, imaginative and worldwide – often independent from one another – efforts to go beyond that stage

This workshop confirms this trend !

I apologize for not being able to cover all talks and give them the merit which they deserve !

My personal collection of some results presented @ the workshop and an outlook at the end...

Atmospheric neutrino experiments

➠ Kajita

There have been many "massive underground detectors" for atmospheric neutrino studies. and proton decay

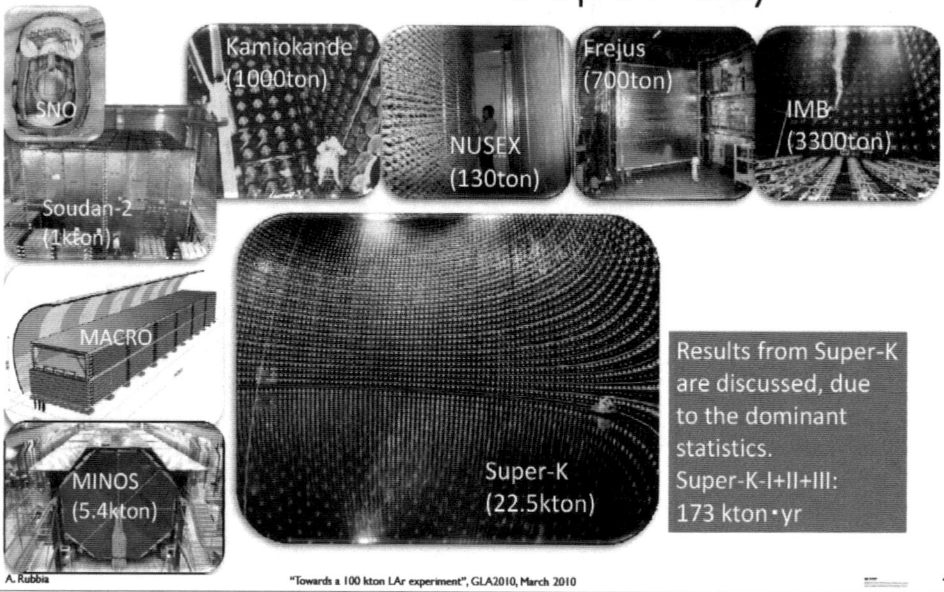

Results from Super-K are discussed, due to the dominant statistics.
Super-K-I+II+III: 173 kton·yr

A. Rubbia "Towards a 100 kton LAr experiment", GLA2010, March 2010 4

Thursday, April 1, 2010 4

Giant !

1,000 m underground

Superkamiokande
50'000 m^3 pure H_2O
Cerenkov imaging
22.5 kton fiducial mass
(Data taking 1996-)

>200 m underground

GLACIER detector
77'000 m^3 ultra pure LAr
Liquid Argon TPC
100 kton fiducial mass
(?)

A. Rubbia "Towards a 100 kton LAr experiment", GLA2010, March 2010 5

Thursday, April 1, 2010 5

Why a 100 kton LAr detector?

Next generation detector, a new way to look at rare events, to significantly go beyond present experiment(s)

- ☑ **Neutrino oscillations and CPV in the lepton sector** (known CPV does not account for matter excess over antimatter)
- ☑ **Grand Unification and proton lifetime** (electron and proton charge cancellation)
- ☑ **Neutrino astrophysics** (stars' & supernovae's inner mechanisms, search for new sources,...)

A. Rubbia "Towards a 100 kton LAr experiment", GLA2010, March 2010 6

Thursday, April 1, 2010 6

3 ordinary ν's oscillations and CPV

atm Δm^2_{31} **Pontecorvo–Maki–Nakagawa–Sakata matrix & neutrino masses** Δm^2_{21} solar

$$U = \begin{pmatrix} 1 & 0 & 0 \\ 0 & c_{23} & s_{23} \\ 0 & -s_{23} & c_{23} \end{pmatrix} \begin{pmatrix} c_{13} & 0 & e^{-i\delta}s_{13} \\ 0 & 1 & 0 \\ -e^{i\delta}s_{13} & 0 & c_{13} \end{pmatrix} \begin{pmatrix} c_{12} & s_{12} & 0 \\ -s_{12} & c_{12} & 0 \\ 0 & 0 & 1 \end{pmatrix}$$

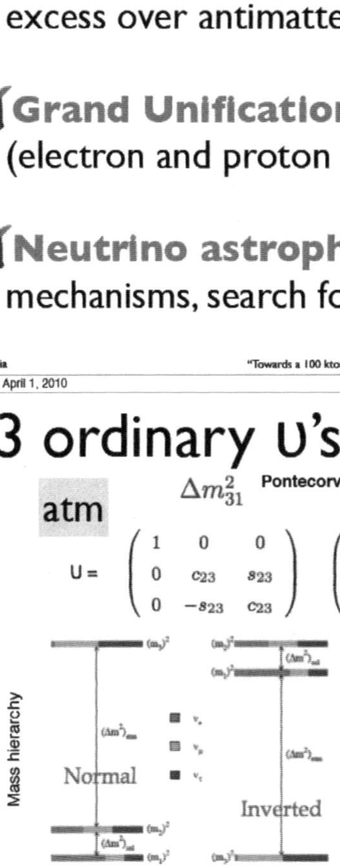

Unknown mass hierarchy

Last unknown angle

Well meas'd parameters:

Schwetz, Tortola, Valle, New J.Phys.10:113011,2008

Δm^2_{21} $(7.65^{+0.23}_{-0.20})\, 10^{-5}\,\mathrm{eV}^2$

$\sin^2\theta_{12}$ $0.304^{+0.022}_{-0.016}$

$|\Delta m^2_{31}|$ $(2.40^{+0.12}_{-0.11})\, 10^{-3}\,\mathrm{eV}^2$

$\sin^2\theta_{23}$ $0.50^{+0.07}_{-0.06}$

$\sin^2\theta_{13} < 0.056$ @ 3σ

Neutrino flavor oscillation probability:

$$P(\nu_\mu \to \nu_e) = \Sigma_{i=1,4} P_i$$

$$P_1 = \sin^2\theta_{23}\sin^2(2\theta_{13})\left(\frac{\Delta_{13}}{B_\pm}\right)^2 \sin^2\frac{B_\pm L}{2}$$ Atmospheric term

$$P_2 = \cos^2\theta_{23}\sin^2(2\theta_{12})\left(\frac{\Delta_{12}}{A}\right)^2 \sin^2\frac{AL}{2}$$ Solar term

$$P_3 = J\cos\delta\left(\frac{\Delta_{12}}{A}\right)\left(\frac{\Delta_{13}}{B_\pm}\right)\cos\frac{\Delta_{13}L}{2}\sin\frac{AL}{2}\sin\frac{B_\pm L}{2}$$

$$P_4 = \mp J\sin\delta\left(\frac{\Delta_{12}}{A}\right)\left(\frac{\Delta_{13}}{B_\pm}\right)\sin\frac{\Delta_{13}L}{2}\sin\frac{AL}{2}\sin\frac{B_\pm L}{2}$$

Interference terms CP-phase

Hints:

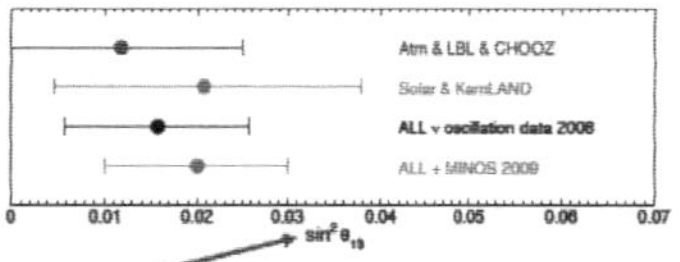

$\sin^2 2\theta_{13} \approx 4\sin^2\theta_{13}$

Fogli et al., J.Phys.Conf.Ser.203:012103,2010

A. Rubbia Towards a 100 kton LAr experiment 7

Thursday, April 1, 2010 7

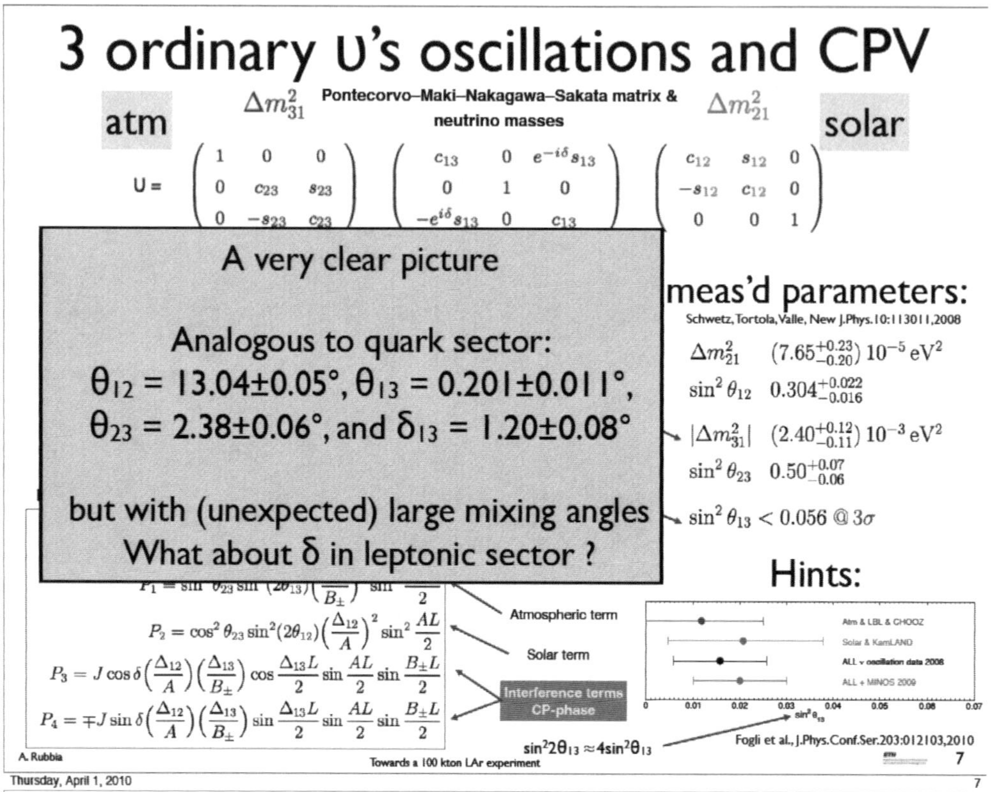

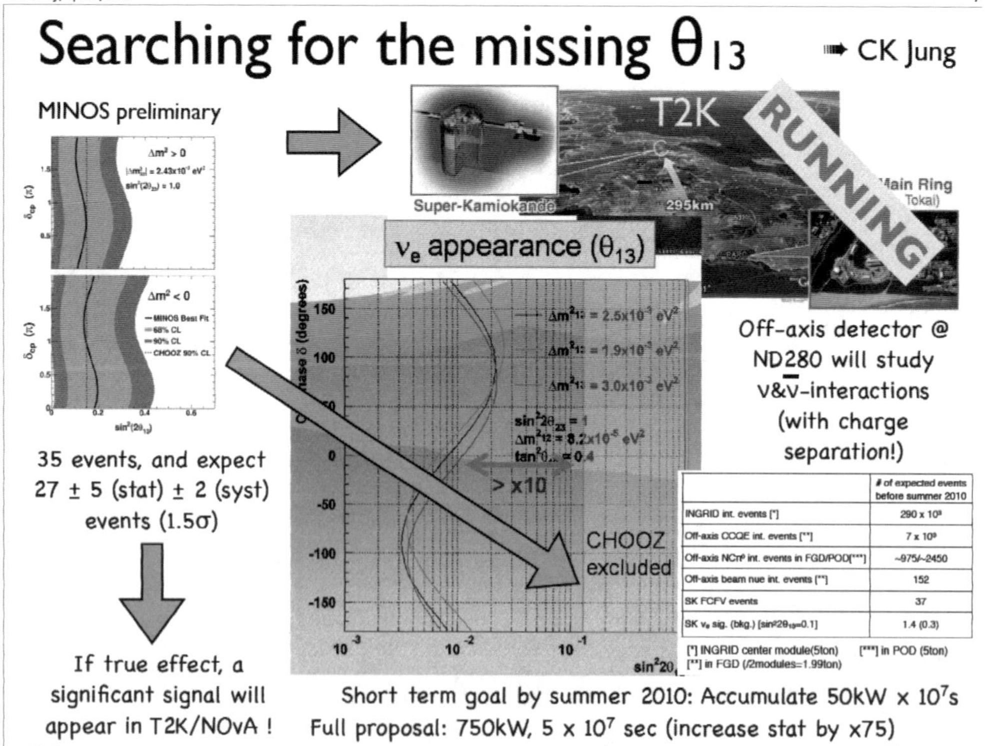

	# of expected events before summer 2010
INGRID int. events [*]	290 x 10³
Off-axis CCQE int. events [**]	7 x 10³
Off-axis NCπ⁰ int. events in FGD/POD[***]	~975/~2450
Off-axis beam nue int. events [**]	152
SK FCFV events	37
SK νe sig. (bkg.) [sin²2θ13=0.1]	1.4 (0.3)

[*] INGRID center module(5ton) [***] in POD (5ton)
[**] in FGD (/2modules=1.99ton)

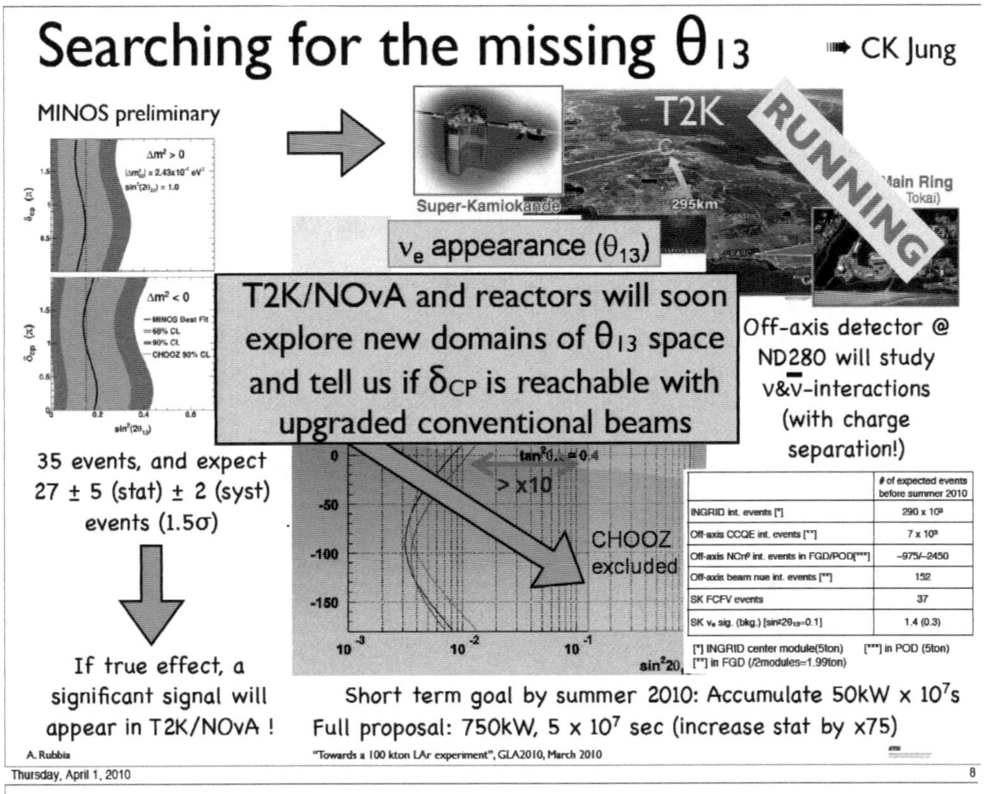
Searching for the missing θ13
CK Jung
MINOS preliminary
T2K
RUNNING
Super-Kamiokande
295km
Main Ring (Tokai)
νe appearance (θ13)
T2K/NOvA and reactors will soon explore new domains of θ13 space and tell us if δCP is reachable with upgraded conventional beams
Off-axis detector @ ND280 will study ν&ν̄-interactions (with charge separation!)
35 events, and expect 27 ± 5 (stat) ± 2 (syst) events (1.5σ)
> x10
CHOOZ excluded
If true effect, a significant signal will appear in T2K/NOvA !
Short term goal by summer 2010: Accumulate 50kW x 10^7s
Full proposal: 750kW, 5 x 10^7 sec (increase stat by x75)
A. Rubbia
"Towards a 100 kton LAr experiment", GLA2010, March 2010
Thursday, April 1, 2010
8

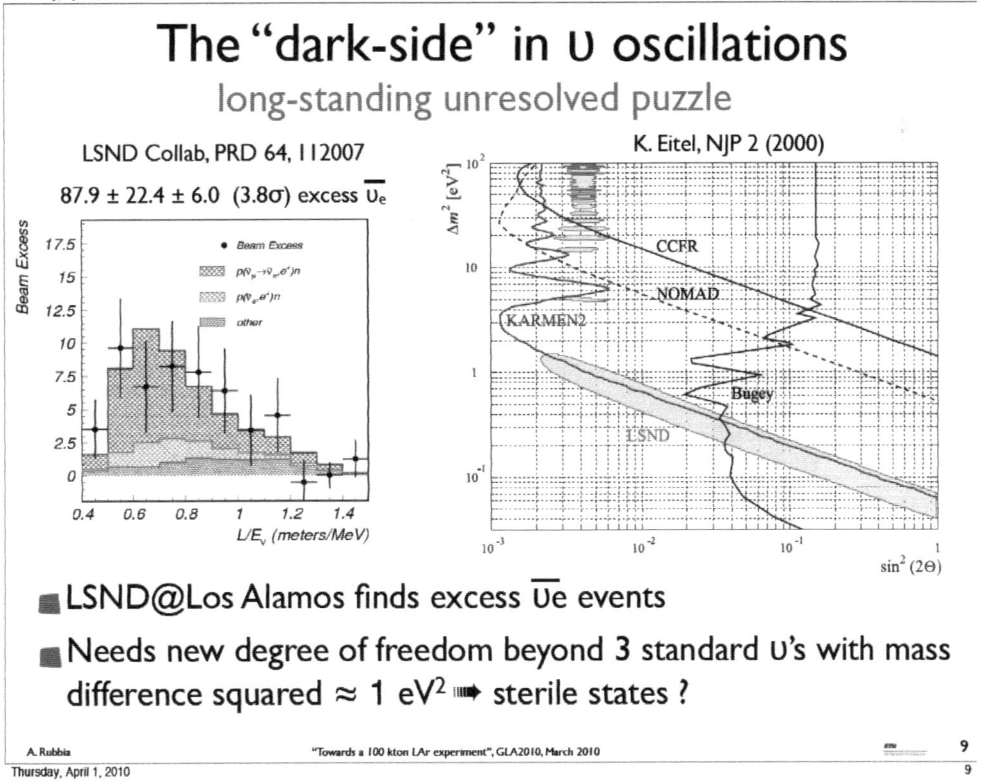
The "dark-side" in ν oscillations
long-standing unresolved puzzle
LSND Collab, PRD 64, 112007
87.9 ± 22.4 ± 6.0 (3.8σ) excess ν̄e
K. Eitel, NJP 2 (2000)
Beam Excess
L/Eν (meters/MeV)
CCFR
NOMAD
KARMEN2
Bugey
LSND
Δm² [eV²]
sin² (2Θ)
LSND@Los Alamos finds excess ν̄e events
Needs new degree of freedom beyond 3 standard ν's with mass difference squared ≈ 1 eV² sterile states ?
A. Rubbia
"Towards a 100 kton LAr experiment", GLA2010, March 2010
9
Thursday, April 1, 2010
9

The "dark-side" in υ oscillations

long-standing unresolved puzzle

LSND Collab, PRD 64, 112007

87.9 ± 22.4 ± 6.0 (3.8σ) excess $\overline{\upsilon}_e$

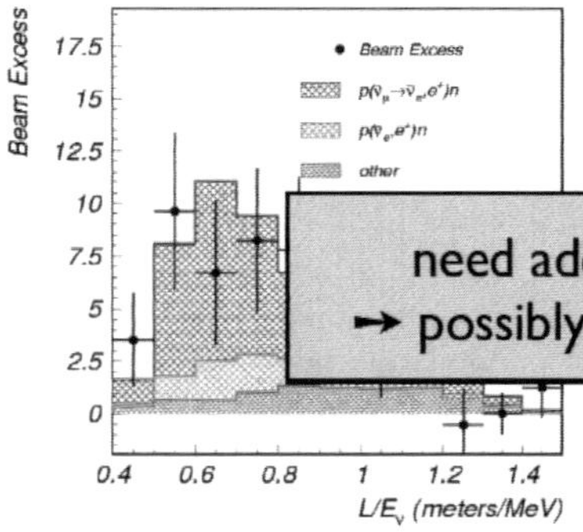

K. Eitel, NJP 2 (2000)

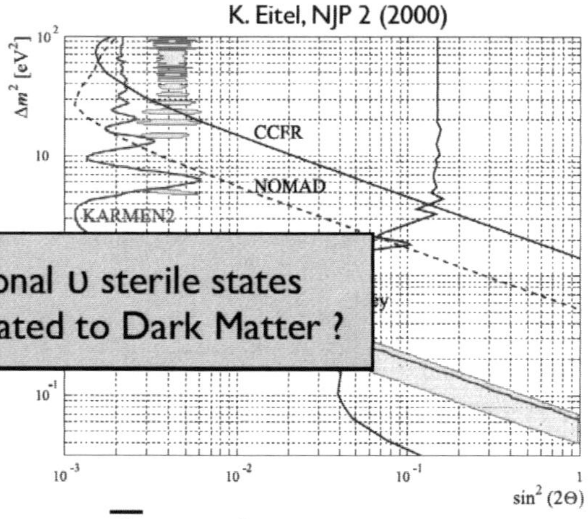

need additional υ sterile states
→ possibly related to Dark Matter ?

- LSND@Los Alamos finds excess $\overline{\upsilon}$e events
- Needs new degree of freedom beyond 3 standard υ's with mass difference squared ≈ 1 eV^2 ➠ sterile states ?

A. Rubbia "Towards a 100 kton LAr experiment", GLA2010, March 2010 9
Thursday, April 1, 2010 9

Unexplained excess of events ?

MiniBOONE Collab, PRL 102, 101802 (2009)
Unexplained excess from 200-475 MeV
= 128.8±20.4±38.3 e-like events

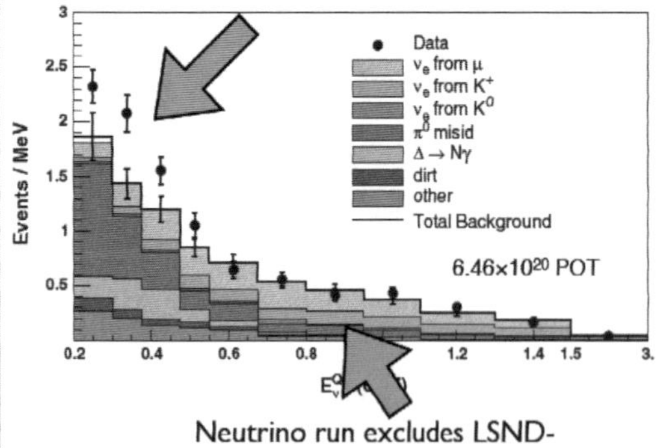

Neutrino run excludes LSND-like flavor oscillation

a new puzzle ?

Antineutrino is inconclusive

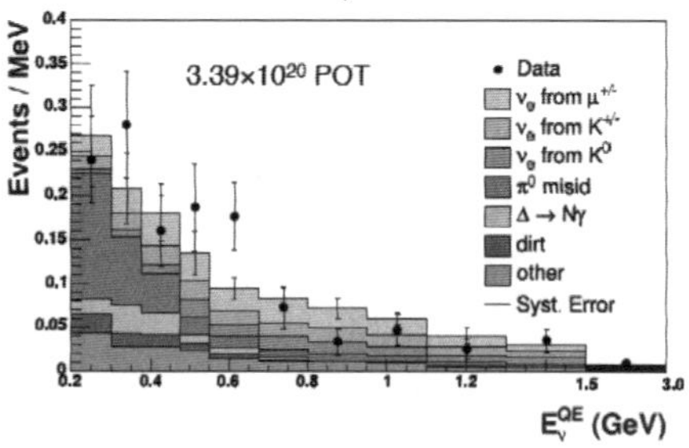

- "MiniBOONE is not able to conclusively determine whether the low-energy excess is due to a Standard Model process (e.g. a NC γ process) or to Physics Beyond the Standard Model (e.g. sterile neutrinos with CP or CPT violation or sterile neutrino decay)" –W.C. Louis, Aspen 2010

A. Rubbia "Towards a 100 kton LAr experiment", GLA2010, March 2010 10
Thursday, April 1, 2010 10

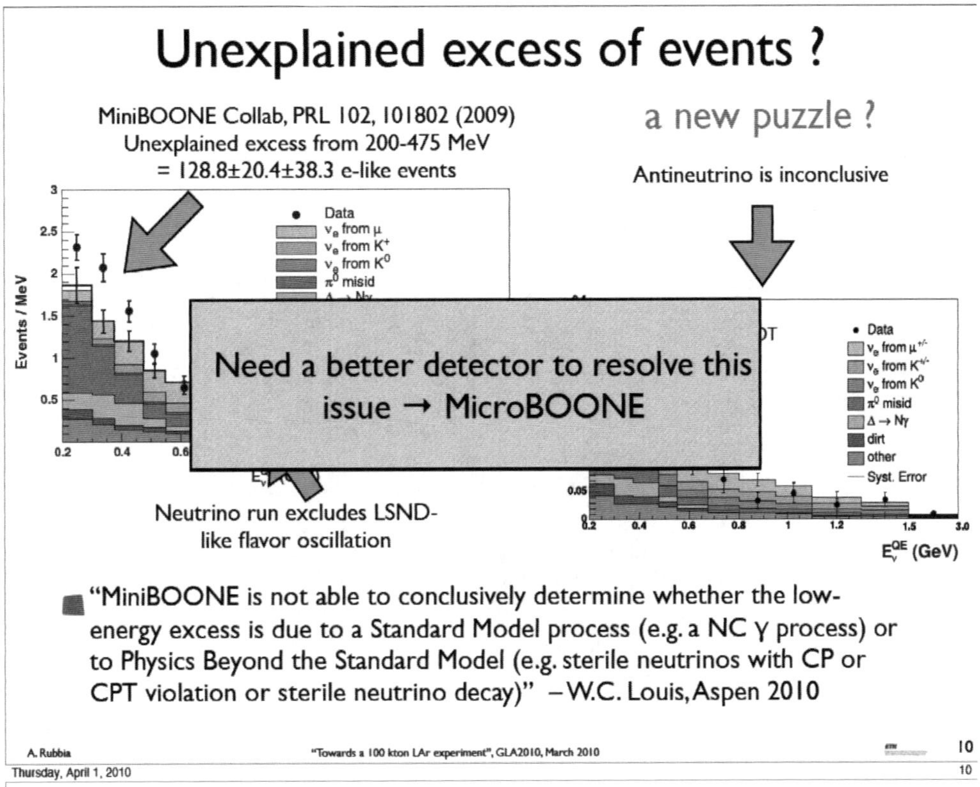
Unexplained excess of events ?
MiniBOONE Collab, PRL 102, 101802 (2009)
Unexplained excess from 200-475 MeV
= 128.8±20.4±38.3 e-like events
a new puzzle ?
Antineutrino is inconclusive
Events / MeV
Data
Need a better detector to resolve this issue → MicroBOONE
Neutrino run excludes LSND-like flavor oscillation
E_ν^{QE} (GeV)
"MiniBOONE is not able to conclusively determine whether the low-energy excess is due to a Standard Model process (e.g. a NC γ process) or to Physics Beyond the Standard Model (e.g. sterile neutrinos with CP or CPT violation or sterile neutrino decay)" –W.C. Louis, Aspen 2010
A. Rubbia
"Towards a 100 kton LAr experiment", GLA2010, March 2010
10
Thursday, April 1, 2010
10

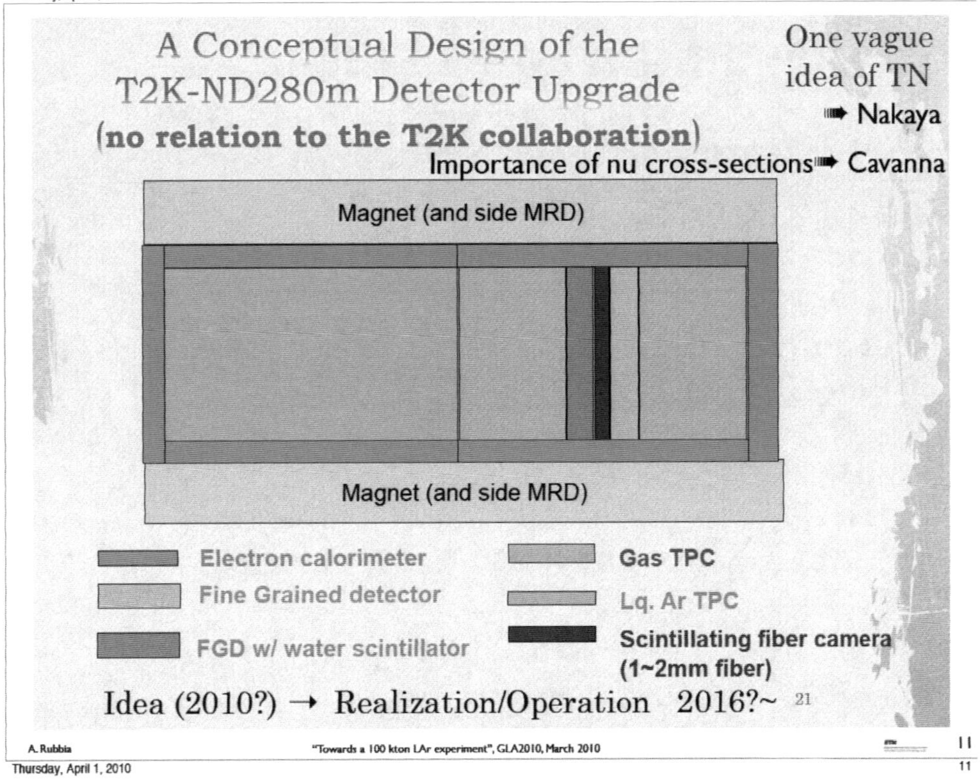
A Conceptual Design of the T2K-ND280m Detector Upgrade
(no relation to the T2K collaboration)
One vague idea of TN
➠ Nakaya
Importance of nu cross-sections ➠ Cavanna
Magnet (and side MRD)
Magnet (and side MRD)
Electron calorimeter
Fine Grained detector
FGD w/ water scintillator
Gas TPC
Lq. Ar TPC
Scintillating fiber camera (1~2mm fiber)
Idea (2010?) → Realization/Operation 2016?~
21
A. Rubbia
"Towards a 100 kton LAr experiment", GLA2010, March 2010
11
Thursday, April 1, 2010
11

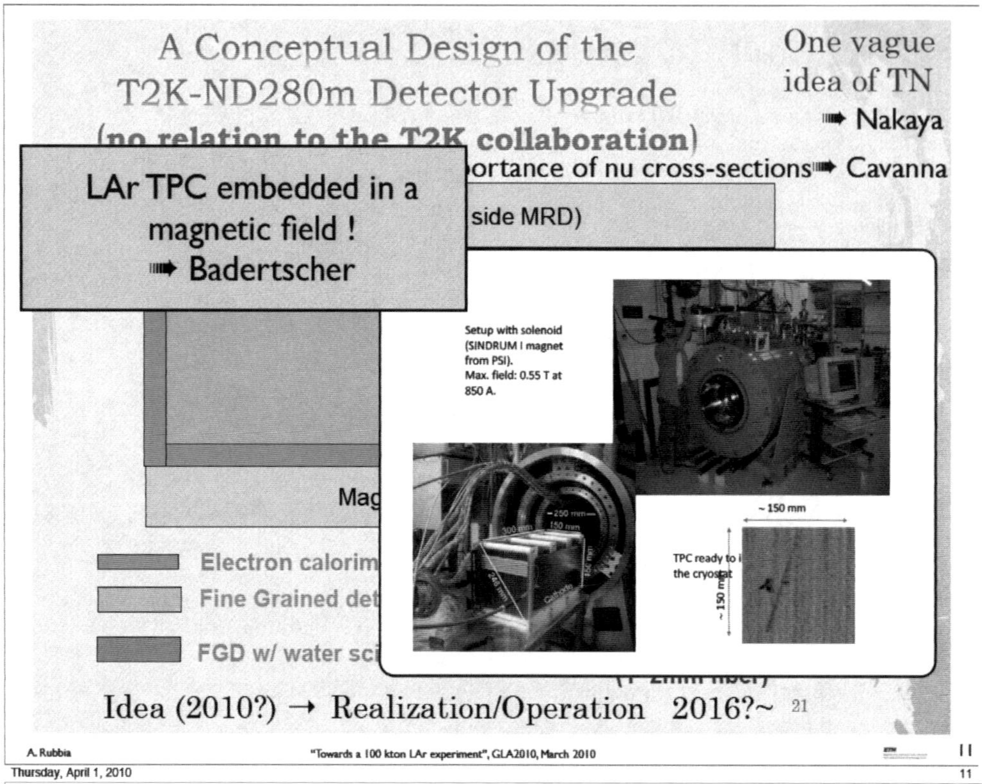

Focusing on the long baseline...

➠ Wark

Several competing projects worldwide

- 1st generation: K2K, MINOS, OPERA
  - ✳ 250 km - 730 km, 1.8-22.5 kton
- 2nd generation: T2K, NOVA
  - ✳ 300 km - 830 km, 750 kW, 15-22.5 kton
- 3rd generation: upgraded JPARC, ProjectX/DUSEL, CERN ? conventional MW-class "super-superbeam" = hyperbeam
  - ✳ **600 - 2500 km ? 100-1000 kton detectors ?**
- 4th generation: **next generation ν beams** (NF/beta)
  - ✳ 2000-7000 km ? 50 kton magnetized ?

A. Rubbia "Towards a 100 kton LAr experiment", GLA2010, March 2010 12
Thursday, April 1, 2010 12

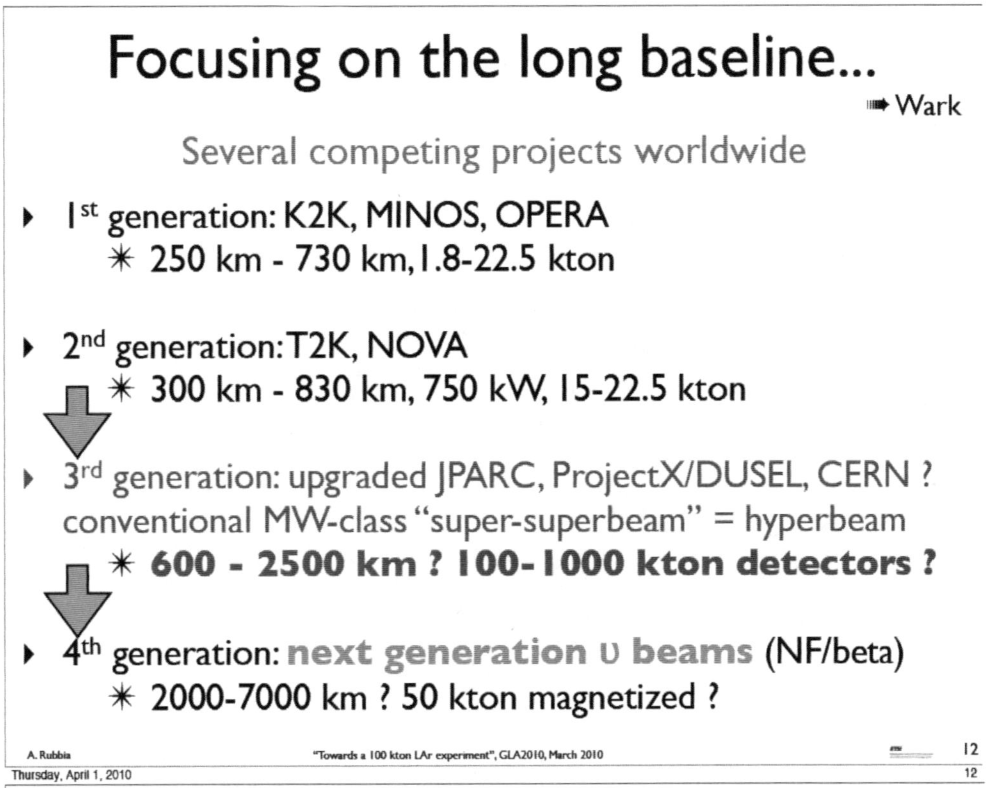
Focusing on the long baseline...
➠ Wark
Several competing projects worldwide
1st generation: K2K, MINOS, OPERA
250 km - 730 km, 1.8-22.5 kton
2nd generation: T2K, NOVA
300 km - 830 km, 750 kW, 15-22.5 kton
3rd generation: upgraded JPARC, ProjectX/DUSEL, CERN ?
conventional MW-class "super-superbeam" = hyperbeam
600 - 2500 km ? 100-1000 kton detectors ?
4th generation: next generation ν beams (NF/beta)
2000-7000 km ? 50 kton magnetized ?
A. Rubbia
"Towards a 100 kton LAr experiment", GLA2010, March 2010
12
Thursday, April 1, 2010
12

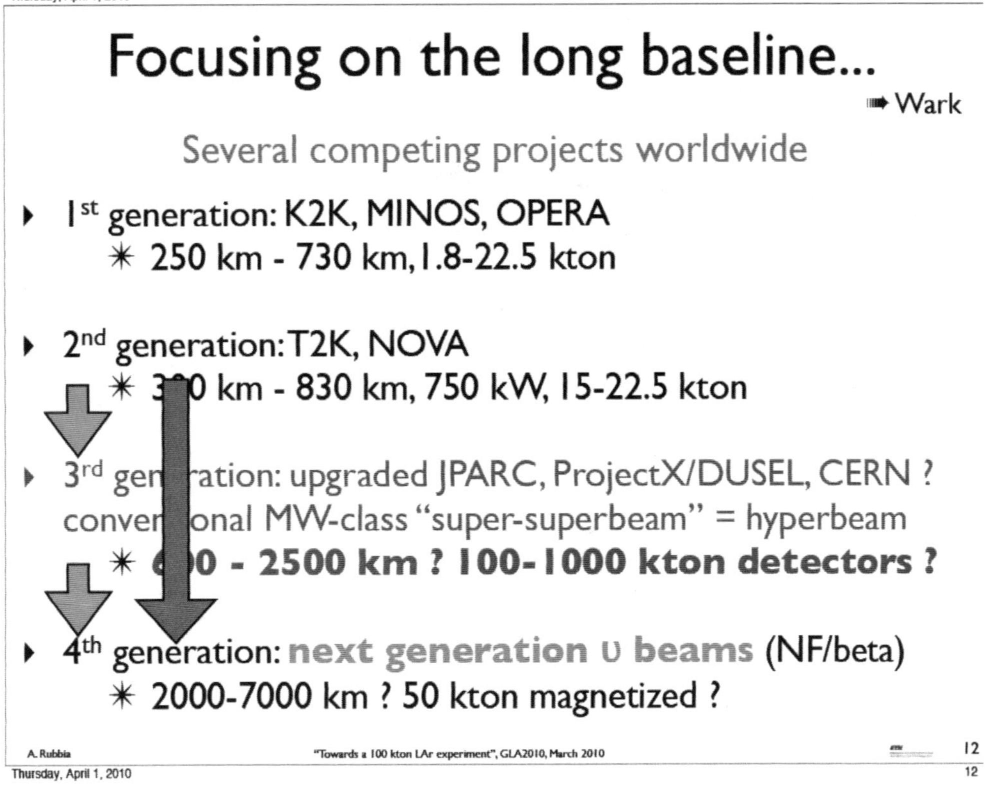
Focusing on the long baseline...
➠ Wark
Several competing projects worldwide
1st generation: K2K, MINOS, OPERA
250 km - 730 km, 1.8-22.5 kton
2nd generation: T2K, NOVA
km - 830 km, 750 kW, 15-22.5 kton
3rd ge ration: upgraded JPARC, ProjectX/DUSEL, CERN ?
conve onal MW-class "super-superbeam" = hyperbeam
0 - 2500 km ? 100-1000 kton detectors ?
4th generation: next generation ν beams (NF/beta)
2000-7000 km ? 50 kton magnetized ?
A. Rubbia
"Towards a 100 kton LAr experiment", GLA2010, March 2010
12
Thursday, April 1, 2010
12

Focusing on the long baseline...

➠ Wark

Several competing projects worldwide

- 1st generation: K2K, MINOS, OPERA
  - ✳ 250 km - 730 km, 1.8-22.5 kton
- 2nd generation: T2K, NOVA
  - ✳ 3 0 km - 830 km, 750 kW, 15-22.5 kton
- 3rd gen ration: upgraded JPARC, ProjectX/DUSEL, CERN ? conver o r-superbeam" = hyperbeam
  - ✳ 6 0 - 2500 km ? 100-1000 kton detectors ?

Skip generation ?

- 4th generation: **next generation ν beams** (NF/beta)
  - ✳ 2000-7000 km ? 50 kton magnetized ?

A. Rubbia "Towards a 100 kton LAr experiment", GLA2010, March 2010 12

Thursday, April 1, 2010 12

Inescapable conclusion

In case θ_{13} is found soon → immediately proceed towards measurement of δ_{CP} phase with high power conventional superbeam

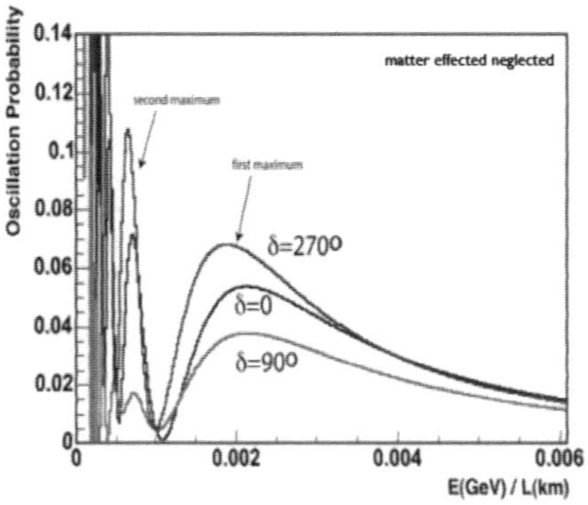

- Effect of CP Phase δ_{CP} appears as
  - ☆ ν_e Appearance Energy Spectrum Shape in Wide Band Beam
    - Peak position and height for 1st, 2nd maximum and minimum
    - Sensitive to all the non-vanishing δ including 180°
    - Investigate CP phase with ν run only, but need WBB
    - Need very good energy resolution
  - ☆ Difference between ν_e and $\bar{\nu}_e$ Appearance Behaviors
    - Also in Narrow Band Beam (off-axis)
    - Need both beam polarities with similar statistics to study effect

A. Rubbia "Towards a 100 kton LAr experiment", GLA2010, March 2010 13

Thursday, April 1, 2010 13

Journal of Physics: Conference Series **308** (2011) 012030 doi:10.1088/1742-6596/308/1/012030

J-PARC to Okinoshima

w/ 100 kton Giant Liquid Argon Charge Imaging Experiment

arXiv:0804.2111

☆ 1.66 MW J-PARC upgraded neutrino beam:

- If $\nu_\mu \rightarrow \nu_e$ signal is found in T2K ($\sin^2 2\theta_{13} > \approx 0.01$) experiment will directly address δ_{CP}
- If $\nu_\mu \rightarrow \nu_e$ signal is not found in T2K, continue the quest with improved sensitivities, down to $\sin^2 2\theta_{13} \approx 0.001$

☆ Underground detector (rock overburden >600 m.w.e.):

- Proton lifetime
- Atmospheric neutrinos
- Astrophysical sources of neutrinos

A. Rubbia "Towards a 100 kton LAr experiment", GLA2010, March 2010 14

Thursday, April 1, 2010 14

J-PARC to Okinoshima

P32 proposal (LAr TPC R&D) recommended by J-PARC PAC (Jan 2010)

Distance = 658 km

Off-axis angle = 0.76°

(2.5° @ SK)

100 kton liquid Argon

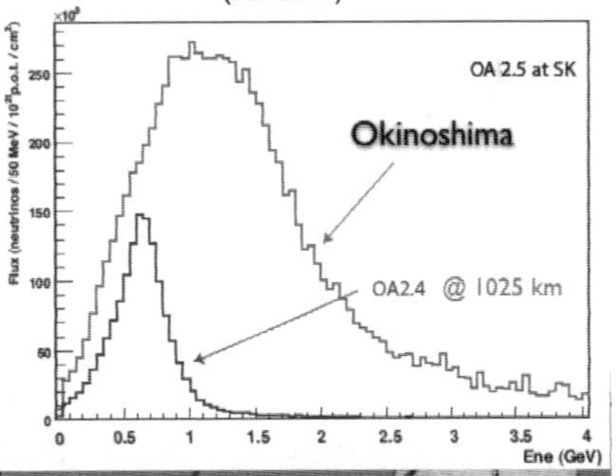

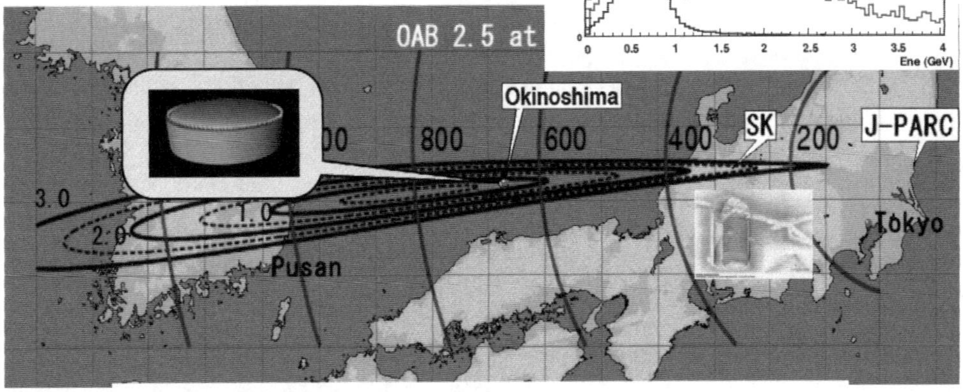

→ Extract δ_{CP} from fit of 1st & 2nd maximum

A. Rubbia 15

Thursday, April 1, 2010 15

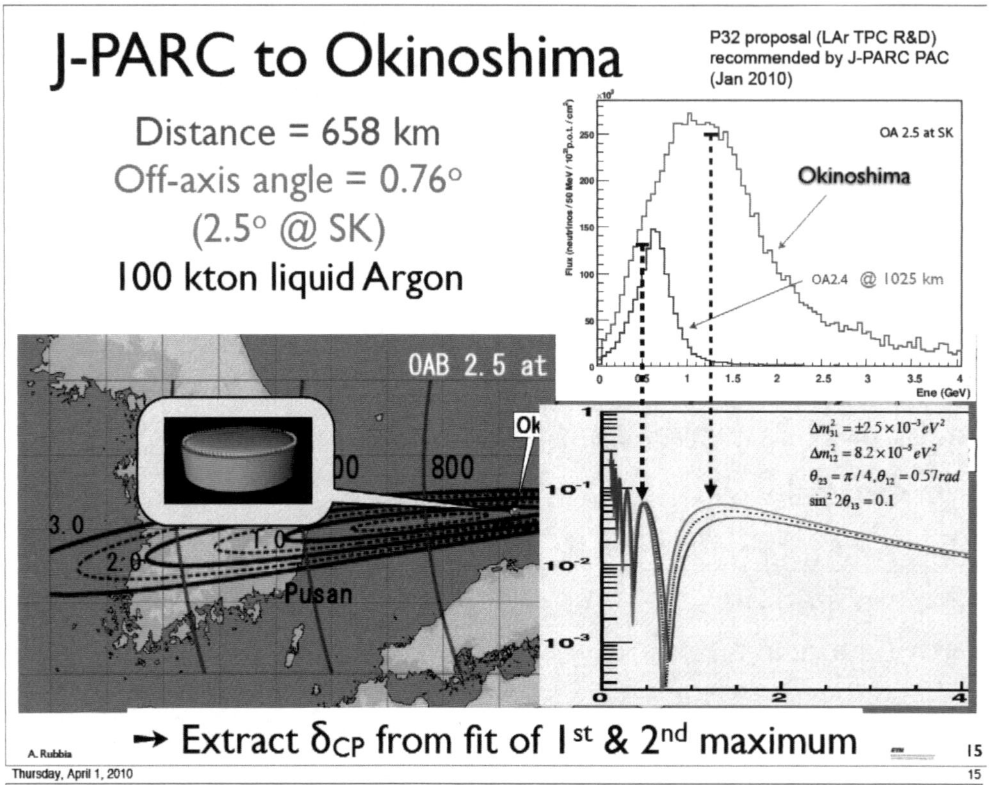
J-PARC to Okinoshima
P32 proposal (LAr TPC R&D) recommended by J-PARC PAC (Jan 2010)
Distance = 658 km
Off-axis angle = 0.76°
(2.5° @ SK)
100 kton liquid Argon
OA 2.5 at SK
Okinoshima
OA2.4 @ 1025 km
Ene (GeV)
OAB 2.5 at
Pusan
800
3.0
2.0
1.0
$\Delta m^2_{31} = \pm 2.5\times10^{-3} eV^2$
$\Delta m^2_{12} = 8.2\times10^{-5} eV^2$
$\theta_{23} = \pi/4, \theta_{12} = 0.57 rad$
$\sin^2 2\theta_{13} = 0.1$
→ Extract δ_{CP} from fit of 1st & 2nd maximum
A. Rubbia
15
Thursday, April 1, 2010

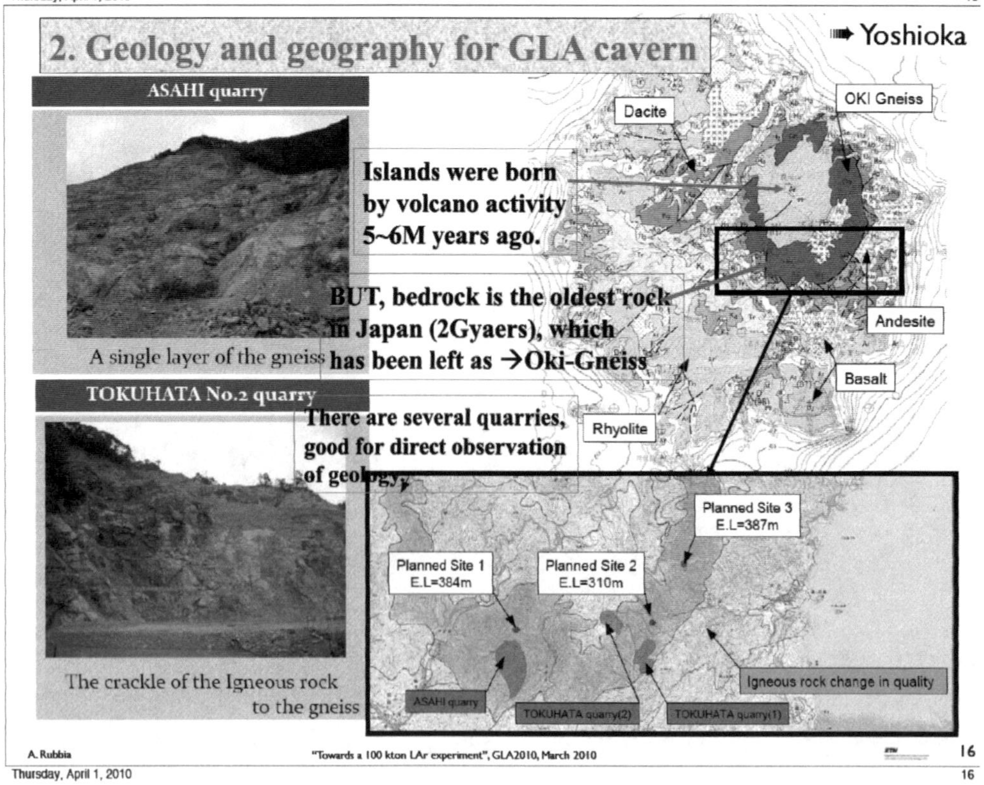
2. Geology and geography for GLA cavern
→ Yoshioka
ASAHI quarry
A single layer of the gneiss
TOKUHATA No.2 quarry
The crackle of the Igneous rock to the gneiss
Islands were born by volcano activity 5~6M years ago.
BUT, bedrock is the oldest rock in Japan (2Gyaers), which has been left as →Oki-Gneiss
There are several quarries, good for direct observation of geology.
Dacite
OKI Gneiss
Andesite
Basalt
Rhyolite
Planned Site 1 E.L=384m
Planned Site 2 E.L=310m
Planned Site 3 E.L=387m
Igneous rock change in quality
ASAHI quarry
TOKUHATA quarry(2)
TOKUHATA quarry(1)
A. Rubbia
"Towards a 100 kton LAr experiment", GLA2010, March 2010
16
Thursday, April 1, 2010

Expected rates at Okinoshima

Events in 100 kton, 658 km, 5 years @ 1.66 MW

	ν_μCC	ν_eCC	$\bar{\nu}_\mu$CC	$\bar{\nu}_e$CC
Beam components (null oscillation)	82000	750	1460	35
	$\nu_\mu \to \nu_e$ oscillations			
$\delta_{CP} =$	0°	90°	180°	270°
$\sin^2 2\theta_{13} = 0.1$	2867	2062	2659	3464
$\sin^2 2\theta_{13} = 0.05$	1489	1119	1342	1908
$\sin^2 2\theta_{13} = 0.03$	942	506	829	1266

→ x10 more statistics than in T2K
→ Electron appearance with high signal efficiency, strong background rejection and good energy resolution:
- $\sin^2 2\theta_{13} < 0.001$ @ 90%CL (if no signal is found)
- measure θ_{13} with $\delta\sin^2 2\theta_{13} \approx \pm 0.01$ and test CP at better than 90%CL if $\sin^2 2\theta_{13} > 0.01$
- CP discovery exceeds 3σ for $\sin^2 2\theta_{13} > 0.02$

J-PARC P32 (Jan2010)

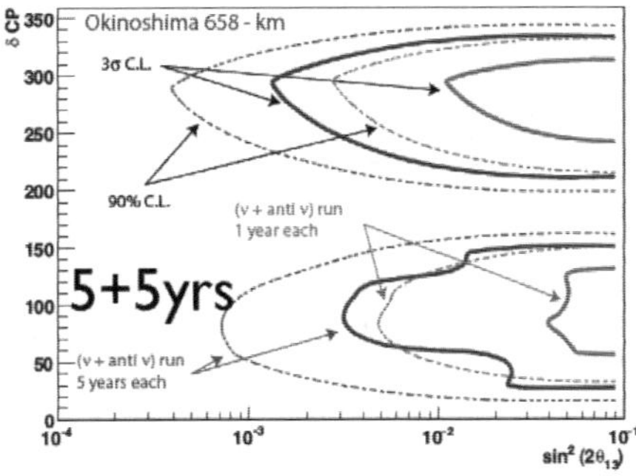

LBNE / LAr20 / DUSEL

➠ see Fleming/ Baller

- Plan to build very massive underground detector at the Deep Underground Science and Engineering Laboratory in the Homestake mine
- New beamline from FNAL to DUSEL (700kW SNUMI →2.4MW with ProjectX)
- Two detector options: WC & LAr20
- CD-0 mission statement in January 2010
- CD-1 foreseen by end 2010 ("technology choice" / aggressive schedule)
  - ☑ integrated plan for LAr
  - ☑ impressive progress in design & engineering effort for LBNE
  - ☑ some aspects of design are voluntarily "conservative" (e.g. 2.5 m drift)
  - ☑ embedded cold electronics *"to save the features of the LAr TPC"* ➠ Radeka
- Looking forward to seeing CD-2 decision process!

LBNE/DUSEL

➡ Baller

how to safeguard the science ?

ProjectX 2.4 MW, LAr50@DUSEL, L=1300km

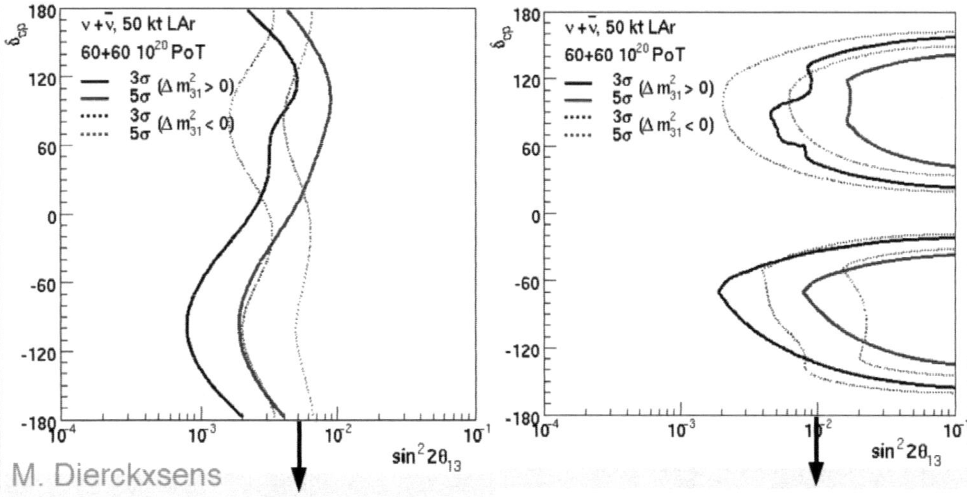

0.005 0.01

A. Rubbia "Towards a 100 kton LAr experiment", GLA2010, March 2010 19

Thursday, April 1, 2010 19

In Europe ➡ LAGUNA

EC Grant Agreement No. 212343 FP7-INFRA-2007-1

- LAGUNA design study addresses feasibility of large underground detectors at 7 potential sites located in Finland (Pyhäsalmi), France (Fréjus), Italy (Umbria region), Poland (Sieroszowice), Romania (Slanic), Spain (Canfranc) and United Kingdom (Boulby).
- Baselines from CERN: 130 km(Fréjus) to 2300 km(Pyhäsalmi)
  ➡ CERN-Slanic & CERN-Pyhäsalmi offer very long baselines not considered elsewhere in the world → unique physics opportunities
- Three detector options: GLACIER, LENA or MEMPHYS
- LAGUNA consortium will prioritize sites by summer 2010, based on technical, scientific, environmental and political criteria
- Next-step: LAGUNA-next to be submitted in December 2010
- **But:** next generation long-baseline program in Europe needs a new CERN high-intensity, high-energy proton source and high power neutrino beam line (arXiv:1003.1921) → ... SPS 400 GeV provides highest power options until new accelerator is built

A. Rubbia "Towards a 100 kton LAr experiment", GLA2010, March 2010 20

Thursday, April 1, 2010 20

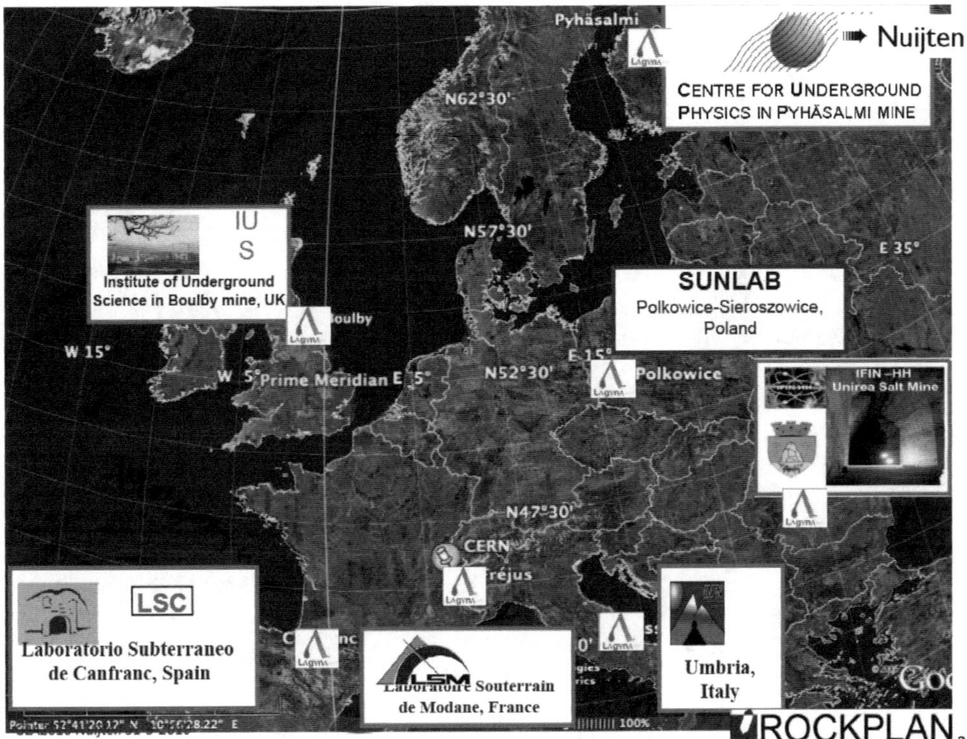

A. Rubbia "Towards a 100 kton LAr experiment", GLA2010, March 2010 21

Thursday, April 1, 2010 21

Nucleon decay searches

JHEP 0704:041,2007

- Profit from excellent tracking, energy resolution, low particle detection threshold, great background discrimination
    - Several (unbiased) decay channels should be accessible with lifetime sensitivities in the range 10^{34}-10^{35} years (in 10 years) with small or quasi-absent backgrounds (discovery at the few events level)
- Simulation conclude that shallow depth (>600 m.w.e.) sufficient to reduce cosmogenic background
- Dominant background is several 100'000's of atmospheric $\nu\&\bar{\nu}$ interactions for 10 year exposure
- Many promising channels: e.g. $p\rightarrow\nu K^+$, $p\rightarrow e^+K^0$, $n\rightarrow e^+K^-$, ...
- Also multi-prong configurations (e.g. $p\rightarrow\mu\pi K$, ...)

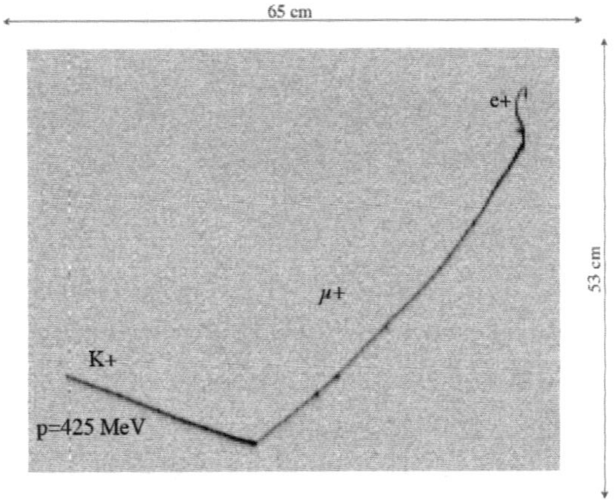

A. Rubbia "Towards a 100 kton LAr experiment", GLA2010, March 2010 22

Thursday, April 1, 2010 22

Proton decay sensitivity curves

Consider two "typical" decay modes (more should be explored):

Curves adapted from E. Kearns, NPo8

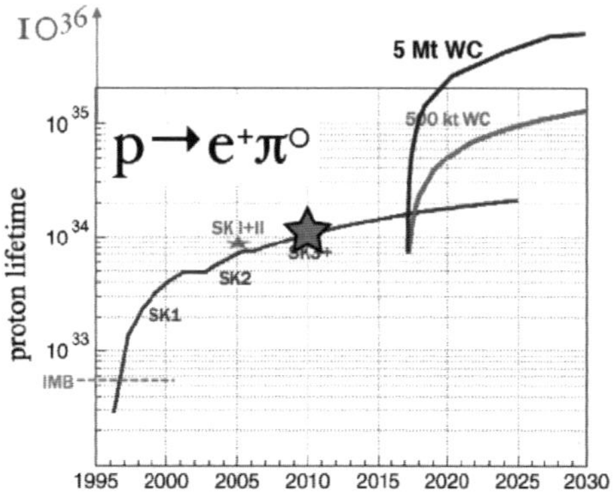

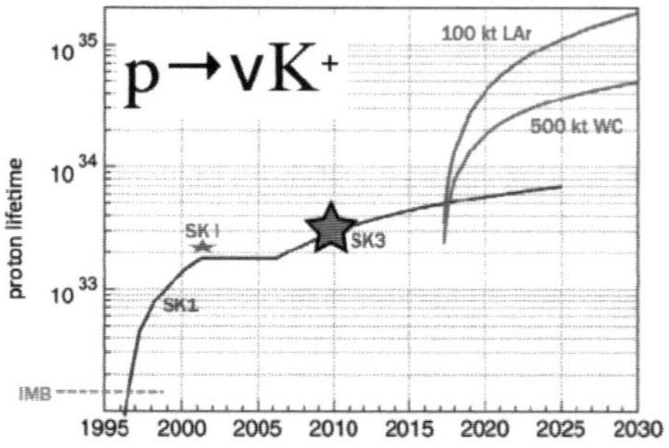

up to 10^{36} yrs ? Theoretical hints: **up to 10^{35} yrs ?**

$$\tau(p \to \pi^0 + e^+) \approx 5 \times 10^{36} \left(\frac{M_X}{3 \times 10^{16}\ \mathrm{GeV}}\right)^4 \left(\frac{0.015\ \mathrm{GeV}^3}{\beta_{lattice}}\right)^2 \text{ years.}$$

$$\tau(p \to K^+ + \bar{\nu}) < \left(\frac{1}{3} - 3\right) \times 10^{34} \left(\frac{0.015\ \mathrm{GeV}^3}{\beta_{lattice}}\right)^2 \text{ years.}$$

Dorsner, Perez, Phys.Lett.B625:88-95,2005

A. Rubbia "Towards a 100 kton LAr experiment", GLA2010, March 2010

Thursday, April 1, 2010 23

All other channels

➠ Kajita, Gil-Botella

Event rates in 100 kton LAr detector

☑ **Atmospheric neutrinos**

- 9000 e/year and 15000 μ/year(no osc) and L/E reconstruction
- 3300 ν NC/year and clear π0 identification
- 100 ντ CC/year from oscillations identified by kinematical analysis
- ν/$\bar{\nu}$ statistically separated in QE events

☑ **Solar neutrinos**

- 200000 events / year @ E>5 MeV

☑ **Supernova neutrinos**

- 25000 νeCC, 30000 νNC, 1000 ES, 540 $\overline{\nu e}$CC events at d=10kpc
- CC rates strongly sensitive to oscillations
- νe/$\overline{\nu e}$/νx flavor separation
- neutronization visible
- relic S/N ≈ 60/30 in 5 yrs (16-40 MeV window)

A. Rubbia "Towards a 100 kton LAr experiment", GLA2010, March 2010 24

Thursday, April 1, 2010 24

The ICARUS LAr TPC

➠ C.Rubbia

- Modern "bubble-chamber"
- Tracking device
  - ✦ Precise event topology
  - ✦ Low particle detection threshold
- Measurement of local energy deposition
  - ✦ Particle ID with dE/dx & range
  - ✦ e/π0 separation
- Full sampling, homogeneous calorimeter
- ✦ Excellent energy reconstruction

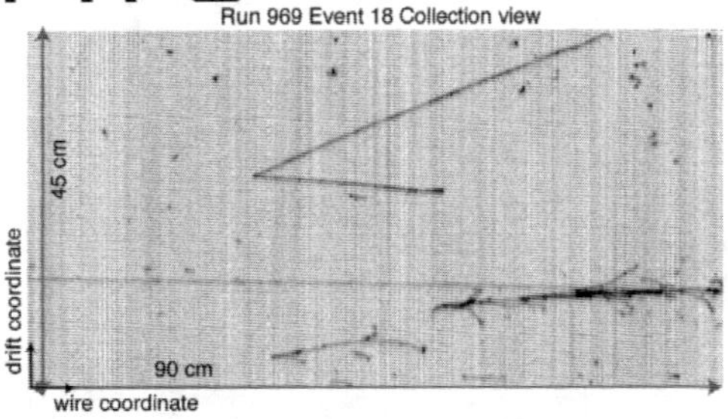

Successful demonstration of T300 on surface in Pavia (2001)

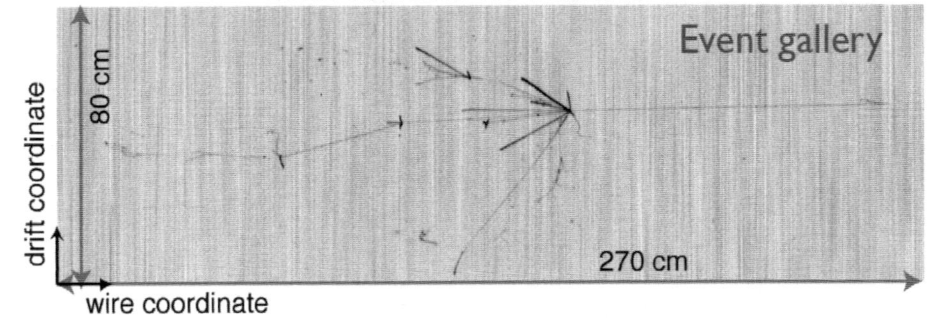

A. Rubbia "Towards a 100 kton LAr experiment", GLA2010, March 2010 25

Thursday, April 1, 2010 25

ICARUS (CNGS2): the first large scale LAr experiment

- ICARUS represents a major milestone in the practical realization of a large scale LAr detector. Successfully operated on surface in Pavia in 2002, will soon be operational in the underground HallB of LNGS.
- The T600 at LNGS will collect simultaneously "bubble chamber like" neutrino events events of different nature
- Cosmic ray events
  - ≈ 100 ev/year of unbiased atmospheric CC neutrinos.
  - Solar neutrino electron rates >5 MeV. ~1-2 ev/day
  - Supernovae neutrinos.
  - A zero background proton decay with 3 x 10^{32} nucleons for "exotic" channels.
- CERN beam associated events: 1200 n_m CC ev/y and 7-8 n_e CC ev/year
  - Observation of neu-tau events in the electron channel (with sensitivity comparable to OPERA
  - A search fo sterile neutrinos
- Other unexpected phenomena

Looking forward to see start of physics programme !

Rubbia, KEK_March2010 Slide: 44

Thursday, April 1, 2010 26

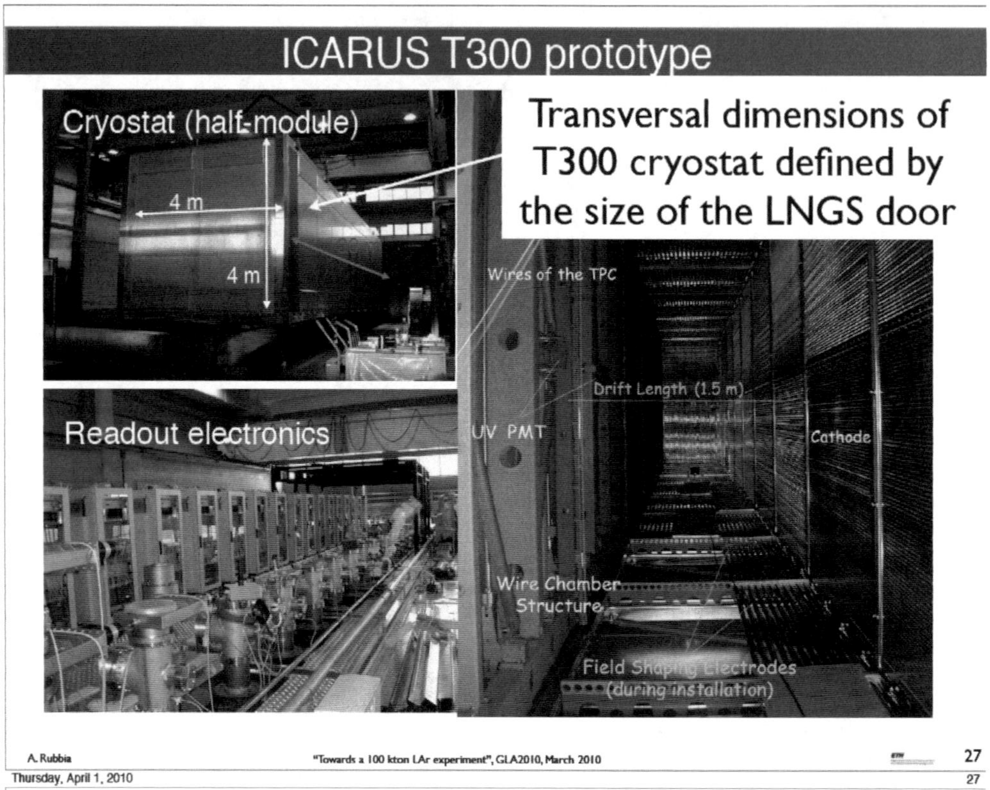
ICARUS T300 prototype
Cryostat (half-module)
4 m
4 m
Transversal dimensions of T300 cryostat defined by the size of the LNGS door
Wires of the TPC
Drift Length (1.5 m)
UV PMT
Cathode
Readout electronics
Wire Chamber Structure
Field Shaping Electrodes (during installation)
A. Rubbia
"Towards a 100 kton LAr experiment", GLA2010, March 2010
27
Thursday, April 1, 2010
27

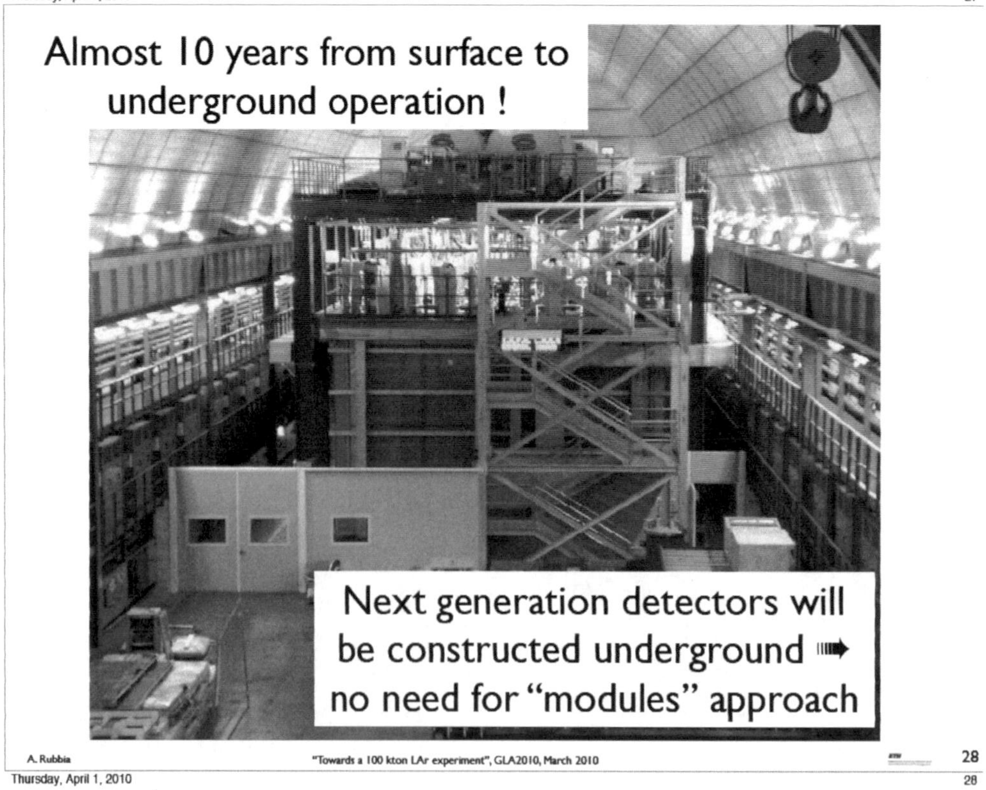
Almost 10 years from surface to underground operation !
Next generation detectors will be constructed underground ➠ no need for "modules" approach
A. Rubbia
"Towards a 100 kton LAr experiment", GLA2010, March 2010
28
Thursday, April 1, 2010
28

Journal of Physics: Conference Series **308** (2011) 012030 doi:10.1088/1742-6596/308/1/012030

Concepts for Giant LAr detectors

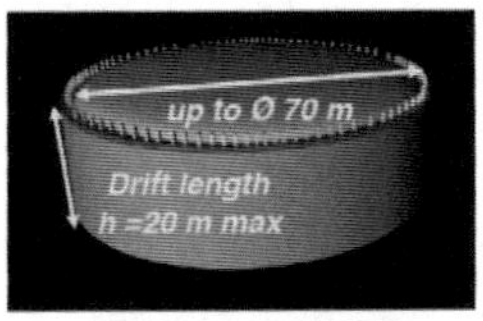

GLACIER
2003
LNG-tank up
to 100 kton

evacuable
LANND
2006

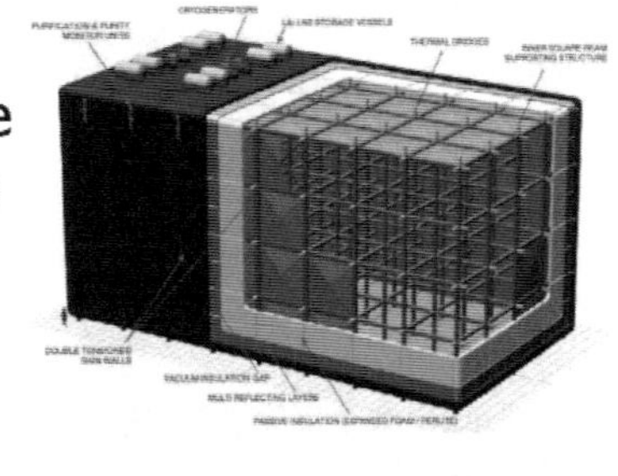

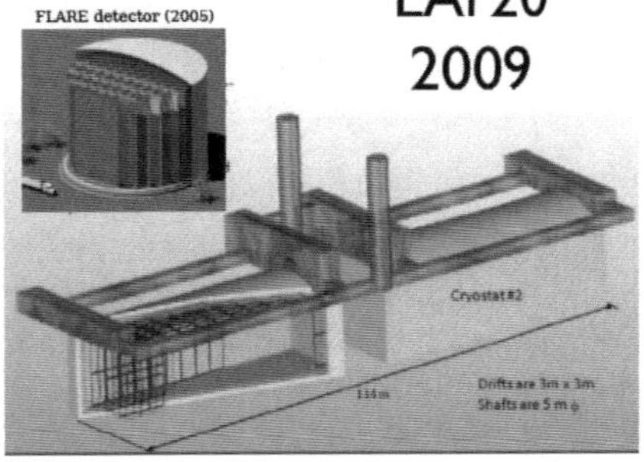

LAr20
2009

MODULAr
2008
10 kton

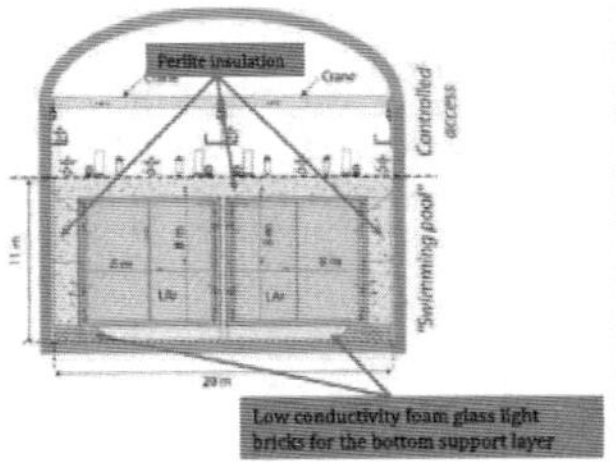

A. Rubbia "Towards a 100 kton LAr experiment", GLA2010, March 2010 29

Thursday, April 1, 2010 29

Giant underground caverns

➠ Nuijten, Yoshioka, Rameika

☑ USA LAr20 @ DUSEL

- CD-1 → CD-2 approval process

☑ Japan Okinoshima

- Site visit; potential location found, feasibility study started and preliminary concept defined (Penta-Ocean Co)

☑ Europe 7 potential sites studied in LAGUNA

- FP7 design study to select among promising locations in Europe: Pyhäsalmi, Slanic, Sieroszowice, Umbria, Canfranc; concepts and prices for main cavern excavation, ancillary infrastructure and access tunnel available

☑ For safety horizontal access is probably easier than vertical !

A. Rubbia Towards a 100 kton LAr experiment 30

Thursday, April 1, 2010 30

GLACIER tank

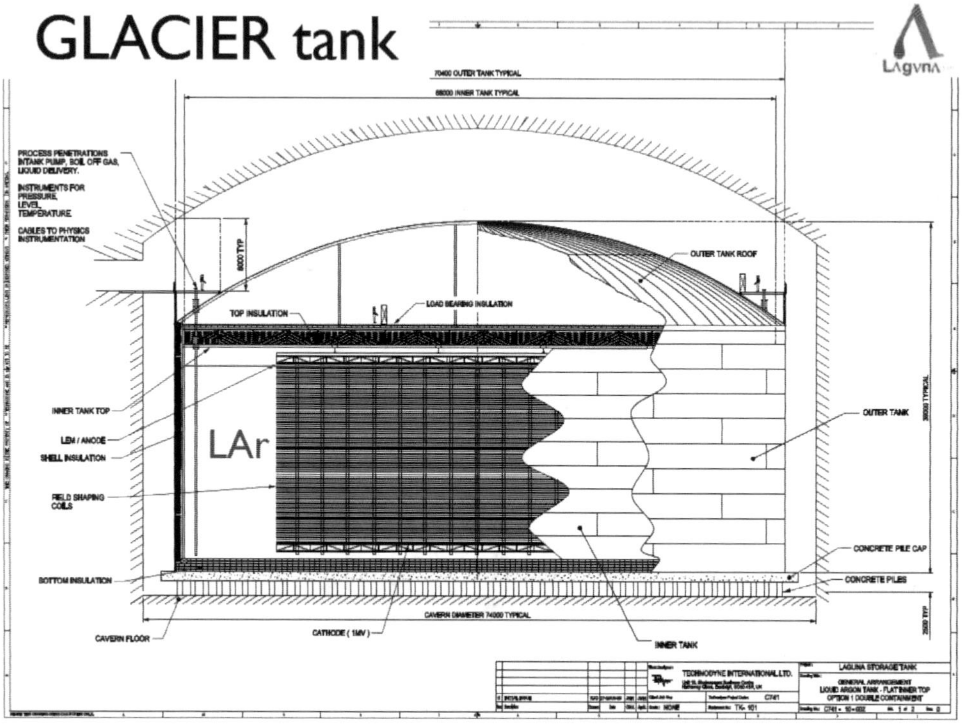

A. Rubbia Towards a 100 kton LAr experiment 31

Thursday, April 1, 2010 31

Style variations

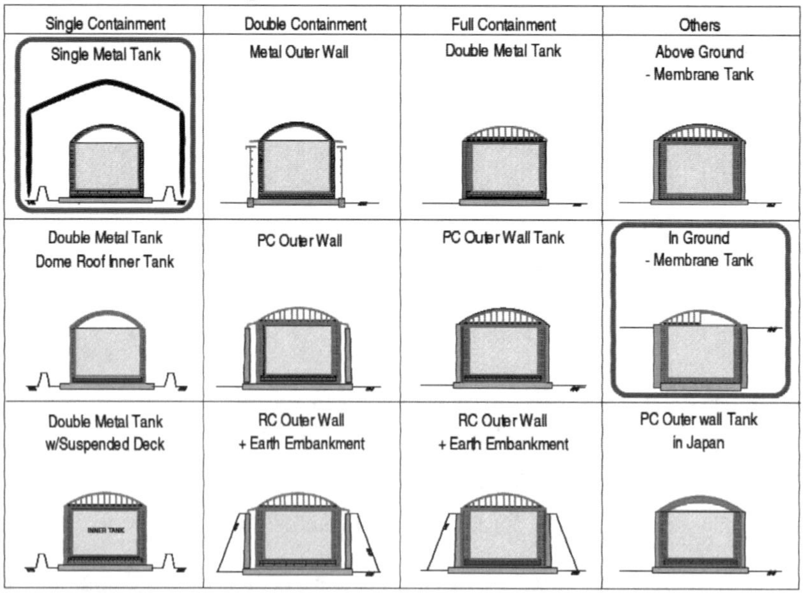

PC = prestressed concrete, RC= reinforced concrete

A. Rubbia Towards a 100 kton LAr experiment 32

Thursday, April 1, 2010 32

DUSEL (LBNE)

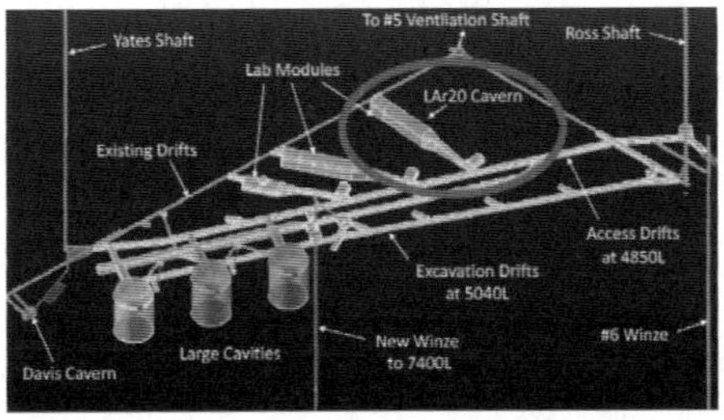

Pyhäsalmi (Rockplan)

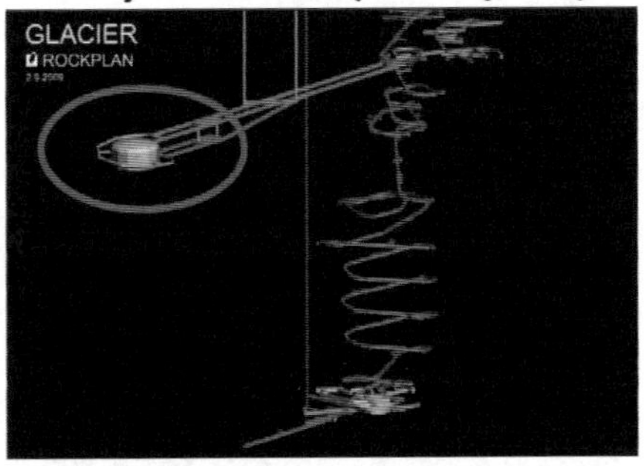

Sieroszwice (KGHM Cuprum)

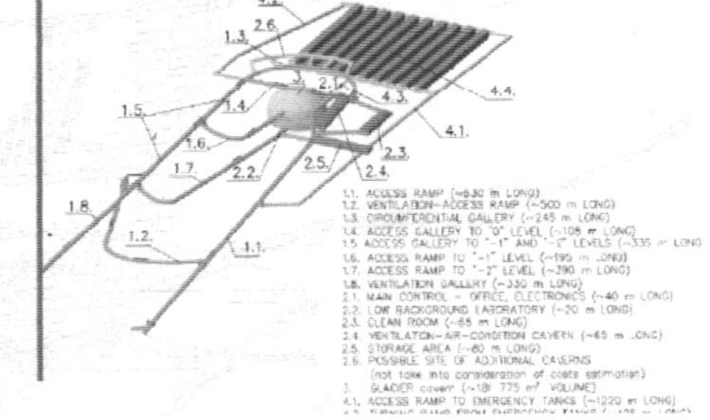

Okinoshima (Penta-Ocean)

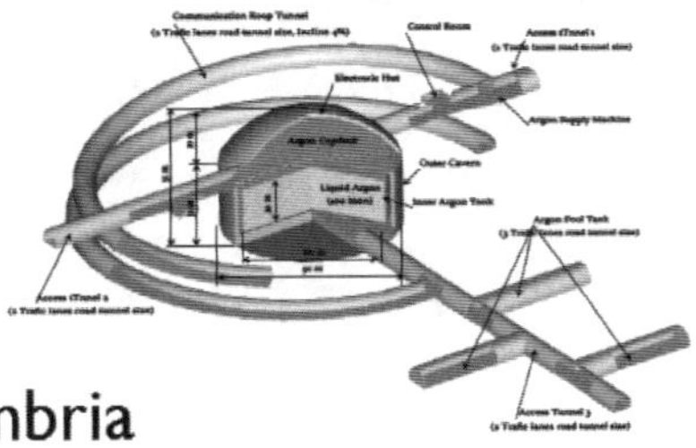

Also Canfranc, Fréjus, Slanic, Umbria

A word on procurement...

➠ Nuijten & Yoshioka

- 100 kton ≈ 4500 trucks ≈ 6 trucks / day × 5/7 days × 3 years ➠ advantageous to procure in parallel to cavern excavation (duration of environmental impact reduced). On island → transport by small tanker
- Estimated LAr yearly production: Japan 300kt/yr, Italy 100kt/yr, ... entirely used by local industries (big user: steel)
- Local liquefaction is solution only if associated nitrogen & oxygen can be commercially exploited ➠ purchase bulk argon
- Need several years to procure LAr ➠ store and purify locally in storage tanks before transfer to main detector tank
- During detector operation boil-off: ≈0.03% / day ≈ 30 tons / day ≈ 60kW @ 87 K re-condensed by local cryocoolers ≈ 6MW electrical power

More small scale R&D efforts

- R&D efforts targeted at developing further the technology towards large scale detectors and assessing performance
  (1) on particle beams (2) but also for DM searches
  - ICARUS 50L ➠Cavanna, ArgoNEUT@FNAL ➠Bromberg, 250L@JPARC ➠Maruyama
  - WArP/ArDM/DEAP/CLEAN but also MEG/XENON/XMASS ➠Montanari, Marchionni, Boulay, Mihara, Giboni, Sekiya...
  - LAPD, 6m3@CERN, ArgonTUBE ➠Rebel, Curioni, Rossi

ArgoNEUT @ FNAL

➠ Fleming, Bromberg

- First neutrino interactions in US LAr TPC !
  - 1/4-ton TPC, 3 wire planes, 2 readout planes (wires ± 30° wrt vertical)
  - 4 mm pitch, 4 mm plane separation, 240 wires/plane
  - Cryo-cooler driven purification system, ~ 500 μs e⁻ lifetime (PM)
  - Bias voltage distribution/decoupling caps in the LAr

~175 lt on NuMI ν beam

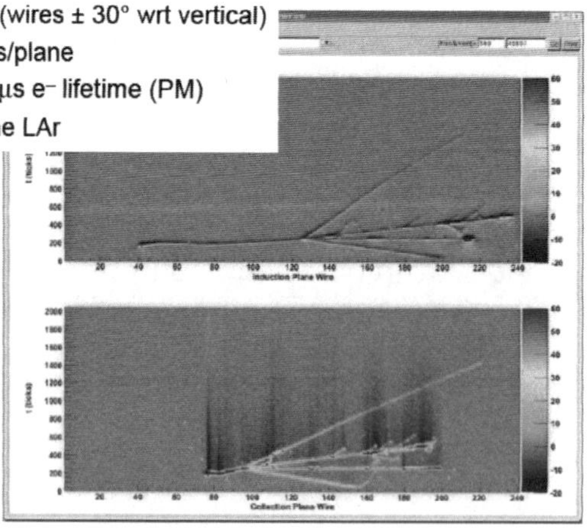

Analysis of several 1000's of υ events in progress !

Journal of Physics: Conference Series **308** (2011) 012030 doi:10.1088/1742-6596/308/1/012030

Bromberg

ArgoNeuT

Raw Data

- Electronics
  - "Long" 3.5 μs peaking time, both planes
  - Unipolar collection, bipolar induction
  - Both planes useful for ionization

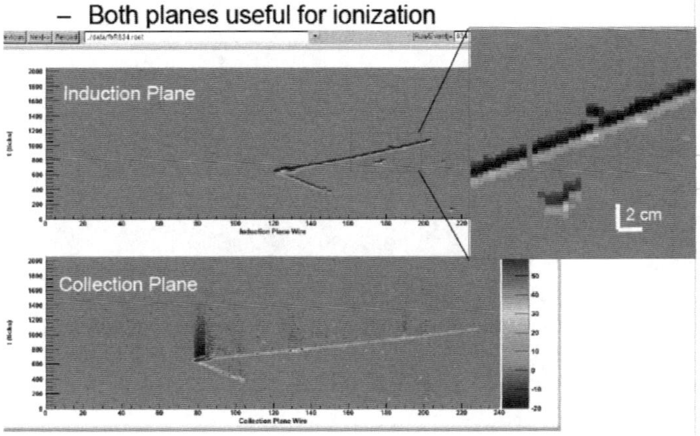

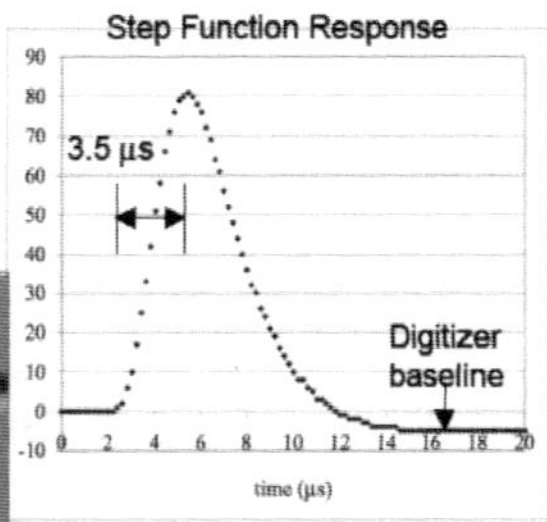

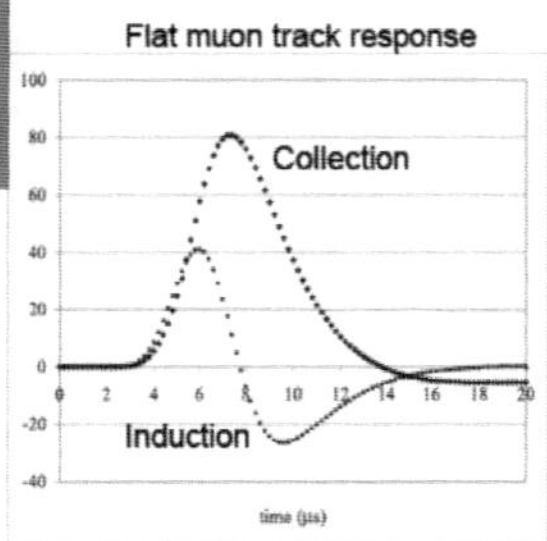

LAr20 Reference Design Parameters

- Fiducial mass = 1/6 of 100kt Water Cherenkov → 16.7kt (LAr20)
  - Size ~20m x ~20m x ~40m
- Drift distance 2.5m
- 2 or 3 wire planes, wire spacing = 3 - 5 mm
  - Cost-benefit decision (ν_e ID efficiency, NCπ^o rejection)
  - ~500k channels for 2 wire planes with 5mm spacing

Cryostat Design: "Warm" vs "Cold" Electronics

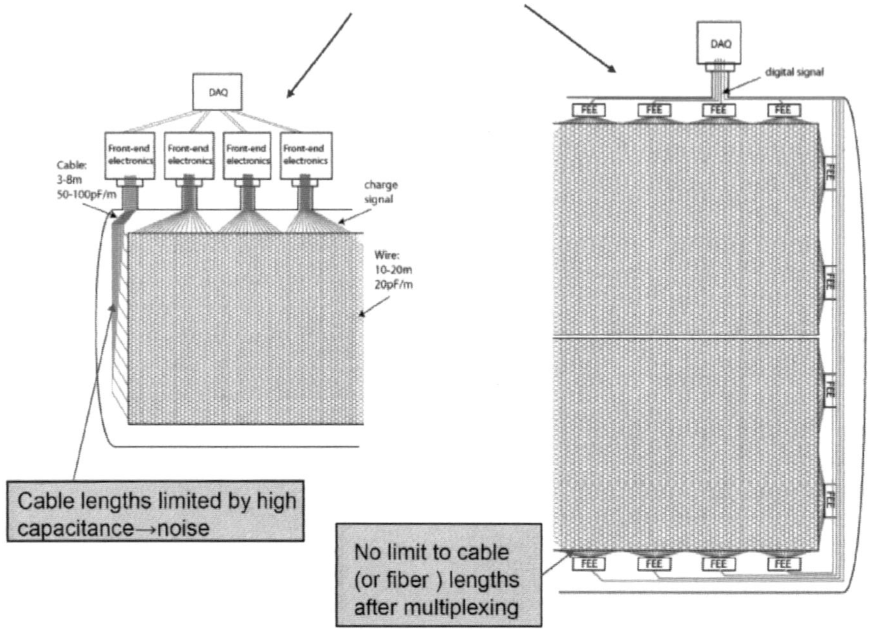

A. Rubbia "Towards a 100 kton LAr experiment", GLA2010, March 2010 39

Thursday, April 1, 2010 39

TPC wire noise considerations

First transistor (wire capacitance, bandwidth) +
wire thermal noise (wire resistivity) +
attenuation along drift path

yields a loss in physics performance

Radeka et al., June 2009

Collection plane S/N for cold electronics with low resistance (copper+gold plated) wire, 3 ohm/m					
		wire spacing >	3	4	5
Wire length	Noise	Signal >	6000	8000	10000
5	500	S/N >	12	16	20
10	650	S/N >	9.2	12.3	15.3
15	800	S/N >	7.5	10	12.5
20	1000	S/N >	6	8	10

Collection plane S/N for warm electronics (+3m cable) with low resistance (copper+gold plated) wire, 3 ohm/m					
		wire spacing >	3	4	5
Wire length	Noise	Signal >	6000	8000	10000
5	1500	S/N >	4	5.3	6.7
10	1750	S/N >	3.4	4.6	5.7
15	2150	S/N >	2.8	3.7	4.7
20	2450	S/N >	2.4	3.3	4.1

A. Rubbia "Towards a 100 kton LAr experiment", GLA2010, March 2010 40

Thursday, April 1, 2010 40

➠ Radeka

Internal cold electronics

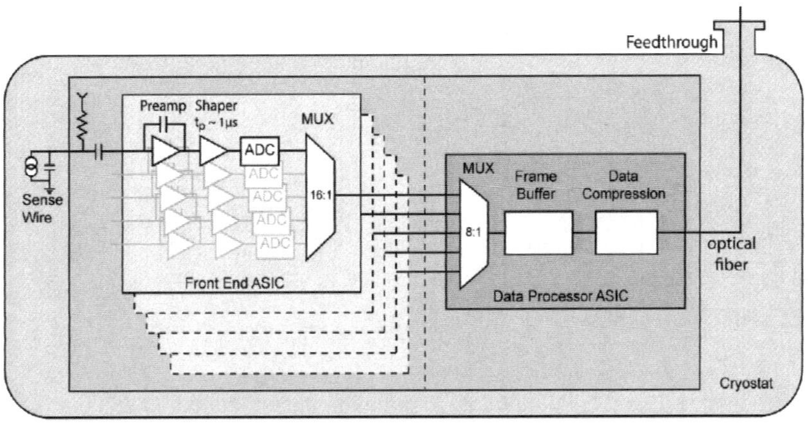

Fig. 8. A Functional Outline of a Multiplexed Readout Chain for Very Large LAr TPCs. Multiplexing will be performed in two steps at appropriate locations within the cryostat. A CMOS, or a BiCMOS technology with circuit design and operating conditions for long term operation in LAr will be used. A preliminary goal is multiplexing in two steps by 16 x 8=128. Power dissipation has been estimated to be ≤10mW/signal wire.

A. Rubbia "Towards a 100 kton LAr experiment", GLA2010, March 2010 41
Thursday, April 1, 2010 41

R&D towards alternative r/o methods

"to save the features of the LAr TPC"

➠ Resnati, Delbart, Tanaka, Spooner, ...

against loss of signal from diffusion, attachment to electronegative impurities, ...

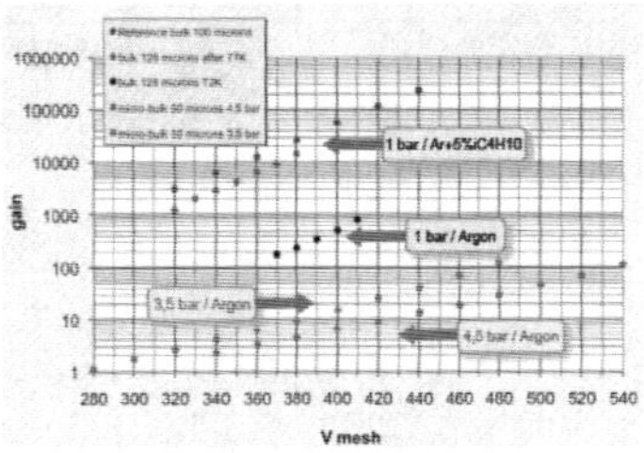

(1) double phase Ar LEM/THGEM TPC

(2) Double phase+ MicroMegas

(3) secondary scintillation from THGEM

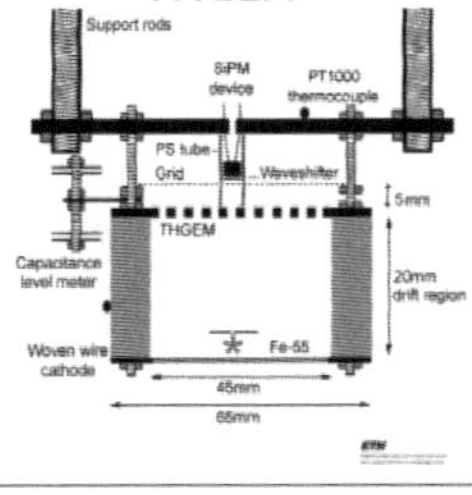

A. Rubbia Towards a 100 kton LAr experiment 42
Thursday, April 1, 2010 42

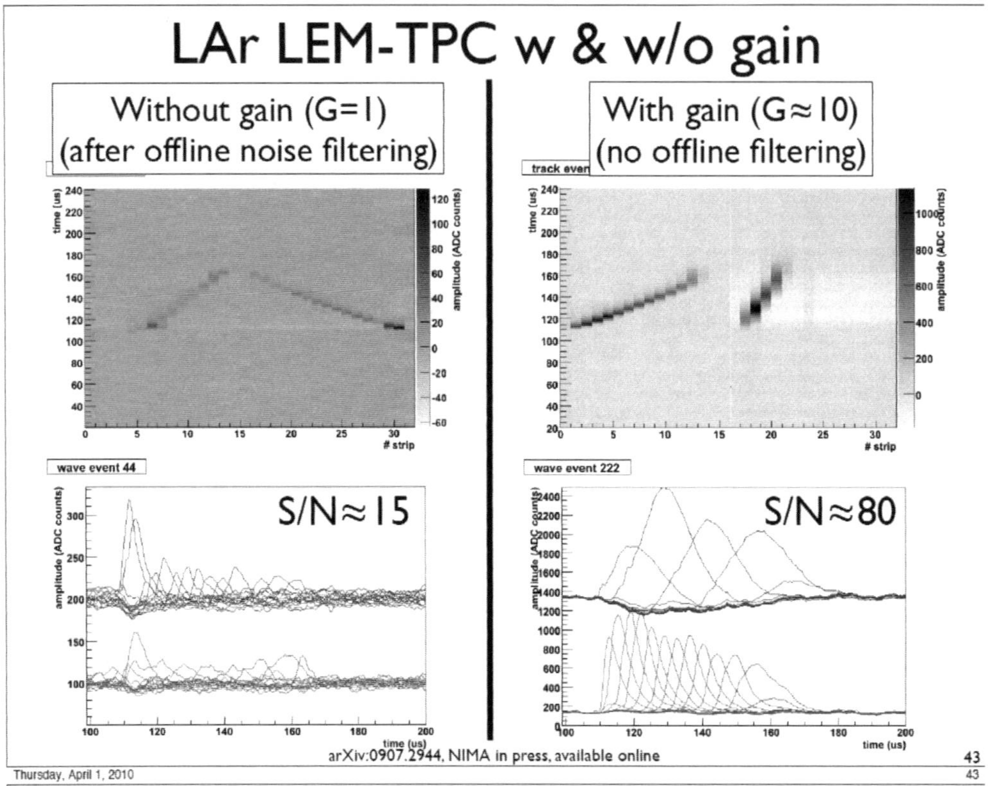
LAr LEM-TPC w & w/o gain
Without gain (G=1)
(after offline noise filtering)
With gain (G≈10)
(no offline filtering)
time (us)
amplitude (ADC counts)
strip
wave event 44
S/N≈15
wave event 222
S/N≈80
time (us)
arXiv:0907.2944, NIMA in press, available online
43
Thursday, April 1, 2010

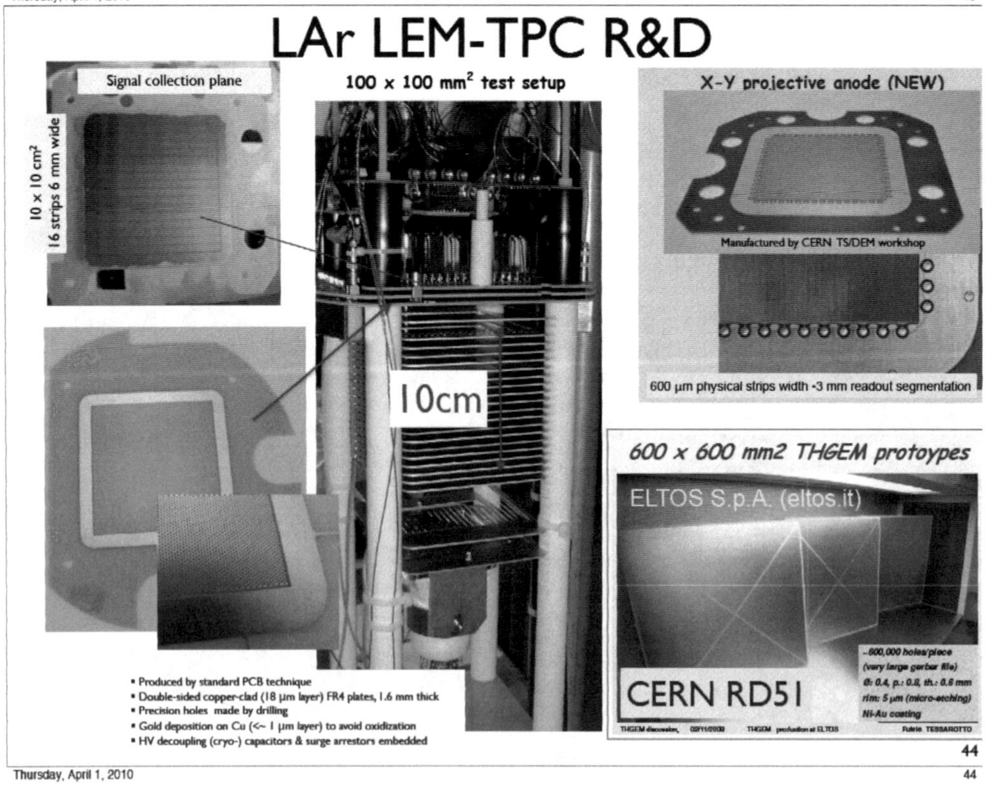
LAr LEM-TPC R&D
Signal collection plane
10 x 10 cm²
16 strips 6 mm wide
100 x 100 mm² test setup
10cm
X-Y projective anode (NEW)
Manufactured by CERN TS/DEM workshop
600 μm physical strips width -3 mm readout segmentation
600 x 600 mm2 THGEM protoypes
ELTOS S.p.A. (eltos.it)
CERN RD51
~600,000 holes/piece
(very large gerber file)
Ø: 0.4, p.: 0.8, th.: 0.6 mm
rim: 5 μm (micro-etching)
Ni-Au coating
▪ Produced by standard PCB technique
▪ Double-sided copper-clad (18 μm layer) FR4 plates, 1.6 mm thick
▪ Precision holes made by drilling
▪ Gold deposition on Cu (<~ 1 μm layer) to avoid oxidization
▪ HV decoupling (cryo-) capacitors & surge arrestors embedded
44
Thursday, April 1, 2010

Giant detectors need long drift paths

Challenge: very long drift paths ➡ 5m ➡ 10m... ➡ 20m...

Drift fields E=0.5,0.75,1,1.25,1.5 kV/cm

Papers on diffusion see e.g.
K. McDonald, LArTPC Document 532-v1

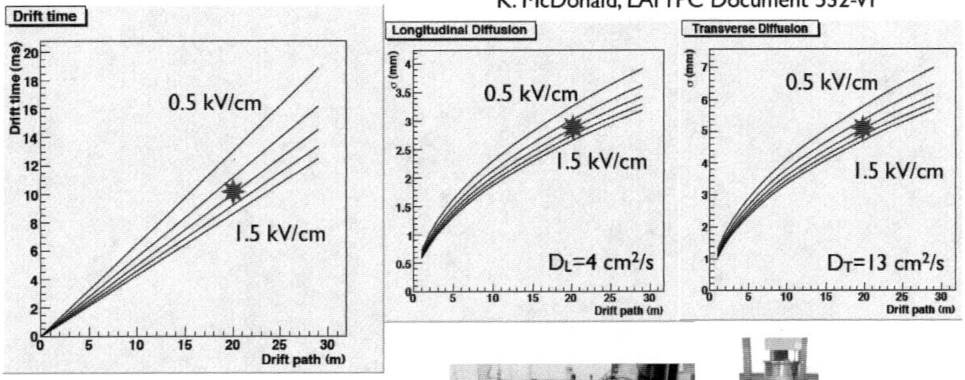

Experimental verification needed:
➡5m drift (F.Sergiampetri), ArgonTube (B. Rossi)

A. Rubbia
Towards a 100 kton LAr experiment
Thursday, April 1, 2010
45

Drift high voltage

➡ Horikawa

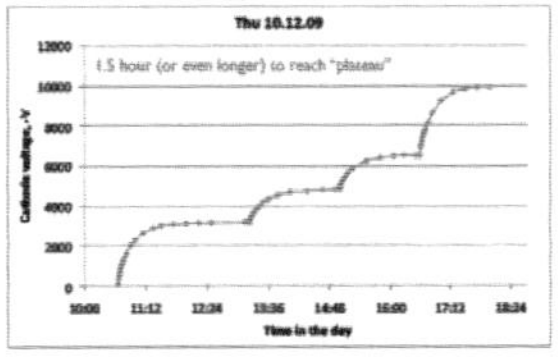

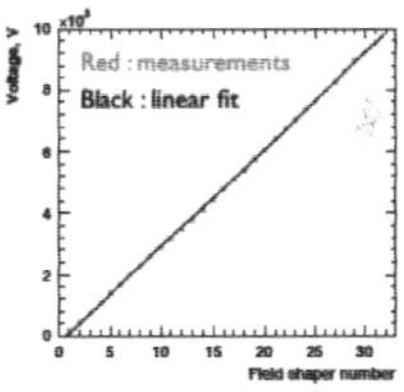

Extrapolation to long drift

Extrapolation of the ArDM design

Changing Cs for fixed Cp = 2.35 pF and Vpp-in = 2E = 2.5 kV

		ArDM				
Drift length	m	1.24	5	10		20
Total output voltage for 1 kV/cm	V	124k	500k	1M		2M
Input voltage Vpp-in = 2E	V	820	2.5k	2.5k	$\times\sqrt{2}$	3.5k
Shunt capacitance, Cp	F	2.35p	2.35p	2.35p	$\times 1/2$	1.18p
Capacitor	F	328/164n	475n	1.90μ		1.90μ
Number of stages, N	–	210	319	638		903
N per 10 cm	–	16.9	6.38	6.38		4.51
Total capacitance	F	125μ	303μ	2.43m		3.43m
Capacitance per 10 cm	F	10.4μ	5.99μ	24.3μ		17.2μ
Total stored energy	J	21.7	948	7.58k		21.5k

Actual ArDM parameters are given just for comparison.
For extrapolation, 2γN = 1.42 is always assumed.
LAr vaporization heat 160 kJ/kg

$$V_{\rm max} = \frac{E}{\gamma},\ \gamma \approx \sqrt{\frac{C_{\rm p}}{C_{\rm s}}}$$

A. Rubbia
Towards a 100 kton LAr experiment
46
Thursday, April 1, 2010
46

Journal of Physics: Conference Series **308** (2011) 012030 doi:10.1088/1742-6596/308/1/012030

And the purity issue...

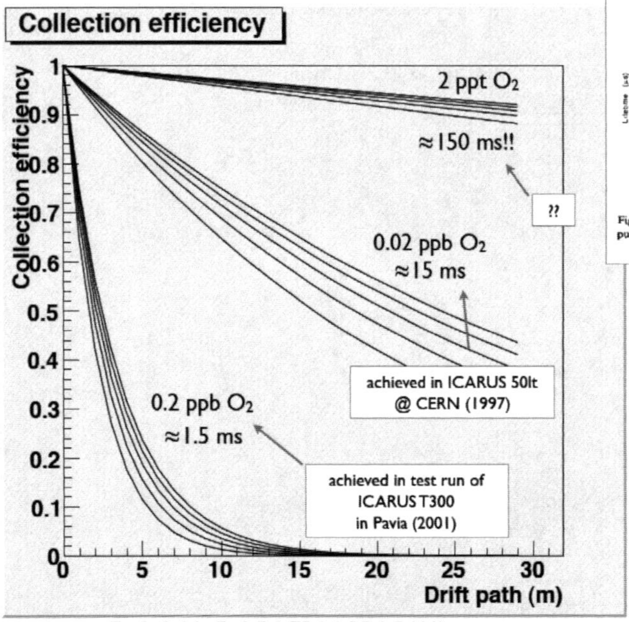

Drift fields E=0.5,0.75,1,1.25,1.5 kV/cm

A. Bettini et al., NIM A305 (1991) 177

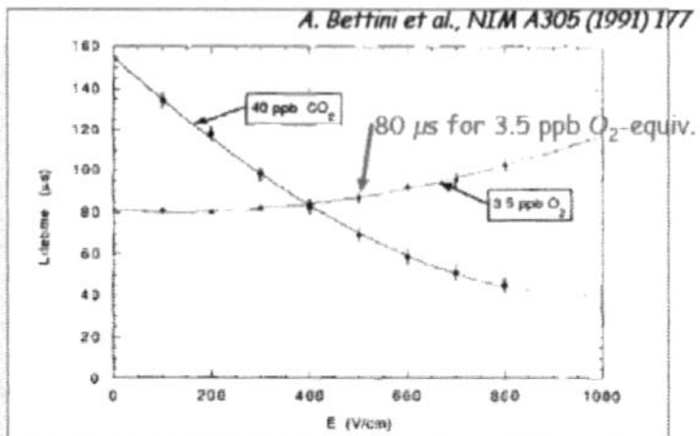

Fig. 12. Electric field dependence of the electron lifetime τ in purified LAr doped with 40 ppb CO_2 and 3.5 ppb O_2. Lines are drawn to guide the eye.

$$\tau \approx \frac{300\,\mu s}{O_2(ppb)}$$

More efforts needed to understand "ultimate" purity on large volumes

A. Rubbia — Towards a 100 kton LAr experiment — 47

Thursday, April 1, 2010 — 47

UV laser beams technique for LAr TPCs ➠ Rossi

The use of UV lasers to produce ionization tracks give some advantages

- lasers are cheap and transportable;
- provide good resolution in space over several meters;
- no production of δ-rays, no multiple scattering, no Landau fluctuation;
- no curvature in magnetic fields.

Ideal to monitor:

1. Possible drift electric field distorsion
2. LAr purity
3. Drift speed
4. Diffusion
5. Stability of operation over large volumes

For "double-phase" TPC:

1. Uniformity of the charge amplification over large volumes
2. Stability of the charge amplification in time

Could be also used to calibrate the detector

Biagio Rossi - University of Bern — GLA2010 - Tsukuba - 30 Mar 2010 — 3

A. Rubbia — Towards a 100 kton LAr experiment — 48

Thursday, April 1, 2010 — 48

➠ Rossi

UV laser beam-induced tracks events gallery

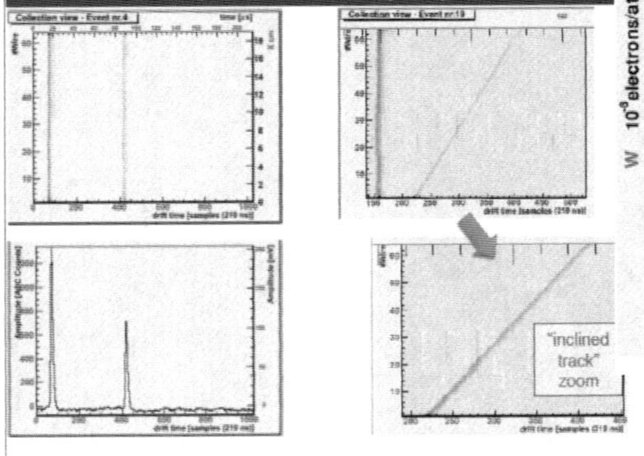

Two-photon absorption cross section

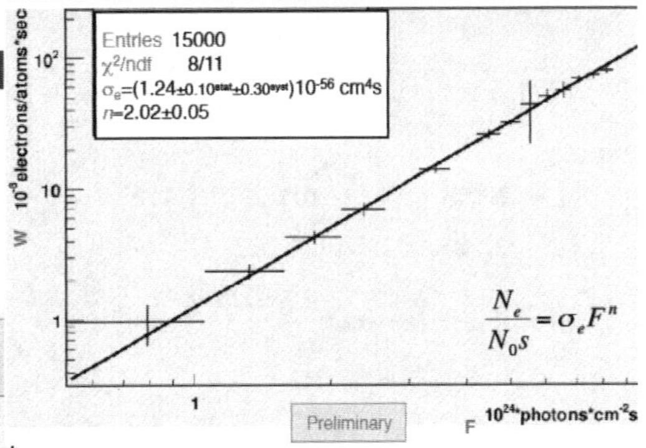

- UV laser beams technique succesfully tested
- Drift velocity and LAr purity measurements performed
- "First measurement" of the two-photon absorption cross-section achieved

A. Rubbia Towards a 100 kton LAr experiment 49

Thursday, April 1, 2010 49

Scintillation light readout

➠ Mavrokoridis, Marchionni

- **LAr emission spectrum peaked at 128 nm**
- **use reflector foils coated with TPB in order to use standard bialkali PMTs**
  - TPB shifts to a mean wavelength of 430 nm

coating achieved by evaporation of TPB is very uniform

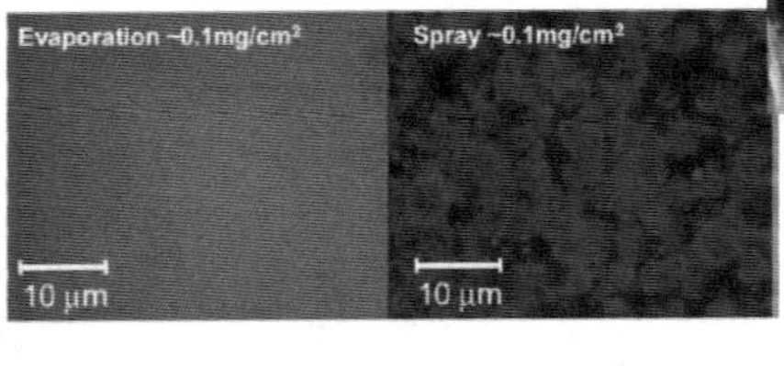

Position	Distance from cathode [cm]	Light yield 7 PMTs [p.e./keVee]
B	10	0.502 ± 0.007
D	30	0.440 ± 0.005
F	50	0.370 ± 0.006
G	60	0.319 ± 0.005

(Error only statistical)

...ith NaI Xtal trigger (1.8 MeV

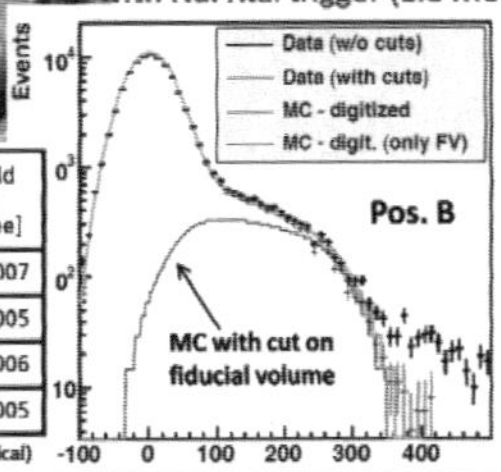

A. Rubbia Towards a 100 kton LAr experim

Thursday, April 1, 2010 50

J-PARC P32

➠ Maruyama

Project to build a small-scale LAr TPC in Japan

- Cryogenic vessel originally built for MEG liquid xenon calorimeter
- Vessel currently at KEK LAr lab
- Ultra-Vacuum established
- Cryocooler and liquid argon filling under investigation
- Liquid argon purification system under procurement
- Chamber being designed and built
- Exposure to low-momentum separated kaon beam @ JPARC

340 MeV/c K+

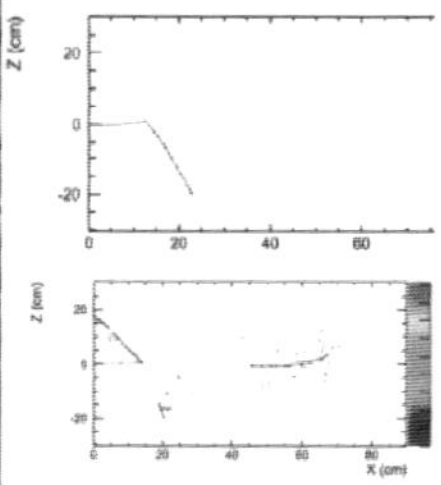

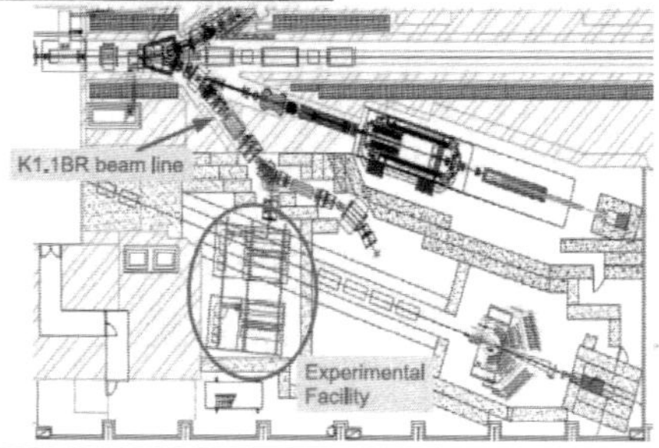

J-PARC PAC recommendation

Beam test of LAr
Charged particle beam

6. PROPOSAL EVALUATIONS

1. **P32:** (Towards a Long Baseline Neutrino and Nucleon Decay Experiment with a next-generation 100 kton Liquid Argon TPC detector)

The PAC acknowledges the high scientific merit of a neutrino oscillation experiment with a baseline longer than T2K. The measurements of the mixing angle θ_{13} and a possible CP violation in the lepton sector are of highest significance.

The specific P32 proposal is to set up and test a 250 Liter LAr prototype TPC in a low-energy charged particle beam at J-PARC, preferentially with kaons from the K1.1BR beamline. **The PAC encourages the team to proceed with this development work and recommends the allocation of beam time of a low intensity charged particle beam at J-PARC for this test.**

KEK's first step toward LAr !

Development of readout electronics

❑ Preamplifier development

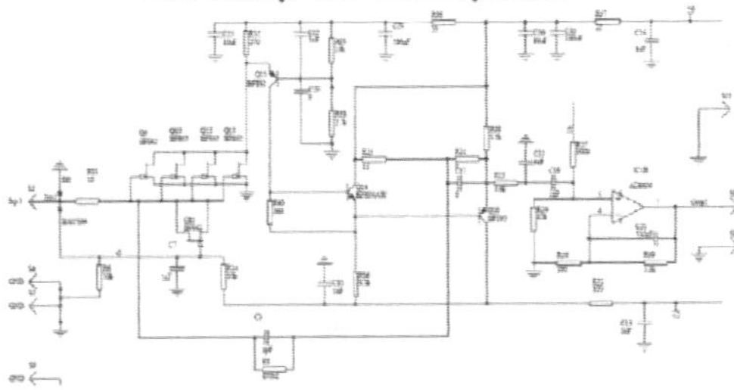

Inspired from C. Boiano et al.
IEEE Trans. Nucl. Sci. 52(2004)1931

ETHZ preamps

~11 mV/fC
S/N =10 @ 1 fC, C_i=200 pF
MIP ~10 fC/cm

❑ Development of LAr TPC electronics for small scale devices

- CAEN, in collaboration with ETHZ, developed A/D and DAQ system
- 12 bit 2.5 MS/s flash ADCs + FPGA

see talk by C. Tintori

A. Rubbia Towards a 100 kton LAr experiment 53
Thursday, April 1, 2010 53

Readout electronics (II)

➠ Autiero

F/E CMOS ASIC

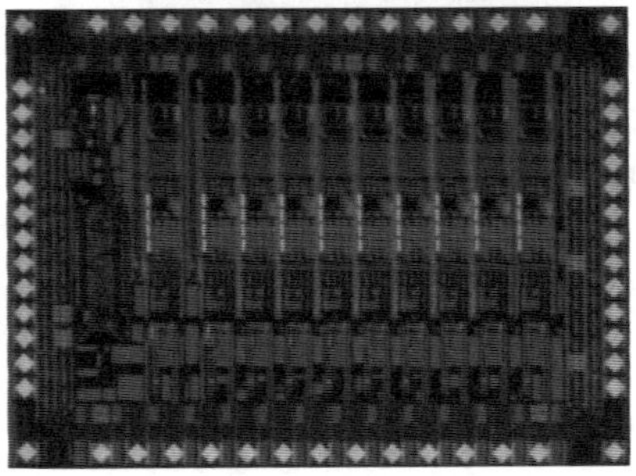

Good stability at cryogenic temperature, phase margin improved
8 channels Preamp + shaper + line driver
+ 2 test channels for individual components characterization
Slow control of all functionalities and components selection with embedded I2C slave
Gain increased to 18 mV/fC
Preliminary results 1500 ENC at the end of the full chain with 250 pF input capacitance @ 110K (-20%)

μTCA network based DAQ architecture

μTCA network based DAQ architecture

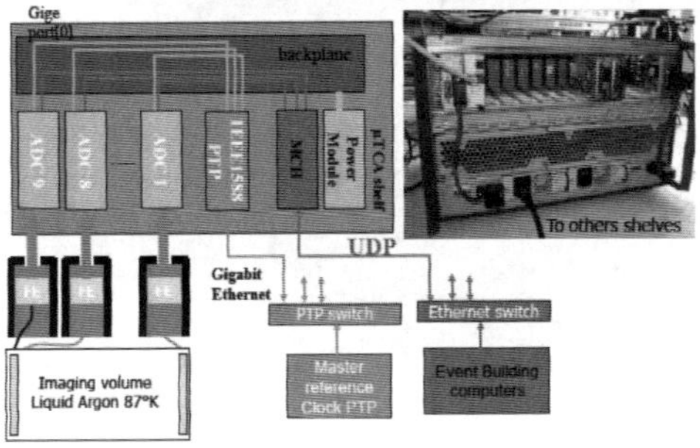

μTCA ⇨

➠ see also talk of Radeka

A. Rubbia Towards a 100 kton LAr experiment 54
Thursday, April 1, 2010 54

Methods to attack the purity problem

➠ Rebel & Curioni

- In future, we will need three very large scale independent systems: (1) **GAr flushing** (2) **GAr recirculation & purification** and (3) **LAr recirculation**
- **GAr flushing** (no evacuation) ➠ Curioni's talk
- **GAr purification**
  - ICARUS 50lt (in 1997) ➠ best result <30 ppt after several weeks
  - 3 lt setup @ CERN (2010) ➠ best result 0.6 ppb at filling
  - 12 m3 GAr/hr (from 6m3@CERN), GAr recirculation pump + SAES getters to be tested soon ➠ Curioni's talk
- **Liquid recirculation** industrial purification system developed by Airliquide for ICARUS (composed of Barber&Nichols pumps + Messers-Griesheim Hydrosorb+Oxysorb)
  ➠ Unit capacity: 2500 lt LAr/hr per T300
  ➠ best result 0.2 ppb after 50 days

A. Rubbia Towards a 100 kton LAr experiment 55
Thursday, April 1, 2010 55

6m3 @ CERN

➠ Curioni

ETHZ-KEK-Liverpool

- R&D towards non evacuated vessels
- First test purging - satisfactory!
- Piston effect seen
- Demonstrated 3ppm O_2 contamination
- Gas recirculation under construction (summer 2010)

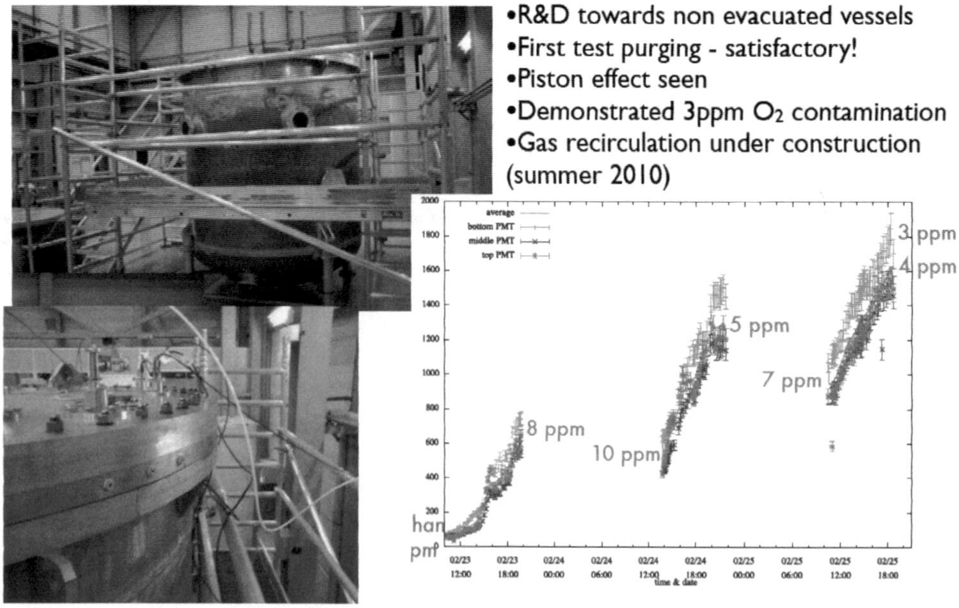

A. Rubbia Towards a 100 kton LAr experiment 56
Thursday, April 1, 2010 56

➠ Rebel

Liquid Argon Purity Demonstrator

- Currently operating systems such as test stands at FNAL and ArgoNeuT use evacuation as the first cleaning step
- Building large vessels that can be evacuated is very expensive - scales the cost by at least a factor of 2 for small vessels, worse for large vessels
- Need to find an alternative to evacuation for large vessels - LAPD is test stand at FNAL to study alternatives
- Vessel holds ~30 t of LAr and is 3/16 inch thick stainless steel
- Makes use of MTS experience in design of system

7

A. Rubbia Towards a 100 kton LAr experiment 57

Thursday, April 1, 2010 57

➠ Rebel

Liquid Argon Purity Demonstrator

- Currently operating systems such as test stands at FNAL and ArgoNeuT

use

step

- Buil eva the sma
- Ne eva test alte
- Ves inc

We seem to have not yet reached the intrinsic limits of commercially available purification cartridges, but the purity is in practice rather determined by the volumetric capacity of recirculating the gas and liquid phases in a given amount of time → 100kton detectors will require at least x100 GAr/LAr recirculation capacity compared to today's state-of-the-art (another synergy with the petro-chemical industry)

- Makes use of MTS experience in design of system

7

A. Rubbia Towards a 100 kton LAr experiment 57

Thursday, April 1, 2010 57

➠ Rebel

System Flow

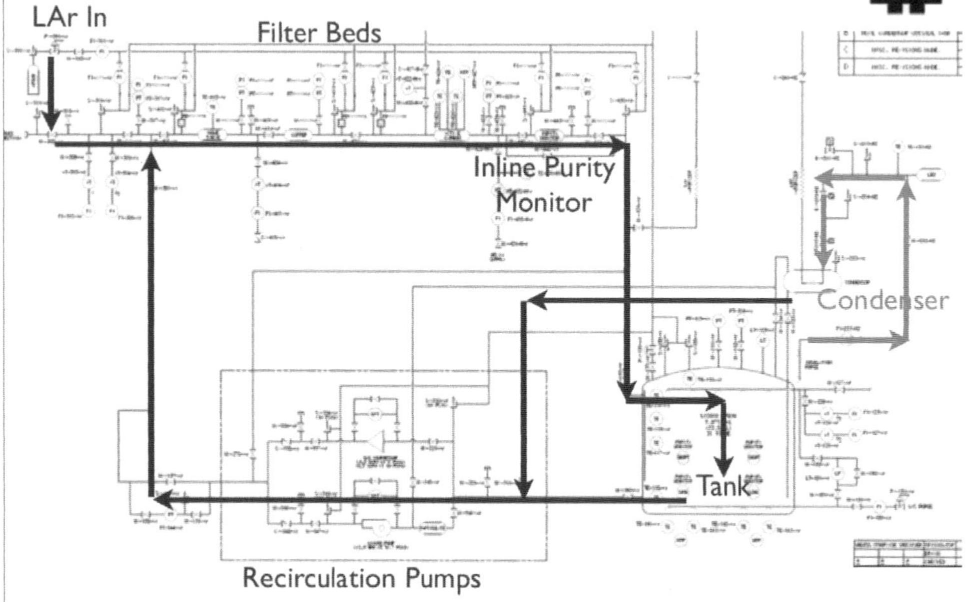

A. Rubbia Towards a 100 kton LAr experiment 58

Thursday, April 1, 2010 58

ArgonTube purification - bellow pump

➠ Rossi

The liquid recirculation system scheme (2/2)

2 asynchronous cryogenic pumps driven by GN_2
400 l/h Maximum flow rate
120 mm Pump diameter
10 mm Rise

Complete recirculation of the inner volume (4500 l) in about 12 hours

Automatic control system to immediately detect possible bellows failure and preventing pure LAr contamination.

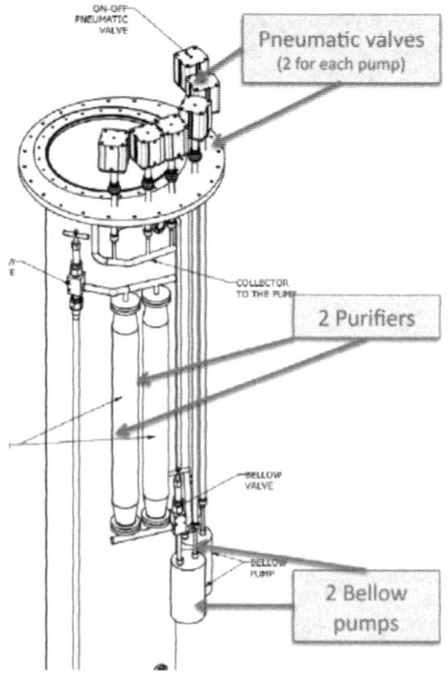

Biagio Rossi - University of Bern GLA2010 - Tsukuba - 30 Mar 2010 26

A. Rubbia Towards a 100 kton LAr experiment 59

Thursday, April 1, 2010 59

So what are the main show stoppers?

- Several groups are now engaged in R&D programs towards very large liquid Argon detector. So what are the "main" reasons for not proposing the 100 kton detector as of today ?
- The most important "show stoppers" seem to be today the following:
  - Is the physics performance of LAr detector fully understood (signal eff., backgrounds) ?
    - need to study $\approx 10^6$ υ & 5×10^5 $\bar{\upsilon}$ interactions on LAr
  - What are the plausible maximum drift length & highest purity ?
    - need experience with long drift detectors of smaller size
  - How to get the funding agencies to commit about ≈500M€ ?
    - need results from T2K/NOvA and LHC... to have very strong arguments for finding CPV & p-decay

A precursor step

A. Rubbia Towards a 100 kton LAr experiment 60

Thursday, April 1, 2010 60

GLACIER roadmap

see J.Phys.Conf.Ser.171:012020,2009

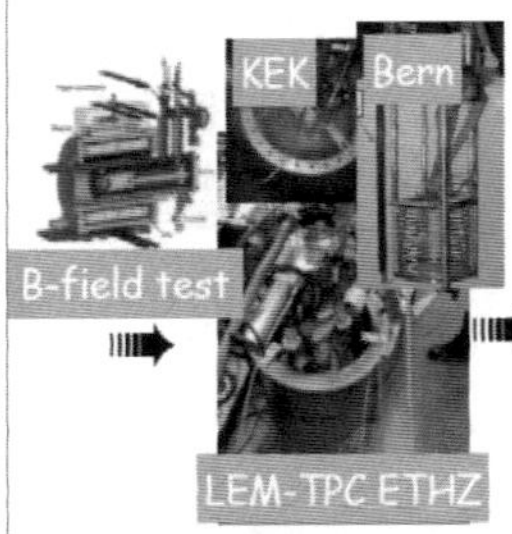

proof of principle double-phase LAr LEM-TPC on 0.1x0.1 m² scale

LEM readout on 1x1 m² scale UHV, cryogenic system at ton scale, cryogenic pump for recirculation, PMT operation in cold, light reflector and collection, very high-voltage systems, feed-throughs, industrial readout electronics, safety (in Collab. with CERN)

direct proof of long drift path up to 5 m

Application of LAr LEM TPC to neutrino physics: particle reconstruction & identification (e.g. 1 GeV e/μ/π/K), optimization of readout and electronics, possibility of neutrino beam exposure

6m3 → CERN NA ?

full engineering demonstrator for larger detectors, acting as near detector for neutrino fluxes and cross-sections measurements, ...

A precursor step

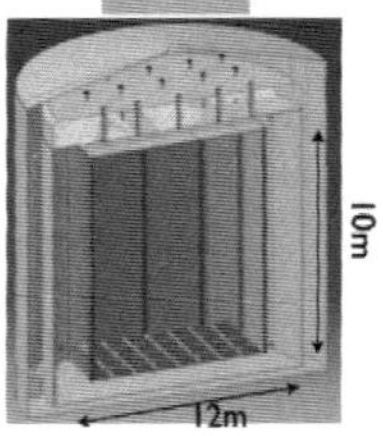

A. Rubbia "Towards a 100 kton LAr experiment", GLA2010, March 2010 61

Thursday, April 1, 2010 61

Strategies at the CERN PS beam

Several scenarios discussed at neutrino detector studies and possible experiment at CERN PS, March 17-18th, CERN, chaired by A. Blondel

- Rebuild CERN-PS nu beam (timescale min. 3 years)
- Build new 1kton far detector in BEBC hall as precursor of future 100 kton experiment
  - LNG-based tank (not evacuable)
  - Very long drift path as play-ground for new readout designs, purity demonstrator, new HV and electronics
- A near detector for near/far oscillation physics
- A "pilot non-accelerator underground experiment"

A. Rubbia "Towards a 100 kton LAr experiment", GLA2010, March 2010 62
Thursday, April 1, 2010 62

Our view on a 1 kton @ CERN PS

A step beyond MicroBoone with 10x statistics

- Search for ν_e & $\bar{\nu}_e$ appearance with very large statistics
  - answer open questions (e.g. LSND oscillations and/or address whether the MiniBOONE low energy excess is due to <u>electrons</u> or <u>photons</u>)
  - unique opportunity to test in realistic beam conditions the LAr TPC performance for future long baseline program
- Statistically precise exclusive neutrino and antineutrino differential cross-section measurements in the GeV range
  - with $>10^6$ interactions with low neutrino beam flux systematics (tdb), with both horn polarities
  - to study backgrounds in future long baseline and future non-accelerator physics measurements with high statistical accuracy (x10 expected backgrounds)
- A technical precursor of a 100 kton underground future neutrino physics and proton decay experiment
  - the largest possible detector to minimize extrapolation to 100 kton
  - the smallest possible detector to minimize cost and timescale for realization

A. Rubbia "Towards a 100 kton LAr experiment", GLA2010, March 2010 63
Thursday, April 1, 2010 63

Comparison event statistics

	MiniBOONE (running)	MicroBOONE (CD1)	1kton @ PS
Det. mass	500 t	70 t	1000 t
Distance L (m)	541	541	850
υμCC energy (peak)	0.8 GeV	0.8 GeV	1.5 GeV
E/L (eV^2)	1.8	1.8	2.2
Proton energy (GeV)	8	8	20
Pots	6.46x10^{20}(υ-run) 5.5x10^{20}(antiυ-run)	6x10^{20}(υ)	1.25x10^{20}(υ)* few 10^{20}(antiυ)
υμ CC(CCQE)	υ: 7x10^5(4.2x10^5) antiυ: 1x10^5	1x10^5(6x10^4)	1.2x10^6(4.8x10^5) antiυ: 5x10^5

(*) 6x10^{19}pot/yr for 2yrs running (R. Steerenberg)
(**) antineutrino horn polarity rate being computed (Ludovici)

A. Rubbia "Towards a 100 kton LAr experiment", GLA2010, March 2010 64

Thursday, April 1, 2010 64

Outlook

- We all look forward to ICARUS T600 (1.5m drift, 50k wires) starting commissioning @ LNGS and produce some physics results
- A 20 kton LAr TPC (2.5m drift, 500k wires) is part of LBNE program @ DUSEL with a deliberately chosen conservative design (e.g. 2.5 m drift) but with some challenging aspects (e.g. fully embedded cold electronics) → indeed a great opportunity to build 1/5 of the 100 kton! currently getting to CD1-review, waiting for CD-2
- In parallel, a smaller technical "precursor step" of a 100 kton detector in the range of the 1 kton mass, which could perform a sensible neutrino physics program in addition to being the playground for future technical solutions, has been considered since a while.
- Its feasibility @ CERN PS, presumably after MicroBOONE startup, could provide an ideal setting, however needs much more detailed studies and assessment of physics performance → an expression of interest ?
- A significant part of this detector (e.g. cryogenic components, purification systems, detectors, electronics, etc..) could be eventually deployed underground as an "pilot underground experiment".
- It looks like LAr technology has very exciting times ahead !

A. Rubbia Towards a 100 kton LAr experiment 65

Thursday, April 1, 2010 65